Commercializing Successful Biomedical Technologies

Second edition

Transform your research into commercial biomedical products with this revised and updated second edition.

Covering drugs, devices, and diagnostics, this book provides a step-by-step introduction to the process of commercialization, and will allow you to create a realistic business plan to successfully develop your ideas into approved biomedical technologies.

This new edition includes:

- Over 25% new material, updates and practical tips on startup creation.
- Tools for starting, growing, and managing a new venture, including business planning and commercial strategy, pitching investors, and managing operations.
- Additional real-world case studies, updated to include emerging technologies such as regulated medical software and artificial intelligence (AI), offer insights into key challenges and illustrate complex points.
- Tips and operational tools from established industry insiders, suitable for graduate students and new biomedical entrepreneurs.

Shreefal S. Mehta is a successful serial entrepreneur and investor, currently Executive Chair and co-founder of Pulmokine Inc, Entrepreneur in Residence at the Severino Center for Technological Entrepreneurship, Rensselaer Polytechnic Institute, and Visiting Faculty at Center of Advanced Studies on Entrepreneurship in BioMedicine (CASE BioMed) Università della Svizzera italiana.

Commercializing Successful Biomedical Technologies

Second edition

SHREEFAL S. MEHTA

CAMBRIDGE
UNIVERSITY PRESS

Shaftesbury Road, Cambridge CB2 8EA, United Kingdom

One Liberty Plaza, 20th Floor, New York, NY 10006, USA

477 Williamstown Road, Port Melbourne, VIC 3207, Australia

314–321, 3rd Floor, Plot 3, Splendor Forum, Jasola District Centre, New Delhi – 110025, India

103 Penang Road, #05–06/07, Visioncrest Commercial, Singapore 238467

Cambridge University Press is part of Cambridge University Press & Assessment,
a department of the University of Cambridge.

We share the University's mission to contribute to society through the pursuit of
education, learning and research at the highest international levels of excellence.

www.cambridge.org
Information on this title: www.cambridge.org/mehta2

DOI: 10.1017/9781108186698

First published 2023

A catalogue record for this publication is available from the British Library.

Library of Congress Cataloging-in-Publication Data
Names: Mehta, Shreefal S., author.
Title: Commercializing successful biomedical technologies / Shreefal Mehta.
Description: 2nd edition. | Cambridge, United Kingdom; New York, NY: Cambridge University Press,
    2022. | Includes bibliographical references and index.
Identifiers: LCCN 2022011773 (print) | LCCN 2022011774 (ebook) | ISBN 9781316510063 (hardback) |
    ISBN 9781108186698 (epub)
Subjects: MESH: Biomedical Technology–economics | Marketing–methods
Classification: LCC HD9999.B442 (print) | LCC HD9999.B442 (ebook) | NLM W 82 |
    DDC 660.6068/8–dc23/eng/20220528
LC record available at https://lccn.loc.gov/2022011773
LC ebook record available at https://lccn.loc.gov/2022011774

ISBN 978-1-316-51006-3 Hardback

To Gauri, whose continuing support and encouragement, and patience and willingness to shoulder my share of parenting when necessary, and more, made the completion of this book possible. Without your love and help, there would have been no book.

For this 2nd edition, I would like to add a dedication to the readers, students, teachers, scientists, engineers, technicians, executives, and entrepreneurs who work with such zeal to generate ideas and develop new biomedical products every day. Your efforts are the inspiration for this book.

Contents

Preface

This book will give you an overview of the steps involved in taking a new biomedical invention and making it into a commercially available product. The products covered are drugs (both small molecules and biologics), medical devices, diagnostics, and their combination products. These product definitions are given by the Food and Drug Administration (FDA) – the regulatory agency responsible for overseeing the world's single largest healthcare market in the United States. The term "biomedical technologies" refers to the collective technologies underlying these FDA-regulated products: biotechnology, various engineering technologies, chemistry, and materials science, etc.

The book's goal is to highlight key issues that will help improve chances of success through the complete commercialization process for biomedical technologies and products. This book aims to help you understand what questions to ask as you go through the planning and processes involved. In addition, the text will highlight issues to expect when you launch your invention from the laboratory into a business for commercialization.

This text started as an expansion of a series of lectures given to students at the Lally School of Management and Technology, Rensselaer Polytechnic Institute in Troy, New York, in a course titled "Commercializing Biomedical Technology." The course filled a gap in biomedical and biotechnology engineering and science education by providing practical information about the process in commercializing the engineered ideas and bringing those solutions to the people that need them.

This content in this book could be used to bring science and engineering students together with business and law students, showing them the benefits of approaching this complex process as a team. Many students who studied the book in courses have found the information useful in securing positions and fitting into the work environment of the biotech industry and its service sectors from day one. In addition, the book helped them better understand the big picture context within which they were working. It turns out that the overview provided by this book is also useful as a quick reference guide for strategic planning or for career transitions for senior executives.

I have attempted to keep a practical perspective in selecting the content, so that scientists and managers in the industry can apply these concepts, issues, and exercises within the context of their job functions in industry. What's more, aspiring entrepreneurs may walk through all the steps and exercises found here, to create a commercialization plan and form a business plan for a new venture (Figure 1). Business models and financial plans vary with the economic or personal context and goals of the founders. However, any business model, to be successful, must come from an understanding of the complete commercialization path for the regulated product.

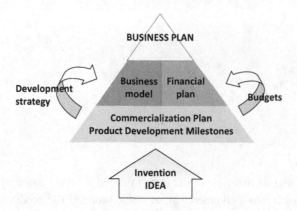

Figure 1 First you have to understand how your idea will be developed into a product and identify key development milestones on the critical path to reach the paying customers. That gives you a budget and financial plan. Then you can choose a business model and prepare a business/financial plan to execute that development strategy. It all starts with thoroughly understanding the product commercialization path.

COMPONENTS OF A PRODUCT COMMERCIALIZATION PLAN

Plan	Position	Pitch	Patent	Product	Pass!	Production	Profits
INDUSTRY CONTEXT	MARKET RESEARCH	STARTUP A NEW VENTURE	INTELLECTUAL PROPERTY RIGHTS	NEW PRODUCT DEVELOPMENT	REGULATORY PLAN	MANUFACTURE	REIMBURSEMENT
Technology positioning and strategy, corporate portfolio strategy, industrial value chain context	Market need, Specific indication of interest, market size and segments, product characteristics	Opportunity recognition, people in the business, business plans, pitch for financing, from scientist to CEO	Intellectual property management and licensing strategy, Patent content for market protection, Business models	Stage gate new product testing and development plan, budget, Gantt chart	Regulatory strategy -working with FDA toward approval	Production Planning	Coverage, Coding, Payment, Distribution, marketing and sales planning

Figure 2 Roadmap to creating a commercialization plan. The linear stages shown here reflect the layout of the book chapters. Figure 3 represents the iterative feedback from various areas that would eventually define and launch an idea into a viable commercial product

The linear roadmap in Figure 2 shows the components that must be assessed to build a sound commercialization plan. The planning is carried out iteratively as you proceed through the chapters, with increased understanding of the needs of each step. The arrows below the specific chapters in Figure 2 illustrate the fact that all these components feed into a successful commercial and product development plan. The planning component on starting a business – Chapter 3 on pitching an opportunity and founding and financing a new venture – is new in this edition. Chapter 3 was written in response to queries from readers who asked in some manner to include a practical guide to ease the transition from scientist to company founder and executive (CEO, CSO, CTO, COO, etc). I had the benefit of also drawing from my own experiences

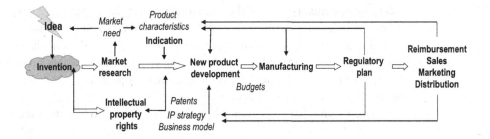

Figure 3 Successful development of new biomedical products for a competitive and regulated marketplace requires a full and thorough understanding of specific issues in the full value chain, discussed in the book. As feedback from various areas is defined for the specific product concept, the commercialization and product development plan will be revised (indicated by thinner feedback arrows above).

and those of many others from the annual BioBusiness course (offered to executives and first-time entrepreneurs in USI [Universita della Svizzera Italiana], Switzerland). Thus, instead of echoing many texts on the generic basics of writing and pitching business plans, I have tried to highlight issues that I have seen arise in science-based startups. Scientist-founders (technopreneurs) often face unique challenges in these science-based ventures due to their highly analytical and technical training backgrounds, which don't always translate well in a sparse-data environment. I hope the points in this chapter paint a clearer picture of the rocky reefs ahead, so founders can navigate better and launch and grow their ideas successfully.

The process of doing science and the process of building commercial entities can be represented as linear, but the practice of both is path-dependent and iterative. Learning and understanding grow by doing each experiment or building each step of a commercialization plan. The schematic in Figure 3 illustrates with arrows the process of feedback between the various components of a commercialization process. As an example, the regulatory path influences the product development plan and defines the markets accessed by the product. Likewise, the scope of intellectual property rights influences the direction of development and access to specific markets. Thus, iterative feedback from evaluating the specific regulatory pathways or intellectual property rights might require reconfiguration of the original invention in its product characteristics or its applications.

The process for planning new product development might for instance follow the steps:

Idea – invention – market research – intellectual property search – define product and indication(s) of interest – plan the key product development steps – check on regulatory strategy – revise product development plan and characteristics – check on reimbursement strategy – revise product characteristics and product development plan.

The result will be a comprehensive product development and commercialization plan with a timeline and budget. The exercises at the end of the chapters will help guide the reader through these steps. Somewhere in this iterative loop of business planning, you decide the opportunity is the right one and launch the new venture, stepping through Chapter 3 in greater detail. Many companies start with a novel technology and a general idea about the market application. While that will not change in this innovation-driven, science-based industry, the steps outlined in this book highlight how important it is to define the specific indication (application) within the context of regulatory and reimbursement gateways.

I hope that this text, in addition to serving as a reference to industry executives and practitioners, continues to be taught at the undergraduate, graduate and executive education levels. Courses that teach this book will, it is hoped, create a more conscious and self-aware breed of scientist and engineer who will use this foresight to better guide their inventions to become useful products in the world. Finally, it is my hope that better thinking and planning in the development of regulated products will help improve the efficiency, success, and quality of biomedical technology commercialization, increasing the number of innovative products that can be delivered to help people.

Visit https://shreefalmehta.com/csbtbook for additional enriching readings around the topics covered in the book, topical updates on the content and for industry viewpoints and news.

Shreefal Mehta PhD MBA
Entrepreneur in Residence, Severino Center for Entrepreneurship at RPI
Executive Chair and co-founder, Pulmokine Inc.
Commercialization advisor, investor, and serial entrepreneur.

Acknowledgments

The contributions and suggestions of friends and colleagues who shared their time, their insights from years of industry experience, their editorial suggestions, and specific case studies have significantly improved this book. The inputs from readers, entrepreneurs, students, and professors who used the 1st edition over the years were very helpful in improving the book and in fact inspired me to continue and update this 2nd edition. The Startx medical group of entrepreneurs was a super resource as always, as were the discussions with executives and entrepreneurs at the BioBusiness course that I teach annually at USI, Switzerland. I would also like to recognize the formative early discussions and exchanges with my colleague and friend Dr. Jan Stegemann during the creation of the eponymous class that we co-taught at Rensselaer Polytechnic Institute (RPI).

Contributors:
Sarvajna Dwivedi, Scientist and serial entrepreneur, co-founder of Angiosafe Inc, and Pearl Therapeutics, Inc., previously Chief Scientist, Pharmaceutical Sciences, BioPharmaceuticals R&D at AstraZeneca
Ulrich Granzer, President of Granzer Regulatory Consulting, GMBH
Ankur Gupta, Ophthalmologist and entrepreneur, founder of Spect Inc.
Christoph Hergersberg, past Global Head of Bioscience Technology at GE and now Executive VP at ThermoFisher
Karl Hermanns, Managing Partner at Seed IP Law Group
Shantala Mehta, Artist, architect-in-training, and my loving daughter, for help with all the figures
Parashar Patel, Past VP of Health Economics and Reimbursement, Boston Scientific
Kim Popovits, Chief Operating Officer and President, Genomic Health
Andrew Radin, CEO & Co-Founder at twoXAR
Raghu Rao, Serial entrepreneur and Board Director at Sonnet Bio, for improvements in Chapter 3.
Dan Recinella, Past VP of Product Development at Angiodynamics
Jonathan Romanowsky, Co-Founder and Chief Business Officer at Inflammatix, Inc
Lawrence Roth, Past VP of Product and Business Development, Percardia, Inc
Eric Schur, CEO at Hepatx Inc.

Jayson Slotnick, past Director of Medicare Reimbursement and Economic Policy at Biotechnology Industry Association (BIO), now partner at Health Policy Strategies, Inc.

Mitchell Sugarman, Past VP of Health Economics, Policy and Payment at Medtronic, now President and CEO at Anchor Bay Consulting, Inc.

Dan VanPlew, Exec VP, General Manager – Industrial Operations and Product Supply at Regeneron

1 The biomedical device and drug industry and their markets

Plan	Position	Pitch	Patent	Product	Pass	Production	Profits
Industry context	Market research	Start a business venture	Intellectual property rights	New product development (NPD)	Regulatory plan	Manufacture	Reimbursement

---

Learning points

- Description of types of regulated biomedical products covered in this book
- Understand the technological base and application for each product type – diagnostics, drugs, and devices
- Understand the industry context for development of new products in healthcare
- Gain an overview of the technology trends and their potential to impact the delivery of healthcare
- Understand the impact of information technology in biomedical product development
- Analysis of industry sector competitiveness by value chain model and Porter's Five Forces analysis

---

1.1 The healthcare industry components and large cycle trends

The healthcare industry and the health markets for services and products differ from the usual free-market industries such as consumer retail or industrial products. For example, while purchasing a retail product or a service in a competitive, free-market economy, the user is the primary customer and makes the purchasing decision from available choices. The user is given all appropriate requested information on the product and the user is then the payer. In the healthcare industry, the end user (patient) usually does not make the purchasing decision (the provider and other intermediary institutions like pharmacy benefit managers make that decision), the patient does not get all the information (the care provider typically gets the detailed briefing and information packages), and the patient is not the direct payer (the payer is the insurance company or government). In many developing countries, while the patient or user is usually the direct payer, the purchasing decision is still made by a more informed decision maker – usually the caregiver, who could be a doctor or nurse, the pharmacist, or a traditional medicine practitioner. In general, patients do not have the knowledge or training necessary to make an informed decision even if information is provided. The law of averages and pattern recognition built from experience is often the basis for selection and success of the therapy, and the reason why patients are partial to caregivers with the most experience.

Additionally, this marketplace is highly regulated, starting from the early product development stages to the preparation and dissemination of marketing information, and including the flow of payments, goods, and information. The government is also the largest single payer organization in the healthcare industry and thus has a strong influence on payment policies and procedures in the industry. Finally, and most significantly for manufacturers, the government and laws and policies enacted by the legislative bodies play a very important role in shaping the marketplace. Companies must be proactive in monitoring and interacting with legislators (elected representatives) in government and with regulatory agencies to monitor changes in policy that impact the market and to proactively educate and inform the drafting of such policy and regulation. Any commercialization plan for a new biomedical technology must be mindful of the context of this regulated and politically charged healthcare marketplace.

The healthcare marketplace (detailed in Figure 1.1) now must be seen in the broader context of how patients and caregivers make decisions, both for access to and purchase of therapies, whether reactive or proactive, prescriptive (driven by caregiver choice), or over the counter (driven by consumer choice).

The dynamics of power in this industry are now being challenged and changed by the recent increases in computing, making information available to the caregiver and

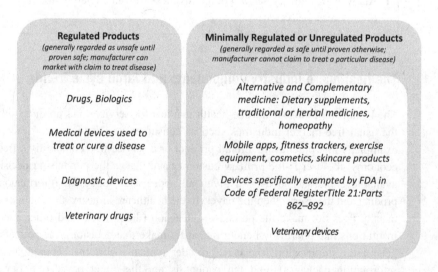

Figure 1.1 Regulated human health products and minimally or unregulated human health products

the patient at the point of service. This information is made available as a synthesis or sum of experience and advice from machine learning or artificial intelligence (ML/AI) software that extracts patterns from large pools of "big data." These same information analysis tools, which are available in more sophisticated and validated form for the caregiver, are making it possible to expand the context for healthcare to be broader than the individual experience, to the community knowledge, and even beyond to bring historical and global knowledge bases to the local point of care.

While social media and internet search engines may have made the patient a "google" doctor, in the end, these new computing methods and technologies may only be shifting some of the decision-power from the caregiver to companies that control the data and analysis. Data are still siloed among providers as a means of control over the patient, and access to potentially useful data is sometimes curbed by legal and business issues.

Advances in computing power are going to continue to impact the healthcare industry at many levels, from the patient to the regulators. In the end, the patient and society have to evaluate whether these and any other new technologies have improved health outcomes, and that will take time.

The rest of this chapter discusses the various product sectors involved in the larger healthcare industry and highlights methods to analyze and better understand their functional structures *from a product development perspective*.

1.2 Biomedical technology – definition and scope, applications

This book covers regulated biomedical products that go through the regulatory process (e.g. USA – Food and Drug Administration or FDA, Europe – European Medicines Agency or EMA, China – China Food and Drug Administration or CFDA, India – Central Drugs Standard Control Organization or CDSCO, Japan – Pharmaceuticals and Medical Devices Agency or PMDA) for marketing approval, including thera-peutic or prophylactic drugs (the term includes small molecule and biologic drugs), diagnostics, and devices. The term *biomedical technology* will be used to refer to companies whose products need regulatory approval to get to market. The "technolo-gies" include engineering and various sciences, including natural sciences (e.g. life sciences or biology) and applied sciences (e.g. materials science, computer science).

Proceeding through these first few chapters, it will become apparent that the terms *biotechnology* and *device* have blurred boundaries today, as an increasing number of leading medical device companies are incorporating drugs or biological therapeutics such as cells, DNA, or proteins, and pharmaceutical companies are tying their products to diagnostic or delivery devices. Such products, which are codependent or intermingled with other technologies, are called "combination products." Some examples of combination products are the drug Herceptin (used to treat breast cancer), which has to be prescribed based on a diagnostic test for the gene *Her2*, drug-eluting stents, bioresorbable sponges with growth factors, skin grafts containing live cells imbedded in a bio-printed matrix, and insulin pumps with blood glucose monitors.

Other examples of the changing technology landscape are the emergence of Software as a Medical Device (SaMD), where stand-alone software programs, apps, and so on are approved as devices for diagnosis or data analysis and decision-support. The following sections in this chapter define the specific product sectors in greater detail.

1.3   Drugs and biotechnology – definition and scope

Today, drugs are developed from one of two distinct technological platforms:

(1) Synthetic organic molecules – *small molecules* (preferred term used in this book) made de novo by synthetic chemistry processes or naturally occurring compounds which have been isolated or resynthesized in the lab. These are interchangeably called small molecules, drugs, or pharmaceuticals. Oligonucleotide-based drugs (RNA or DNA; composed of nucleic acids) made using synthetic processes are also included in this classification of small molecule drugs, as they have more in common with small molecule drugs than the large molecule biologic proteins.

(2) Biological molecules made by living organisms – using cells or other living organisms to produce therapeutic proteins or biological molecules. These are interchangeably called drugs, biotech drugs, biopharmaceuticals, large-molecule drugs, or *biologics* (preferred term used in this book).

Therefore, the term *drugs* will incorporate both biologics and small molecule pharmaceuticals in common usage and throughout this book. The US Food and Drug Administration (as per section 201(g) of the Federal Food Drug and Cosmetic Act) defines a drug rather broadly as "a substance recognized by an official pharma-copoeia or formulary, intended for use in the diagnosis, cure, mitigation, treatment, or prevention of disease, and is a substance (other than food) *intended to affect the structure or any function of the body* [emphasis added]."

The term *biotechnology industry* was used early on to refer to the growing biologics segment of the drug industry, but today is used rather broadly to refer to small-sized bio-pharmaceutical firms that are developing drugs (whether small molecules or biologics) or molecular diagnostics, as most of them are founded based on key inventions or discoveries in the life sciences. It is also important to note that biotechnology companies also develop products for other (non-health related) applications and industries (see Box 1.1). The definition of *biotechnology* (as per the *Encyclopedia Britannica*) is, in fact, "the use of cellular and molecular processes to solve problems or make products."

Among the therapies produced by biological production processes (produced in cells or bacteria), the various classes of biotech human therapeutics (biologics) being developed for a large variety of diseases are:

Vaccines, another class of human therapeutics and prophylactics, are produced in biological systems such as chicken eggs or engineered cell lines.

Biologic drugs are based on large-molecular proteins or complex biological molecules such as growth hormones, enzymes, etc. Examples are insulin, growth hormone, enzymes, immunoglobulins.

Box 1.1 Diverse applications of biotechnology

While "biotechnology" in this text focuses on life sciences–based products commercialized in the healthcare industries (needing US FDA approval), it is important to remember that many other applications of biotechnology also have great commercial value. In the popular media, the term *biotechnology industry* is also used to loosely refer to activities that may be based on a range of technologies unrelated to the life sciences, such as laboratory equipment manufacture, device manufacture, lab automation, reagent production, and synthetic chemistry with small molecules. Therefore, it is important to always understand the specific context in which the term *biotechnology* is being used.

The use of biotechnology processes at the organism, cellular, and molecular levels has many diverse applications, some of which are described briefly below but not covered any further in this book (e.g. even though biotechnology food products are regulated, they are not in the same market and approval paths as other biomedical products discussed here). A common technology base of tools and processes for manipulation and analysis of cells, DNA, and proteins ties all these diverse applications together across these different industries.

Healthcare

Discussed in main text

Environmental biotechnology

Engineered microbes and enzymes can efficiently clean up pollution, and application of the life sciences to this process is called *bioremediation*. Environmental applications also include biobleaching, biodesulfurization (removal of sulfur from oil and gas), biofiltration, biopulping, etc.

Industrial biotechnology

Engineered microbes and enzymes can be used as highly efficient components in many industrial chemical synthesis processes. Various industrial applications of biotechnology include the efficient use of enzymes to convert sugars to ethanol (transportation fuel), to make polymers such as polylactic acid (PLA) for consumer plastics production, and to improve processes in the production of fine chemicals, bulk chemicals, and commodity chemicals. Currently, efforts are underway to convert cellulose to sugars (and ethanol) on a large scale, thus harnessing biomass that would otherwise be discarded as waste products of food and grain processing.

Agriculture

Biotechnology has been used to engineer new plant and crop varieties that are pathogen-resistant, have greater yield, or add new nutritional benefits to existing crops. Some specific applications are in the development of new genetically modified plant and seed varieties, improved processing of grain products, and the development of biofertilizers. Basic biotechnologies are also used to improve livestock for food production and to provide new treatments for veterinary

> **Box 1.1** (*cont.*)
>
> medicine. Genetically modified foods are already in widespread use in the US food supply. Agricultural biotechnology is arguably the oldest continuing application of life sciences and includes the manipulation of plants and microorganisms to enhance yield, add in new characteristics such as increased nutrition or taste, and reduce the use of toxic pesticides or fertilizers; these are all key goals of biotechnology in agriculture and in the food-processing industry.

Erythropoietin (sold under Epogen and other brand names) was one of the first blockbuster biologic drugs, with over $7 billion of sales in 2015. Biologics are postulated to replace a large portion of the current small molecules due to higher efficacy in many diseases. These biologic drugs are most efficiently produced by cells or within other living organisms. Biopharmaceutical companies use bioreactors, where cells engineered to produce a specific type of protein are grown in large quantities. The proteins are then purified and most are formulated for intravenous delivery. For example, the human gene that makes insulin is inserted into yeast cells, which then produce insulin molecules similar to the ones humans make.

A *monoclonal antibody (mAb)*, a particularly significant type of biologic drug, is a highly specific, purified antibody (protein) that is derived from only one clone of cells and recognizes only one antigen. Monoclonal antibodies (one class of biologics) are an ideally targeted therapy that will only affect the specific protein target against which this antibody is made. The wave of biologics was driven by mABs like Johnson & Johnson's Remicade (infliximab); Roche/ Genentech's Avastin (bevacizumab), Herceptin (trastuzumab), and Rituxan/MabThera (rituximab); Bristol-Myers Squibb's Erbitux (cetuximab); and Abbott's Humira (adalimumab). In 2006, there were 18 mAB products on the market, and by December 2019 over 79 were approved (Lu et al., 2020), with over $300 billion in global sales projected by 2026 (Fortune Business Insights, 2021). Monoclonal antibodies, like most biologics, cannot be given orally (degraded by digestive enzymes) and hence are infused intravenously. New drug delivery technologies are also in development to allow oral administration.

Next-generation antibodies are already on the market (3 approved) with over 50 in clinical development in 2021. These next-generation antibodies (antibody-drug-conjugates or ADCs) combine the unique targeting capabilities of biologics with the cancer-cell-killing specificity of chemotherapeutics, radioactive isotopes, cytotoxins, or cytokines. The antibodies are directed against antigens that are differentially over-expressed in tumor cells. Potent cancer drugs are chemically linked to these antibodies, giving these antibody-conjugates a superior pharmacological efficiency with minimized side effects.

Cell-based therapies and tissue engineering are used for tissue and organ replacement or functional augmentation. The market for regenerative medicine worldwide is in the billions of dollars, primarily using autologous cells. Gene therapy holds many promises but has been hampered by limitations in delivery vehicles and side effects in some patients. In particular, cell-based therapies are attracting a great deal of attention for the truly regenerative potential of stem cells (embryonic and adult). The Japan and Asia-Pacific region is the fastest growing market for regenerative medicine.

Nucleic acid therapy is a particularly interesting and emerging class of drugs that mostly uses synthetic production processes but is usually included under the biologics sector due to the large size of the molecules:

Nucleic acid therapies include gene therapy, which is the introduction of specific genes or segments of nucleic acids appropriately into the body to modify tissues and the production of proteins that may be lacking or malfunctioning in the disease state. These therapies lie somewhere between small molecule and biologic drugs in size of molecules, with specific considerations for development driven by their larger size and limited uptake into the targeted tissue. Many different nucleic acid therapies are in development, with antisense therapeutics being the first approved in the USA. Other nucleic acid technologies, such as ribozymes, antisense oligonucleotides, siRNA (short interfering RNA, or ribonucleic acid, molecules), microRNA inhibitors, and triplex and chimeric endonucleases, have tremendous current commercial and scientific interest as seen by the awarding of the 2006 Nobel Prize to the discoverers (Andrew Fire and Craig Mello) of gene silencing by double-stranded RNA. This short interfering RNA (siRNA) interferes with gene expression and uses the cell's own control mechanism for controlling production of specific proteins. While extremely promising for their targeted approaches, these nucleic technologies have complex development challenges – as seen by Merck's acquisition of siRNA Therapeutics for over $1 billion in 2006 and subsequent sale a number of years later to Alnylam for $150 million, and Alynlam's Phase III–level failure in 2016 as a result of seeing more deaths in the drug group than the control population.

Recent discovery of highly specific mechanism for gene editing and repair called CRISPR/Cas-9 [https://en.wikipedia.org/wiki/CRISPR] promises to yield a new class of biologics directed at modifying gene expression using transcription factors (proteins that bind to specific DNA sequences).

The biotechnology/biologics segment of the pharmaceutical industry is about 40 years old (since early 1980s) and has seen its revenues grow at an average of 16% per year over the past two decades, to reach over $132 billion in global revenues in 2015 (Ernst & Young Annual Biotechnology Industry Reports, 2015–2020).

Biologics are a rapidly growing portion of the overall pharmaceutical industry, accounting for over 30% of the total pharmaceutical sales compared to 23% in 2014 (www.statistica.com, "Evaluate Biotech and Medtech 2020 in Review" report, 2021). The growth rate and strong product pipeline of the biologic drugs have attracted interest from investors and from the traditional pharmaceutical companies themselves. In particular, the biotech impact on the pharmaceutical industry has led to the industry naming itself the "biopharmaceutical industry," as more large pharmaceutical firms (e.g. J&J, Novartis, Wyeth) have adopted biotechnology manufacturing platforms to make drugs.

The interest in the biotechnology sector lies in the future impact of this technology, as more and more biologic drugs are coming through the pipeline, with over 2,500 biotechnology drugs in the clinical development pipeline in 2021, for a variety of human diseases (www.bio.org/fda-approvals-clinical-development-pipeline). Another component of the interest in biotechnology (life sciences as a more general science platform) today is in the promise of forthcoming new discoveries, like the recent CRISPR mechanism discovery, which will lead to an even better understanding of normal and pathological (disease) processes in the human body, as discussed later in this chapter. The hope is that these new tools and discoveries will lead to new therapies that will truly aim to cure disease instead of merely offering palliative treatment or temporary symptomatic relief.

It is important to mention that a significant portion of the biotechnology industry is composed of companies that provide services or make nonregulated products such as research tools, reagents, bioinformatics programs or services, biomaterials, etc. that are sold to the drug or diagnostic companies or to the research community in general.

The business models, product development cycles, and financial and investment profiles of these companies are very different from most of the companies discussed here. Examples of providers of tools and technologies are Thermo Fisher Scientific (Invitrogen, Applied Biosystems), Qiagen, Perkin Elmer, and Illumina.

1.4   Devices and diagnostics – definition and scope

1.4.1   Medical devices industry

Devices are defined by the US FDA as "an instrument, apparatus, implement, machine, contrivance, implant, in vitro reagent, ... *which does not achieve any of its primary intended purposes through chemical action within or on the body of man or other animals and which is not dependent upon being metabolized for the achievement of any of its primary intended purposes*" [emphasis added] (www.fda.gov/medical-devices/classify-your-medical-device/how-determine-if-your-product-medical-device), i.e. achieve their purpose by mechanical action or placement, or now by data analysis if software. Medical device companies use traditional materials such as metals, plastics, or ceramic and advanced materials such as composites to produce devices that work by providing electrical, mechanical, or physical (not chemical) support and interaction with the human body. Some of these devices are implanted (defibrillators), some noninvasive (e.g. EKG monitors) and others minimally invasive (e.g. catheters). These companies have shorter product cycles and thus are more dynamic in product introductions than biotechnology companies.

Medical device products can be classified by two distinct types of markets – commodity products and innovative medical device products. The former are typically made by large mature companies like Johnson and Johnson, Becton Dickson, Welch Allyn, and others and feature a broad portfolio of products sold to clinics and hospitals. These products have a long life cycle in the market, and their development is marked by incremental innovations that do not change the product mix, merely adding specific features to the design. Profit margins for these products are typically low, as customers have high price sensitivity.

Conversely, innovative medical products such as implantable devices, minimally invasive surgical devices, and new imaging devices are made by both large and small companies, such as Medtronic, Guidant (bought by Boston Scientific and Abbott), Bard, Stryker, and many others. These innovative devices have a short product life cycle, with the next generation entering advanced development even as the first generation enters the market. Innovative medical devices command high profit margins by delivering greater life-saving benefits directly to the patient, but also require high investment in research and development (R&D) for continued improvement and incorporation of new technologies.

The medical device industry's gross revenues grew to $371 billion in 2017. The industry is composed of a few large players that hold market access and brand name, and many small companies (80% have fewer than 50 employees) that have found

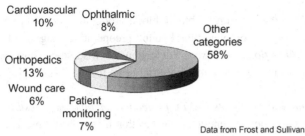

Figure 1.2 US medical device sales by clinical category (Frost and Sullivan 2022a)

niche markets in the device industry. The industry sales, broken into the various therapeutic and clinical areas, are summarized in Figure 1.2. Orthopedics and cardiovascular are the two largest device application areas, but others are growing too, as the population demographics shift.

1.4.2   Diagnostics – IVD industry

The diagnostics market is segmented broadly into the in vitro diagnostics (IVD) (*in vitro* means in the test tube, laboratory, or outside the organism) and in vivo diagnostics businesses (*in vivo* means within a living organism). This book will focus mainly on IVD, which are classified and regulated as medical devices by the US FDA.

In vivo diagnostics is a specialty market, with the key players being large instrument manufacturers of imaging or other instrumentation technology (GE, Phillips, Siemens). Examples of in vivo diagnostics are blood pressure screening, magnetic resonance imaging (MRI), thermometer, ultrasound, x-ray, and computed tomography (CT) scan. The development, sales cycles, and regulatory issues (e.g. radiation safety issues) are quite different from most of the products discussed here. However, it is important to keep in mind that most of these large companies (GE, Siemens, and Phillips) have all launched initiatives in molecular imaging diagnostics (which will be regulated as imaging agents or drugs). Thus, this exclusion (from the book) is on the basis of a specialty market segment, not an exclusion of specific companies.

In vitro diagnostics products are largely regulated as devices by the US FDA. There are two types of IVD products: devices (analyzers for samples such as blood, serum, urine, tissue) and reagents (chemicals used to mark or recognize specific components in the samples). All devices and reagents perform tests on samples taken from the body, and the applications can be divided into five broad types of IVD testing:

(1) *General clinical chemistry* – measurements of base compounds in the body, e.g. blood chemistry, cholesterol tests, serum iron tests, fasting glucose tests, urinalysis.

(2) *Immunochemistry* – matching antibody-antigen to indicate the presence or level of a protein, e.g. testing for allergen reactions, prostate-specific antigen (PSA) tests, HIV antibody tests.

(3) *Hematology/cytology* – study of blood, blood-producing organs, and blood cells, e.g. CD4 cell counts, complete blood count, preoperative coagulation tests.

(4) *Microbiology/infectious disease* – detection of disease-causing agents, e.g. streptococcal testing, urine culture/bacterial urine testing, West Nile virus blood screening.

(5) *Molecular, nucleic acid tests (NAT), proteomic, metabolomic testing* – study of DNA and RNA to detect genetic sequences that may indicate presence or susceptibility to disease, e.g. *HER2*/neu overexpression testing in breast cancer, fluorescence in situ hybridization (FISH) tests for prenatal abnormality testing, HIV viral load assays.

In vitro diagnostics companies are primarily one of four types:

(1) Large pharma with diagnostic divisions
(2) Diagnostic companies, which focus on manufacture, distribution, and marketing of diagnostic test kits (reagents) and devices
(3) Biotechnology (smaller startup) companies, which focus on discovery of technology devices/reagents for novel diagnostic methods or tests for specific diseases (e.g. a marker for cervical cancer)
(4) Clinical sample analysis laboratory services companies

In vitro diagnostics is a mature market with a high volume of clinical tests using immunoassays and simple blood tests that have not changed in decades. More than 20 billion blood tests are performed annually worldwide. The overall estimated IVD market was $65 billion in 2017. Industry segments by sales are shown in Figure 1.3.

A rapidly growing segment of IVD markets is in vitro molecular diagnostics, or nucleic acid testing (NAT, or genetic testing), which analyzes DNA or RNA from a patient to identify a pathogen, a disease, or the predisposition of a disease. These genetic tests also have applications in the area of in vivo diagnostics in the emerging applications of molecular imaging and in the development of new drugs. Biotechnology processes are used to make NAT diagnostic reagents such as nucleic acid probes.

The lab testing industry in the USA has larger companies such as Quintiles, LabCorp, Covance, Roche, J&J, Abbott, Bayer, and others dominating market access, along with large independent companies such as Bio-Rad, Guerbert, bioMerieux, and Idexx. In terms of lab service revenues, the largest market share of about 60% is captured by hospital labs, while independent labs (also called reference labs) hold about 30% market share and 10% is with physician offices. Most small private companies either find a niche or get acquired, as they are typically unable to attain the market reach of the big players to sustain growth. Product sales are dominated by industry-leading companies such as Roche, Illumina, Thermo-Fisher, and Danaher.

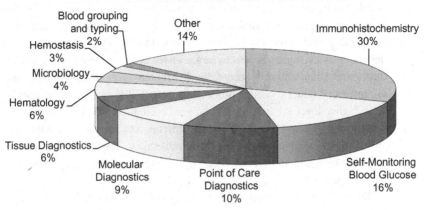

Figure caption within image:
In Vitro Diagnostics Sales $64.9 billion (2017)

Blood grouping and typing 2%
Hemostasis 3%
Microbiology 4%
Hematology 6%
Tissue Diagnostics 6%
Molecular Diagnostics 9%
Point of Care Diagnostics 10%
Self-Monitoring Blood Glucose 16%
Other 14%
Immunohistochemistry 30%

Data from Frost and Sullivan

Figure 1.3 In vitro diagnostics global sales by clinical category (adapted from Frost and Sullivan 2022b).

1.4.3 Healthcare IT and digital therapeutics

The increased prevalence of computational tools (information technology, IT) with multiple interfaces – voice, visual, text, moving images, touch screen, etc. – for accessing larger databases, with automatically curated and contextualized access to information, makes it imperative to discuss the role of IT in biomedical product development and integration into the delivery of healthcare. The ability of faster and high-power computing cycles to find patterns in larger databases that could parse a variety of inputs, including audio, images, videos, and text, is called *cognitive computing*. Easier and cheaper access to this computing power is being made available for many healthcare applications through software such as IBM's Watson platform and other machine learning algorithms being developed in open-source environments. The advent of the wireless internet with higher bandwidths and always-connected mobile devices, such as smartphones, tablets, and wireless sensors integrated into bracelets or clothing items, are all changing the way diseases can be managed and diagnosed. Previously, the hardware was the center, with software playing a minor role, but today, those roles are often reversed. Medical software applications are designed to give devices a range of functionalities. Medical software applications are also becoming increasingly independent of hardware (Software as Medical Device, digital therapeutics). The same software application can run on many different devices and have them perform the same functions. These applications can be downloaded from the Internet onto any connected device/computer/smartphone, making the same device able to do different tasks. The impact of these always-connected, readily programmable, multi-tasking mobile computers and sensors is driving development of new medical products and new business models for healthcare.

Drug product development is a completely new field of computational genomics developed in the late twentieth century with the sequencing of the human genome and expansion of proteomic analysis, creating tools to compile and search vast databases, accelerating the discovery of new targets. In the same time frame, "rational drug design" was driven by access to large, lower cost supercomputers and faster internet throughput that allowed access to distributed computing (carried out by people volunteering unused central processing unit [CPU] cycles on personal computers). This higher-level computing power was used for modeling every atom in a large protein and the mechanics of its movements and interactions with its neighboring atoms, water molecules, and new drug molecules. These advances and their application to biomedical science and technology spawned a discipline known as *bioinformatics*.

Today, IT is used at almost every stage of drug development, in target discovery from searching digitized scientific literature and genomic and proteomic databases, to automated image analysis of large numbers of cell cultures in drug-screening platforms, to compilation of large documents and analysis of large datasets, submitted electronically to the regulatory agencies for approval.

In *medical device product development*, the integration of software and hardware to make "intelligent or smart" devices continues, as higher-density, faster computing chips become more compact as seen in connected devices, augmented reality eyeglasses, retinal implants, implantable defibrillators, cameras in a pill, and other such devices. Information technology, in the form of software or firmware, runs microprocessors in the electronic components of devices, enables medical imaging networks, controls patient electronic records, etc. Devices such as a Halter monitor or a defibrillator are now connected to smartphones or the Internet all the time, making them useful as real-time monitors of conditions or patterns that signal or correct problems before they occur. This impact of IT on device design and functionality is profound, enabling or leading to significant new products such as a defibrillator that also measures physical activity levels and adapts its pacing or a retinal implant that processes images in situ and sends data directly into the brain or the optic nerve.

In diagnostics product development, automated and low-cost genomic sequencing has been made possible due to rapid and exponentially increasing speed, storage, and other advances in computing. Scientists are now able to rapidly analyze larger amounts of data and analytically simulate behaviour of not only individual molecules in a protein (e.g. drug discovery modeling) but also the many biochemical pathways in a cell as networks of dynamic connected systems. This capability makes possible new insights into diseases and personalized or precision medicine. Diagnostic tools go beyond the compiling of IVD or blood or cellular assay results, with larger companies now offering decision support, through machine learning and other IT tools that can help reduce errors and improve diagnostic outcome. Diagnostic sensors continuously sensing various body functions (e.g. wearables and smartwatches) can now be analyzed locally by connecting the data feed to mobile computing devices like smartphones or tablets, or uploaded to the Internet for discovery of patterns in the data that are predictive of various conditions; alerts can be sent out to care providers on the wireless Internet and to patients. The use of the Internet and access to large databases can obtain differential diagnosis by programs that look beyond the standard diagnostic assays and into multiple variables like social status, geography, life incidents, social media connections, emotional responses, or other indicators that may have an impact on disease health and may be relevant for assigning treatment options. The opportunities in this century are truly revolutionary with advances and insights coming faster and faster and regulators and caregivers are challenged to keep up.

1.5 Industrial value chains and industry analysis

There are many ways to analyze an industry with some of the more common methods discussed here. The questions addressed in this section are:

(1) How can you understand opportunity areas in an industry and set strategy to rise above the competition? Porter's Five Forces analysis gives us a method to look at industry competitiveness through forces exerted by suppliers, buyers, substitute products, barriers to entry, and intrinsic industry rivalry. These five forces govern competitive advantage in an industry.
(2) What is the value chain for development of new products or services and the steps and elements that make up the industrial system?

1.5.1 Industry analysis databases

An important tool for evaluating economic metrics for an industry in North America is the NAICS (North American Industrial Classification System, also known previously as SIC – Standard Industrial Classification) codes. These codes are useful for accessing labor and economic trade databases and statistics by region or state. The codes for the biomedical industry are listed in Appendix 1.1.

1.5.2 What is an industrial value chain?

A value chain is a high-level model of the various steps involved in converting raw materials to finished products that are used by customers, as shown in Figure 1.4 and in the description, below. Looking at the value chain of an industry or product is a very useful way to understand the dynamics of the industry and to understand relationships between the company status and its competitive strategy. The individual product development stages and processes are discussed in greater detail in Chapter 5.

As a product moves from basic R&D to market, each step increases the value of the work in progress, with the product reaching maximum value when it is finally sold in the marketplace to the end user. A value chain schematic is used to describe the steps in the development process and also to give an overview of the entire process of taking a concept to market. A supply chain, a common term in industry, is a part of the overall value chain. The supply chain model focuses on activities that get raw

Biomedical Product Industry Value Chain

INPUT VALUE CHAIN

Exploratory/discovery, research and development   **Product Development**

OUTPUT VALUE CHAIN

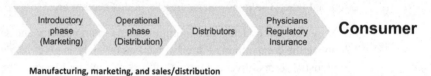

Manufacturing, marketing, and sales/distribution

Figure 1.4 Typical biomedical industry value chain. Competencies and functions are added as the company grows toward commercialization of its first product.

materials and components into a manufacturing operation and carries out specific operations on the raw materials to eventually convert them into a finished product. The goal is to deliver maximum value to the end user for the least possible cost, to analyze the specific functions of the company, and to define strategic advantage. Supply chain management is therefore a subset of the value chain analysis.

The value chain concept is useful in analyzing the specific primary and secondary activities the organization performs, and thus understanding how the organization can use technology better in specific areas, or reduce costs, or reconfigure operations to add value. A value chain can be used to compare a firm's position with its competitive strategy to assess any strategic gaps. Competitive advantage is gained by product differentiation or cost advantage. A cost advantage for a firm can come from either reducing the cost of individual value chain activities or reconfiguring the value chain to its advantage. Product differentiation can be achieved by changing individual value chain activities to increase uniqueness in the product or by reconfiguring the value chain. As a startup company in the life sciences goes through product development in the life sciences, it reaches further down the value chain, and the primary and secondary functions start to grow with the addition of procurement, production, and distribution tasks. The value chain analysis can also help in reviewing the business model, wherein a specific organization's business model can be analyzed by its current and planned location in the value chain.

In a knowledge-based or science-based industry like that of medical devices, drugs, and diagnostics, the product is much more than the fabrication of the physical product (i.e. information and knowledge about its effects and side effects on customers are the key to why regulators will approve and customers will buy). The primary functions in the value chain thus focus on product (knowledge) development, compared to the typical primary functions in a value chain in other industries, which are (1) inbound logistics – which manages the incoming supply chain and vendors and includes warehousing and inventory functions; (2) operations – such as manufacturing, fabrication, packaging, and other steps to convert the raw materials into product; (3) outbound logistics or distribution – such as processing orders, warehousing finished goods, and delivery; (4) marketing and sales – such as identification of customer needs and sales functions; and (5) after-sales service – maintaining product after sales.

Secondary or support activities that a company has to perform and decide on whether those are important to build in-house or buy are: (1) general management – such as business processes, organizational structure, etc., which can be a strong source of competitive advantage; (2) human resource management – identifying the right needs, people, and training development and compensation activities; (3) technology development – improving the product and process through all parts of the company; and (4) procurement – purchasing inputs for raw materials, equipment, or labor. These primary and secondary activities and their interactions with one another are systematically reviewed in a value chain analysis to assess their significance in providing customer value, contributing to profit margin, and maintaining strategic advantage.

Examining the industry's functional segments through this value chain perspective also reveals a view of the business models prevalent. Some companies focus on the supply of specialty raw materials, others specialize in manufacturing the active pharmaceutical ingredient (API), others may only do filling of capsules or injections or contract manufacturing work for regulated products, while some work on value-added distribution services for injectable drugs. Some areas of the value chain – discovery research, for example – are very fragmented, with universities, startups, and contract research organizations (CROs), while others have high barriers to entry and thus see few large players – manufacturing of biologics, for example. Quantifying various measures (e.g. profit margins, return on investment) of these companies and of the incremental value built into the value chain would allow one to understand the highest value-added component of the value chain and the dynamics of each process that involve multiple stakeholders.

The following general descriptions represent some typical value chains in the various segments of the biotechnology industry, acknowledging that there can be specific products and developments that take radically different routes. For example, some companies can license in technologies at one point in the value chain and sell them out at another point, capturing the incremental value represented by that intermediate development step.

1.5.3   Drug development process

The drug development value chain shown in Figure 1.5 (discussed in greater detail in Chapter 5) begins with a discovery or hypothesis of a potential drug target's key involvement in a disease. The target is usually a protein, an enzyme, or a receptor in a

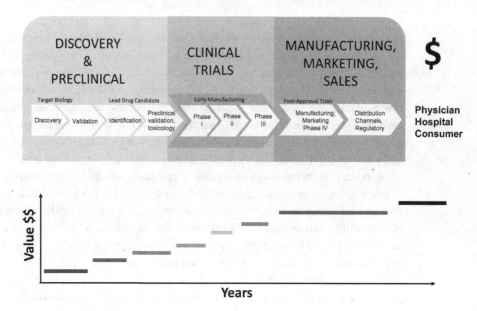

Figure 1.5 Drug discovery process and value chain

cell or tissue that has been discovered to play a central role in the development of a disease or its symptoms. The drug can be a synthetic chemical small molecule that binds to the target and inhibits or activates its function, or it can be a biological molecule that replaces or corrects the faulty or defective enzyme or protein. A large part of the effort in preclinical research work is to verify the validity of the target (to verify that interventions aimed at the target will have the desired effect on the system) and to develop a molecule that can become a drug compound. This preclinical research stage then ends with the key milestones or gates (see Chapter 5 for more details) passed: (1) validation of a therapeutic effect of the drug in animal models of the human disease, (2) satisfactory safety profile as seen by clearance of formal toxicology and other (absorption, distribution, metabolism and excretion profile, other in vivo behavior) testing, and (3) validation of reliable, repeatable manufacturability at scale meeting regulatory guidelines.

The clinical trial process is carried out in specific development steps, Phase I–IV clinical trials, each with specific goals:

Phase I   Toxicity in normal healthy volunteers (usually) and behavior of drug in humans (pharmacokinetics and pharmacodynamics)

Phase II   Establish that the drug works to treat the disease (efficacy, dosage)

Phase III Establish efficacy in larger population (statistical validity of drug effects)

Once the clinical trials are complete, the results are analyzed and submitted to the FDA for approval to market the drug. The review by the FDA can take up to two years. Post-approval clinical studies (Phase IV) may be required or may be conducted for new indications once the drug is on the market.

Phase IV   Post-marketing surveillance (usually required by the FDA after approval, to further validate efficacy or safety with longer-term or broader population exposure to drug), or may be conducted to expand use of drug to new indications/diseases or population (e.g. children).

The entire process of drug development can take from 12 to 16 years and hundreds of millions of dollars. New drugs have a high failure rate in chemical compound development (slightly lower for biologics), with only an estimated 1% of compounds that enter early preclinical screening successfully becoming drugs for a given disease. An estimated cost of developing a new drug, published by Tufts Center for Study of Drug Development in 2014, is $2.6 billion, which includes out-of-pocket costs of other drug molecules dropped at various stages of development and the cost of lost returns on alternate investments that could have been made with that capital. Another estimate, published by the London School of Economics, puts the cost of new drug development at a more reasonable $59 million. In either case, companies that can find ways to reduce costs by reconfiguring their business models and value chain components can find a significant edge over others. For example, some companies choose to focus on generic drugs, which have a much shorter and cheaper clinical trial period; others differentiate themselves by developing more complex generic drugs that require specialized distribution (e.g. generic biologic drugs or biosimilars) or manufacturing skills (e.g. cytotoxins for chemotherapy).

1.5.4   Medical device and diagnostic development process

The medical device and diagnostics industry value chain, represented schematically in Figure 1.6, typically starts with an R&D project where a concept is developed around some core innovative technology or biological or physiological insight. A project team then develops a design, which is then used to make a working proof of concept prototype, iterating the design process with insights gained at each step. For IVDs, assay development takes place at this stage and a prototype assay protocol and/or reagents are developed. Prototype testing at this stage is typically in vitro or laboratory testing. In IVDs, at this point, the test is placed in the context of usage and a test principle is chosen (the technology platform for the specific assay is chosen). Feasibility testing for IVDs is typically done in cells or in archived human clinical samples to which the company has access. At this point, the IVD company can start to generate revenues by the sale of specific reagents for "research use only" to a selected group of certified laboratories.

The final medical device prototype is then refined for manufacturability and scalability (sometimes in parallel with the design iterations). A refined prototype is then tested in animal models or possibly on humans, as appropriate, for proof of concept of therapeutic benefit. This feasibility study is usually accompanied or closely followed by various laboratory-based or in vivo animal-based safety tests. Human testing follows with pilot and then pivotal clinical trials. New IVD tests are typically first validated retrospectively in clinical trials using any available tissue or blood samples and then

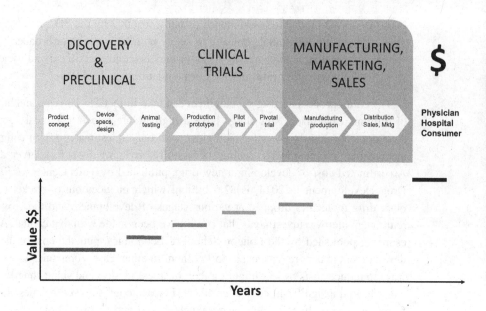

Figure 1.6 Biomedical device value chain

more rigorously through prospective clinical trials. False positive and false negative outcomes are a concern in the safety evaluation of a new IVD. The final medical device product, with manufacturing processes and designs locked in, is then put through final steps along all of the above functions. The results of safety and efficacy testing are submitted to the FDA, and on approval, the device can be distributed and marketed. This entire process can take from two to six years and from a few million to tens or hundreds of millions of dollars (time and costs vary widely due to the diverse nature of products in this industry). Product development stages are discussed in greater detail in Chapter 5. Specific steps for product development in the diagnostics industry are shown in Figure 1.7. Diagnostics offer several intermediate steps for commercialization as the industry has a large market for nonregulated supplies – hence the product development value chain for diagnostics is shown in a different format here.

1.6   Competitive analysis of an industry or sector with Porter's Five Forces model

In Michael Porter's Five Forces model of industry analysis (Porter, 1985), the five dominant forces of supplier power, barriers to entry, buyer power, threat of substitutes, and industry rivalry can be analyzed to understand the best way to gain competitive advantage in that industry. This method is a commonly used strategic planning and analysis tool and is summarized briefly below and shown as a schematic in Figure 1.8.

For a given industry, gather market data and analyze various inputs to determine:

Supplier power – How much influence do suppliers have in the industry and how is it exerted? Are there many smaller suppliers with a standard set by an industry group

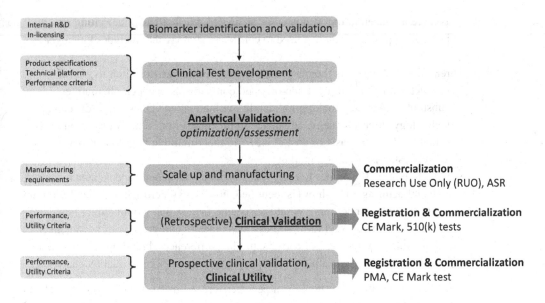

Figure 1.7 Diagnostics commercialization value chain. CE Mark is the label carried on devices approved by the European Commission and European Medicines Agency; 510(k) is one of the FDA medical device approval pathways; PMA = premarket approval, which is the FDA authorization to market novel medical devices; ASR = analyte specific reagents.

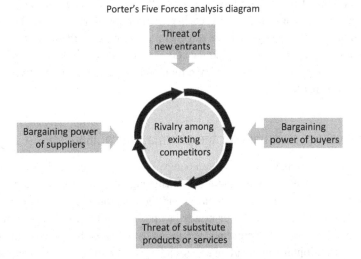

Figure 1.8 Porter's Five Forces analysis methodology. When applied to an industry, it provides a perspective as to strategic positioning and attractiveness for a new entrant.

or are there a few large suppliers that establish a de facto standard by volume? If the latter, is there a need to consider a strategy that includes the supplier as a partner? Buyer power – How much influence does an individual buyer have in the industry and how is it exerted? What is the price sensitivity among various buyer groups?

Is there a concentration of buyers or an aggregator that has price setting or bargaining power? Is there a need to consider a strategy that includes the buyer as a partner?

Threat of substitutes – Is there a switching cost for buyers to switch to a rival's products and what are the trade-offs and comparisons between alternatives/ substitutes? Are they looking for stability or innovation? Does a new differentiated technology offer a strategic advantage in this market or is the competition fairly established on the basis of cost alone? Are there multiple "me-too" products already established in the market?

Barriers to entry – If a barrier to entry (patents, large investment, specialized materials, or manufacturing know-how) is identified, how can you cross it and then keep it up to slow down competitors? What are specific factors that would attract new entrants? Can you excel in or address those factors and grow your advantage?

Industry rivalry – What are exit barriers, product differences, brand power, growth rate in industry, fixed costs among firms, concentration of firms in market share, etc.?

Unfair competitive advantage is a perspective used by many investors to evaluate the prospects of a new enterprise or project along each of the above components of competitiveness in an industry. This unfair advantage can come through technology, exclusive partnerships, business model, geography, government contract, patent position, etc. By going through each segment of strategic competitiveness in an industry and addressing the dynamics and specific issues in that segment, a picture of the industry and direction for development of competitive advantage in the industry can be created. A summary analysis for each product type (device, drug, diagnostic) is presented here. These analyses serve as general overviews for the industry. A specific analysis around an innovative product allows one to focus strategic attention and resources on the primary basis of competition and the specific competitive advantage in the product market of interest.

1.6.1   Competitiveness summary for the pharmaceutical industry

Suppliers to the biopharmaceutical companies for small molecule drugs are typically bulk chemical processing manufacturers. Switching costs are low as there are many well-established bulk chemical producers and these suppliers have low power in this industry. On the other hand, buyer power has been steadily increasing, with consolidation among payers or increasing regulation and oversight on cost increases in healthcare. Distributors in retail markets and specialty dispensation companies are concentrated geographically or over many markets and thus have a sway over the decisions on which drugs to offer as priority for their patients and market partners. State and federal governments (buyers) are also placing tremendous pricing pressure on the larger pharmaceutical industry. Substitute products are typically generics; generics manufacturers enter markets when a patent expires but have recently been using legal mechanisms to enter markets before anticipated patent expiration, reducing profits to innovator drug companies. Biosimilars are also coming to market, and competition will continue to grow even in this highly complex drug product area, as this segment of the drug market has grown significantly in the past two decades. Lastly, some governments in developing countries have started to void global patents, licensing local manufacturers to bring out new medicines at a lower price. Hence, substitute power is high in this industry. While the long and expensive product development cycle is a barrier to entry, companies also have time to create follow-ons and me-too products to take market share after the primary product launch. The long and expensive development path leads to alliances among biopharmaceutical firms and brings larger pharmaceutical companies to access innovation developed more inexpensively at smaller innovator biotech firms. In addition, many large pharmas have started partnering with buyers to reduce pricing pressure, tying the value of the new drugs to clinical benefits in the population served. Merger and acquisition activity in this sector continues, with larger players capturing development pipelines and market share. The competition within the industry is fierce and follow-on products to a new innovation or drug target emerge rapidly (18 months or less). However, the attrition rate for new products in development is high at all stages. In summary, industry rivalry is a strong (high) competitive force.

The pharmaceutical industry is thus under tremendous pressures from many inter-faces (forces). The increased sophistication of external contract research organizations (CROs) could give rise to a stabilizing factor, building scalable discovery engines while using contracts to reduce their risk, and potentially reducing the cost and time for development of a single drug. Even with a few large players owning the path to commercialization, smaller firms can still survive and succeed through innovation and intellectual property capture; niche drugs can allow smaller companies to address focused markets. Small innovative companies will continue to play a role in this industry as generators of new technology and translators of innovation from academia to industry. Access to capital remains a differentiator and a barrier to entry among smaller firms as product development is very long and expensive.

Biologics face similar pressures from these various forces (as compared to overall small molecule drug industry), with a few key differences – the supplier power for raw materials in this sector of the drug industry is low. The manufacture of proteins and monoclonal antibodies requires specific cell lines, and maintaining consistent prod-uctivity and product requires a specialized degree of process know-how. The power of substitutes (biosimilars, generics) is rather low at this point due to a poorly defined regulatory path forward for biogenerics and a more complex and expensive set of studies required for biosimilars compared to small molecules, but biosimilars have entered the market and in 2016, there were about 50 seeking approval from the FDA. Barriers to entry are high, requiring investment in specialized production facilities and access to established cell lines. Biosimilar alternatives see an approximately 15% price drop from the original drug, rather than the typical 80–90% price drop typically seen with small molecule generics, a sign of the higher cost to bring a biologic to market.

The Porter's Five Forces analysis can also be carried out at the company level, from the perspective of either a large pharmaceutical company or a small biotechnology company, not focused around a specific product, but focused around the company. Each of these perspectives will yield variations in the analysis and outcome.

1.6.2   Competitiveness summary for the biomedical device industry

Due to the diversity of firms and technologies in the device industry, a general analysis is presented here, with a focus on products that are innovative, implanted devices. Box 1.2 contains a specific example of strategic competitiveness assessment using Porter's Five Forces methodology.

Buyer power tends to be medium, since larger purchases by hospitals or group purchasing organizations can sometimes be offset by individual physician preferences at a hospital. However, buyer power is very high in the case of commodity products (such as syringes). For new innovative products the manufacturer may have substantial negotiating power due to the limited market monopoly that the patent provides. Device firms typically buy relatively common parts and materials and transform them with knowledge and processes to provide extensive value added. Consequently, supplier importance and power are generally relatively low. The multi-year, multimillion-dollar process to take a product to market through FDA approval creates a barrier to entry in the industry, but the path through FDA approval can be relatively short in many instances. Patent protection reduces competition for many new products and first-mover advantage has been noted in many medical device markets. Consequently, a firm that is first to market and/or temporarily controls a market due to patent protection is well placed to dominate the market with brand recognition on expiry of the patent. However, there has been a tendency for established products to become commodities in the device industry. These commodity product markets are highly competitive, low-margin markets with a focus on reducing manufacturing costs.

Box 1.2 A competitive analysis for a novel medical device using Porter's Five Forces

The example product is a vena cava filter, a metal filter placed in the large vein near the heart to block an embolus (blood clot) from going to the brain or lungs where it could cause death. The following analysis identifies the key competitive forces in this market, using Porter's Five Forces model.

Supplier power – Supplier bargaining power is a **weak** competitive force, as the device companies are taking up commodity materials and adding high value processing to make the filters.

Buyer power – Buyer bargaining power is a **strong** competitive force with high impact in this industry due to the small number of decision makers (physicians) at each purchasing hospital. Therefore, the firms all compete for the attention of these physicians, and the buyers can exert significant force in the sales process. Buyers will become more powerful as the type and number of filters increases. The mix of payers and buyers is different in emerging economies such as in Brazil, Indonesia, or India, where physicians and patients make the purchasing decision due to out-of-pocket payment rather than through third-party payers.

Threat of substitute products – Substitute products are a weak force, as the only other option to the filter is a blood-thinning drug. Many people cannot take blood

Box 1.2 (*cont.*)

thinners for long periods of time, and in fact blood thinners are a complementary product. There are no other known innovations in development at this time. Competition from substitutes is likely to be very **low**.

Barriers to entry (or new entrants) – New entrants are a weak force in this industry, as brand recognition, limited access to decision makers (physicians), and high regulatory requirements and long development times combined with high development costs of clinical trials keep new entrants away. Current products have good clinical outcomes with high efficacy, and thus new innovations do not have much room for improving on the existing solutions, making it more difficult to enter the market

Rivalry – Rivalry among competitors is very strong as each competitor fights for market share in a mature market that has seen no significant growth. A combination of incremental innovations in product and aggressive sales methods is used to compete for market share. Rivalry will increase in the future.

In summary, the main competitive forces in the vena cava filter market are thus rivalry among competitors and buyer influence on purchasing decision. Rivalry is likely to grow and gaining competitive advantage will continue to hinge on product innovations that show significant clinical utility and positive clinical outcome.

A competitive analysis for in vitro cancer diagnostics using Porter's Five Forces

*– Adapted from a source material project originally written
by Roger Kemp for Simon Fraser University EMBA*

The five-year survival rate for people who are diagnosed with late- stage non-small-cell lung cancer is less than 15%. Early stages of lung cancer are usually asymptomatic, and no good diagnostic is currently available. The survival rates for advanced lung cancer have not changed much in the past 40 years. One company developing an early stage diagnostic assay is assessed here.

Supplier power – Components for diagnostic assays and the platform used to carry out the diagnostic assay are highly specialized, and these suppliers are typically large companies with specialized product lines. There is a high switching cost for constituent components of the test, once qualified and validated. Many new diagnostics will be based on some established reference platform technology, such as PCR, which large equipment manufacturers control. Hence suppliers have **high** power over the smaller diagnostics companies.

Buyer power – Customers are usually governments and large healthcare management organizations or insurance companies who decide on reimbursement and coverage for specific tests. If not reimbursed, patients must pay out of pocket and then must be convinced by the physicians, who become the target of marketing from the company. Thus, buyer power is **high**, creating a greater burden of proof of cost savings or clinical validation from the diagnostic assay.

Threat of substitutes – Computer tomographic or magnetic resonance imaging, nuclear antibody tagged imaging, and other advanced medical imaging techniques are available at a relatively high cost, and this used only

Box 1.2 (*cont.*)

when there is reasonable suspicion of lung cancer prevalence. In addition, there are many false positives as screening protocols are not yet clearly established. The option of "no diagnosis" must be considered here, especially if it is perceived that there is no effective treatment anyway. Hence, unless adopted as early stage screening assay, there is **moderate** threat of substitutes with established imaging techniques.

Barriers to entry (or new entrants) – Many universities, large diagnostic imaging companies, and medical institutes are engaged in developing various methods for earlier detection of cancer, such as antibody-tagged imaging methods and detection of blood-circulating cancer cells. There is continued strong government support for cancer research in lung cancer with discoveries made at universities being spun out into companies making competitive diagnostic products. This R&D spending by a private company cannot create a barrier to entry. Regulatory barriers exist, and other companies must build quality systems to address manufacturing consistency and reliability requirements. Cancer diagnostics are classified as Class III risk products in Europe, Canada, and the United States and require clinical trials. Thus, bringing a cancer diagnostic to market takes significant time and money, creating a barrier to entry for all firms. Access to clinical validation specimens, patents filed by universities and other companies, and setting up channels for the distribution of diagnostic assays are all further complexities and barriers to entry. In summary there are **moderate to high** barriers to new entrants in this industry.

Rivalry – Rivalry among competitors is **low to moderate**. Companies are pursuing many different approaches, so it may seem like high rivalry, but the market is large enough and has room for growth for effective products. Firms pursue approaches such as immunohistochemistry, cytometry, image analysis, blood markers, protein markers in urine, and breath analysis. Very different solutions could emerge for various aspects of the types and stages of lung cancer. Market growth in the diagnostics industry continues to remain strong. There are high switching costs between diagnostic platforms and tests, as test labs and physicians need to be convinced of validity and clinical outcome by developing experience with the specific assays.

In summary, the cancer diagnostics industry does not look attractive. Although rivalry is low and moderate, the other four of Porter's Five Forces are high. From this objective review, cancer diagnostics development is a high-risk proposition. The effort is extremely expensive, taking a long time with significant risk of newer technology or methods replacing the diagnostic. Patent landscape is likely highly occupied and, even if a test reaches market, there are enormous difficulties and hurdles in distributing it and having it adopted as part of standard practice. On the other hand, the market need is significant and will continue to be a large market opportunity, looking attractive to the firm that believes it has significantly better technology with effective patent protection.

1.6.3   Competitiveness summary for the diagnostics market

The diverse nature of this product type also limits this segment to a more generalized review and analysis of this section of the industry. Thus, more qualitative discussion is presented here to cover various issues in the diagnostics industry.

The customers (buyers; hospitals, central labs, and clinics) have been gaining bargaining power over the past few decades due to the formation of hospital buying groups and large health maintenance organizations (HMOs) that use the power of scale to choose specific tests and reimbursement levels. This buyer power is also high since most approved diagnostic tests are done in a centralized core lab with selection of specific tests. Buyer power is medium to high in the diagnostics segment of the biomedical industry. Another buyer is the insurance company or government health insurance program – the decision to accept a new diagnostic test for reimbursement coverage is not made easily, with the onus on the test provider to show clinical utility in the coverage population.

Supplier power is medium to low, depending on the type of reagent (monoclonal antibodies are specialized products; basic chemical reagents are not) or device used. For example, a new diagnostic test may have to be run on a platform from another established supplier in the centralized core lab.

Several large players in the IVD industry have established technology platform standards and control distribution channels. For example, Bayer/Chiron are market leaders in blood testing and Roche still controls a large part of the nucleic acid testing (NAT; or molecular diagnostics) markets due to its proprietary position and established standard base of the polymerase chain reaction (PCR) technology. A pioneer in molecular diagnostics, Gen-Probe, which was an industry-leading developer of molecular diagnostics tests (sold to Hologic in 2012), had to establish distribution and sales collaborations with Chiron, bioMérieux, and Bayer. Companies that have diversified product menus and strong commercialization infrastructure (channel access and established technology platforms) are positioned for the long term to capitalize on the opportunities in the diagnostics markets. Competition is intense at the market level and is focused on cost in the clinical diagnostics area. Industry rivalry is high, and barriers to entry into the traditional markets (central labs or physician clinic labs) for a young company are high, as market access is controlled by a few standard-setting large firms. However, in the molecular diagnostics market segment, patent rights on innovative tests or new techniques of genetic sequencing may allow smaller companies to establish themselves. A lowering of the regulatory bar also lowers the barrier to entry, and these firms can start earning early revenues by selling their tests for "research use only" as specific reagents directly to the clinical laboratories. The emergence of molecular diagnostics tests puts emphasis on innovative content in the IVD market. In particular, about half (49%) of the industry is composed of small companies, with fewer than 20 employees. Another 17% have fewer than 100 employees. Smaller firms are usually focused on specific disease areas or even on single diseases, but larger companies have a diverse portfolio and account for a large part of the revenues. For example, in the molecular diagnostics market ($9.3 billion in

2020 [Molecular diagnostics market – growth, trends, COVID-19 impact, and Forecasts Report (2022–2027), report published by Mordor Intelligence (2021)]), the top 10 companies account for almost 89% of the revenues, but there are many smaller startups with active merger and acquisition activity.

Manufacturers of device platforms (devices that analyze specimens) also command a significant margin in this industry, giving rise to strong marketing power for established platforms on which multiple different assays can be run.

Most IVD tests are used in reference labs (national centers with high volume), centralized labs in hospitals (accounting for 60% of IVD industry revenue) or nursing homes, or physician practice labs. Access to these customers requires building a sales force or partnering to gain access to markets, which limit routes for rapid and successful commercialization of IVD tests. Innovative proprietary tests, which are based on the many emerging insights and discoveries into the human genome and proteome, will always command a premium and interest in the market, but could take time to reach commercial success as described below. A growing area of application is point-of-care (POC) diagnostics, where a simpler version (requiring no training to carry out or interpret) of a diagnostics test is sold for use at home or in a physician's office or emergency room setting.

Significant barriers to widespread adoption of molecular diagnostics exist, lock-in by specific test platforms, reimbursement issues (Chapter 7), changing regulations, education and awareness of the clinical utility of a test, the inability to fully interpret test data, and the fact that (in some cases) gene patents are hindering adoption of the tests by routine clinical laboratories and also preventing competitive development that would be good for increased market development. Unclear or changing regulatory environments and reimbursement practices that create disincentives for innovation, particularly for the new molecular diagnostics tests, remain as key impediments to successful commercialization of new IVD tests. Acceptance of a new test by a few leading academic research clinical centers may be rapid, but adoption in the larger volume markets typically takes time. These market hurdles could be overcome as more biomarkers are being clinically validated and familiarity with molecular diagnostics testing is growing through the new tests that are being launched. For example, new tests that result in improved outcomes in diseases such as cancer by helping identify genetic or other biomarkers of the disease to select optimal treatments are seeing substantial market pull due to demonstrable clinical utility.

1.7   Technology trends in biomedical device and drug development

In-depth information in an area of biology builds momentum, as multiple iterations are made to better understand a phenomenon or a technology, ultimately leading to better tools and new applications and products. These new applications, tools, or products eventually lead to new information that enters the cycle shown in Figure 1.9. The spark of curiosity of humans and the intensified, globally competitive research activities of this century are the drivers for innovations, new technologies, and

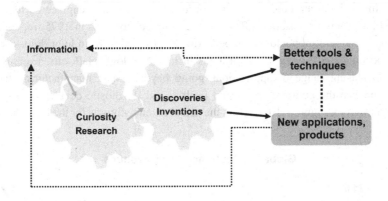

Figure 1.9 Technologies link curiosity, discoveries, and new applications in a cycle of innovation

applications entering the market. New technology breakthroughs in adjacent fields also play a role in offering new or better tools, enabling innovations that were not possible before. An example is the development of high-speed computing chips in more and more compact form with greater processing power available at distributed sites, not just at a central large computer. This breakthrough and ongoing momentum in denser, more efficient, computing capabilities have transformed diagnostics (e.g. camera in a pill for ease of endoscopic diagnosis), medical devices (e.g. implanted cardiac defibrillators that adaptively adjust to activity of the patient), drug delivery (implanted insulin pumps with active sensing of glucose and calculation of insulin dose), and many aspects of the delivery of medical care.

1.7.1   Drug development technology trends

Technology has played an important part in drug development and discovery over the years, either by discovering new targets for better treatments or by speeding up the process of developing drugs. Most early drugs were derived as extracts from natural sources. After their clinical utility was recognized, the components of these extracts, when purified, were identified and synthesized using chemical synthesis methods to yield a reproducible active compound. While small molecule discovery essentially derives from trial and error and serendipity (despite the claims of "rational drug discovery,") drug product design is carried out by medicinal chemists who take some guidance from molecular modeling of drug–protein interactions. Additional technology gains have been made in the field of biomarkers and advanced sequencing (proteins, RNA, DNA) to better identify and confirm the mode of action of a compound. However, the biggest change in drug development in the industry was the advent of biotechnology drugs (biologics) and the totally different methods of discovery, development, and production used for these new drugs.

These biotechnology drugs, typically proteins that are enzymes or antibodies (monoclonal antibodies), are produced using genetically engineered living cells. The biotech industry started off with two basic technologies in 1975, recombinant DNA (rDNA) and monoclonal antibody (mAb) production from hybridomas, and has now accumulated several breakthroughs in its technology platforms, leading to an ever-increasing range of applications going beyond basic manufacturing techniques to enhance the entire supply chain in drug and diagnostic development (Figures 1.10 and 1.11). Figure 1.10 overlays the actual revenue figures for the biotechnology drugs-

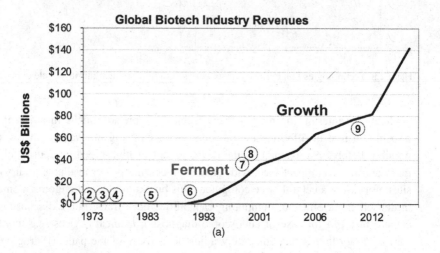

(a)

#	KEY GROWTH STEPS FOR INDUSTRY
1	1953 – DNA structure solved by Crick and Watson
2	1973 – Cohen and Boyer perfect recombinant DNA techniques 1975 – Kohler and Milstein produce mABs from hybridomas
3	1976 – Genentech founded – first commercial life sciences company
4	1980 – Diamond vs Chakrabarty – Supreme Court Case approves principle of patenting genetically modified organisms
5	1983 – PCR technique was developed
6	1990 – The International Human Genome Project was launched
7	1998 – RNA interference phenomenon was published
8	2000 – Human Genome Project 1st draft completed – genomics and technology stock markets spike and fall.
9	2012 – Next-generation sequencing, new therapeutic insights

(b)

Figure 1.10 Biotechnology industry development technical milestones

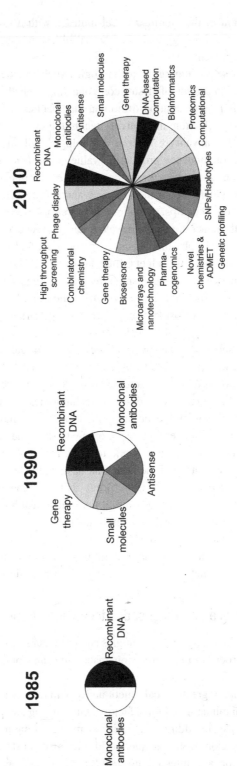

Figure 1.11 Technologies used in the drug discovery and development industry are increasing and the process is more complex. (Data adapted from Alta Partners slide presentation.)

based segment of the pharmaceutical industry with a few selected technology mile-stones. These technology and commercial milestones are meant to be representative and not comprehensive.

The next set of emerging technologies includes stem cells, tissue engineering (including 3D printing of tissue components), gene therapy, siRNA, and in silico biology. This new generation of human therapeutics will likely require the development of new production and delivery technologies.

Additionally, advanced material technologies will also influence the pharmaceutical sector throughout the production value chain, from R&D and drug discovery to manufacturing and packaging. New emerging applications, which include nano-structured polymers (dendrimers) for advanced drug delivery, analytical life sciences instrumentation, biochips, membranes, bioreactor design, coatings, and fine chemicals, will all impact the future development of new classes and types of drugs.

In 2019, the US pharmaceutical industry spent $83 billion dollars on R&D. (data from US Congressional Budget Office, https://www.cbo.gov/publication/57126), with a record number of 59 new small molecule drugs approved in 2018. While the industry has improved its productivity over the last 14 years (28 new drugs approved in 2004), the average cost to bring a new drug to market is three times more than what it was in 2003. One possible explanation for the increased cost is that the explosion of information and new targets with a paucity of historical data on their validity over the past four decades (since 1980s – see Figure 1.11) has led to a net loss of productivity in drug development due to increased complexity and need to assess more and more parameters. Paradoxically, individual technologies promise better productivity. IT systems integration and data integration processes have increased computational intensity in drug discovery and development. The general feeling is that pharmaceutical companies continuously work to integrate a large number of emerging new technologies, insights, and methods that have been shown to individually have great promise in discovering targets or drugs that can lead to a cure for diseases. It is possible that the expected outcome of these new insights will not emerge in the form of real curative medications for some years hence. However, the increasing number of targeted therapies (monoclonal antibodies, drugs that target specific cell-signaling mechanisms) and therapies that require pre-selection of patients using genomic diagnostics indicates that a new and exciting era of precision medicine is emerging.

1.7.2   Medical device and diagnostics technology trends

Medical products rely on technologies such as metallurgy, materials science, electronics and microelectronics, precision machining, and others. The general technology trend has been one of increased miniaturization of devices over time. Additionally, there is a trend of greater local functionality with rapidly advancing computing density and power in miniaturized form factors, with sensing and processing now imbedded in a single device. In addition, ubiquity of mobile computing platforms has brought a further integration of the outside world with medical devices by allowing far greater communication and interactions between the patient and caregiver or between the device and manufacturer or other functions of the healthcare system after it has been implanted. Concerns of privacy and security are now increasingly dominant as data flows more rapidly and ubiquitously over networks.

Another emerging generation of devices has adopted new materials that are biologically active and serve to enhance the therapeutic effect of the device. These emerging combination devices are being developed with multiple mixed types of materials and technologies, such as electronics and microfluidics, cell-encapsulation (material + biologic), metals with protein coatings, nanomaterials, bio-compatible electronic circuits that can flex, and tissue-engineering scaffolding biodegradable materials that can also be printed. Three-dimensional printing of various metals, biological proteins, and other materials has the promise of building highly customized, personalized tissue regeneration or artificial organs or limbs as novel devices. When combined with specific epi-genetic modification methods and stem cells, these could be amazing advances for medical devices.

Additional advances in materials and computing analytics are leading to dramatic improvements in the neural computing interface, making it possible for intelligence of the neural system to be co-opted for better function of prostheses or augmentation of body functions. This combination of technologies that blurs boundaries between traditional science and engineering disciplines (truly multidisciplinary technologies) holds great promise for improved devices with computation to be increasingly integrated into our body function with minimal disruption and maximal benefit. The diversity of devices (sizes, applications, materials types) precludes a succinct summary of technologies and market trends for medical devices.

Technology development in the traditional IVD (traditional blood analysis clinical diagnostics) market has slowed, with continuing incremental innovation in automation and simplification of existing IVD tests, and a focus on increased throughput, automation, and cost reduction. However, accelerating technology developments are seen mainly in the NAT or DNA-based genomic diagnostics market and reflected in the rapidly increasing revenues of this diagnostics market segment. These enabling technologies include genetic sequencing, PCR technologies, and DNA microarray devices, which are leading to new discoveries linking human genetic code and disease.

1.7.3   Emerging technologies and materials in the nucleic acid diagnostics field

- *Next-generation sequencing (NGS) with alternate amplification technologies –* New techniques for copying genetic material for increased sensitivity or speed of testing (compared to PCR) have been developed. The first point of care rapid (<15 minutes) test for influenza A and B virus, using these new methods, was approved by the US FDA in 2015.
- *Bio-chips and lab-on-a chip –* DNA, protein, glycosaccharides, and lipid array chips with multiple probes arrayed on a chip can provide large amounts of information from a single sample. Microfluidic technology and micro-electromechanical system (MEMs) technologies combined with standardized semiconductor industry silicon chip-fabrication and opto-electronic technologies

have made possible the creation of various versions of a lab-on-a chip. Materials that have been used so far include silicon substrate, rubberized silicone, gallium nitride and other electronic-industry based materials. Additionally, nanoscale materials (quantum dots, etc.) are being developed as markers and readouts for various assays. Technologies such as these and NGS are shifting the industry from the prevalent techniques of detection of single analytes to the large-scale, parallel testing of tens, hundreds, or perhaps ultimately thousands of genes and/or proteins in the same multi-analyte test. Together with increased automation, these new lab-on-a-chip devices will continue to bring more testing from central labs to testing at the patient bedside without losing accuracy of the diagnosis.

- *Breath testing* – The first breath test for detection of *Helicobacter pylori* (a bacterium that causes stomach ulcers) was approved by the US FDA over a decade ago.
- *Other multi-probe technologies* – These are novel techniques such as mass spectrometry with simultaneous probes to identify a "molecular signature" that is indicative of disease, rather than a single molecule at a time.

1.7.4　Robotics, 3D printing, connected devices, virtual and augmented reality, mobile technologies, artificial intelligence, and others

The fields of micro- and nano-electronics, 3D printing, machine intelligence, real-time speech recognition and feedback, rapid prototyping, and robotics have been progressing through the past few decades, with costs coming down and capabilities and reliability improving. Mobile computing through smartphones or tablets, integration of virtual or augmented reality on those mobile platforms, location awareness in devices that are connected have enhanced our ways of communicating with the world around us and among us. In addition to enhanced communications between caregivers and patients (whether remote or real-time), the new technologies promise to improve productivity in communications and the delivery of healthcare. These recent and ongoing technical advances and ease of access through wideband Internet connections in the past few years have primed the landscape for increased use of robotic, electronic. and mechanically assistive technologies in medicine and in our daily lives. These applications range from enhanced soldiers with exoskeletons, to nano-bots that can surgically repair tissues from the inside out, 3D-printed casts and prosthetics that can be customized on site for a better fit, neural interfaces that allow for direct mental control of a prosthesis or gloves with electromyography feedback loops for reducing hand tremors, computation interfaces for the human mind, artificial retinas that interface with the optic nerve directly simulating vision for the blind, and many other technological wonders that promise to improve clinical outcomes.

Mobile computing platforms, through the now ubiquitous smartphones, are monitoring our steps, our heartbeats, and our temperature, among other metrics. They have apps that act as an on-demand yoga or fitness trainer, a nutritional consultant, and even a nurse or doctor. These mobile computing devices can now enable patients to be connected to caregivers 24-7 with unprecedented ease of access with video feeds for rapid monitoring, diagnosis, and even remote treatment. Patients can now easily take advantage of earlier intervention to prevent major problems. While these constantly-on devices are constantly feeding information into massive databases, companies are using advanced artificial intelligence tools to gain new understanding. Detailed behavioral studies can easily be performed on large populations to better guide development of new preventive and interventional therapies. For example, artificial intelligence programs can be used to process historical movement patterns and records of individuals with specific symptoms such as high fever or flu, and then used identify specific areas with high risk for viral outbreaks before disease can spread, and an intelligent app on phones can be used to convey early warnings and alerts to individuals when they go near high-risk areas or if their plans include going to those specific areas. The one major concern that will continue to arise with these increasingly ubiquitous, always-on, data portals is the protection of an individual's privacy and security. Ubiquitous data collection could be misused to target individuals with the primary motive of corporate profits, which may not be in the best interests of society. New technologies like this will often be the impetus to create new rules and laws, steering actions that are beneficial to the larger community. A society with open and free dialogue can hope to eventually reach a suitable balance of sharing individual information, protected from misuse, for providing better health outcomes, convenience, and safety while also creating profitable business models.

1.8   Convergence of technologies in biotechnology

In the biotechnology industry, the level of information that is now available at the molecular level is increasing rapidly, and that knowledge is spreading rapidly at all scales of study of biotechnology processes.

As shown schematically in Figure 1.12, early observations many centuries ago (shown to the left of the figure) were made at the phenotype level usually for individual traits. Detailed observations, even though they may have been made at tissue level or greater detail, were either not well recorded or were not carried out with sufficient scientific controls in a rigorous and systematic manner. Emerging from the processes of European philosophical and scientific inquiry, the reductionist approach to biology took shape, where individual elements were studied in isolation from their

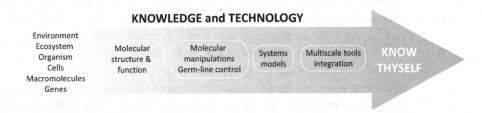

Figure 1.12 Convergence of knowledge base and various technologies toward personalized medicine and a more thorough predictive model of the entire physiology

organism or system to determine the parameters of functional activity and interaction. For example, blood was first studied as a system, then at the level of isolated vessels, cells, and individual serum components to understand how they each changed and behaved in healthy or diseased states. Further reductionist approaches led to isolation of individual receptors on cell membranes, elaboration of intracellular signal transduction pathways, genes, and DNA codons. The goal was always to take this information and knowledge and put the individual pieces back together to be able to understand the complex organism and system – like a child with a box of delicate gears trying to build the mechanism of a complex clock. However, this approach has seen limited success until the recent technological developments leading to much faster computational throughput, cheaper data storage ,and the development of algorithms that can combine disparate types of data.

By putting together computational simulations derived from this increased knowledge, the ultimate goal would be to develop a multiscale model of behavior of any given biological system. Knowing and predicting, at once, the activity of the organism and the molecular or cellular drivers of that particular state of activity (healthy or disease) will allow the true advent of personalized medicine, where a disease state is recognized, understood at the level of specific protein/gene dysfunction, and a therapy developed for the individual, considering all of their systemic unique nuances. This personalized therapy then specifically treats the root cause of the disease in that individual while realigning the systemic processes toward health and well-being. This level of understanding still remains many years in the future for Western medicine and almost calls for a union of Western allopathic medicine and much more individualized and different perspectives offered by traditional systems of treatment such as Ayurveda or Chinese medicine.

Thus, ultimately, basic biomedical research is driven by the dictum "know thyself" toward a future where personalized medicine and individualized therapy are the norm for each of us, made possible by integrating technologies at various scales to achieve detailed knowledge of the inner workings of each individual. This is what all of us would eventually like to have, a complete knowledge of conscious and unconscious, macroscopic and microscopic body processes that allow us to address the root causes of ill health and also make the best decisions about our health.

Treatment of aging as a disorder is one area of medicine in which interest is growing. If the inherent regenerative powers of the human body (stem cells) can be called upon to prevent organs from malfunctioning and to keep joints and mechanical parts working smoothly, aging is now graceful, full of joy, and not dreaded. The body can live out its life span without any chronic ill health issues, making aging a positive experience instead of a debilitating one. Aging can thus be separated from the stage of ill health that we inevitably connect with it. Aging can thus be seen as a disease today, and curing it can assume that we can realign our metabolic systems and enhance our regenerative powers to heal ourselves from within. This is a radical concept but a perspective that is being pursued by many.

In order to truly gain that broad and deep understanding of life's processes, current research is examining the organism through a variety of different perspectives and techniques that include technologies more closely related to

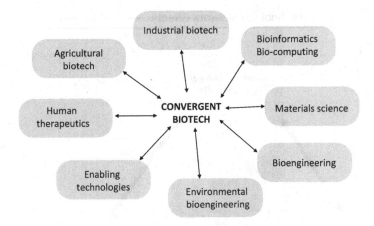

Figure 1.13 Technologies from various disciplines interact in biotechnology

traditional engineering disciplines and "industrialized science." For example, we are using robotics, self-learning adaptive software, and data analytics to analyze genetic coding differences among individuals. The increase in information and knowledge in biological sciences is now critically integrated with technological improvements in basic engineering disciplines and in physical sciences.

Additional convergence of technologies is seen across traditional industrial boundaries (depicted in Figure 1.13), when new tools developed for human medicine find applications in other areas and may in fact be enhanced in interactions with traditional engineering processes in other fields.

Applications of biotechnology that include considerable interaction with other emerging or advanced technologies such as advanced materials and nanotechnology, include some of the following areas:

- Biomass renewable energies, e.g. biodiesel
- Biosensors
- Diagnostic medical devices
- Bioremediation
- Digitization of healthcare information
- Bioinformatics
- Biotherapeutics
- Biometrics
- Biosecurity

Continuing along this trend is an increase in multidisciplinary research and development, with advances in other disciplines impacting developments in energy and biotechnology. As represented in Figure 1.14, the integration of various emerging technologies into the field of energy and biotechnology makes it likely that breakthroughs in performance or reduction in cost, or both, will result. One example of the integration of synthetic nano-biomaterials, information sciences such as advanced data analytics, semiconductor nanotechnology, and energy storage is a device that will

Technology Network Effects

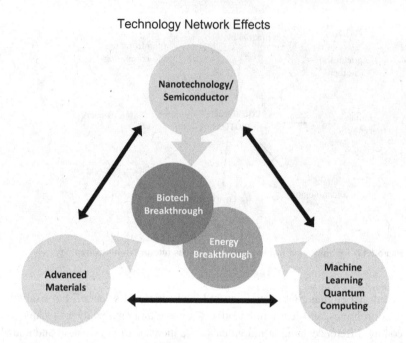

Figure 1.14 Convergence across industries and interdisciplinary studies will result in breakthrough applications at the interfaces, in the focus areas of energy and healthcare

make a neuronal- or human–machine interface possible. This type of electronic device may make it possible to run two-way communications directly with the nervous system of a human and an electronic device or computing system. Current commercial devices, such as artificial retinal implants, have already started this trend, making devices that seemed like science fiction (as in the *Star Trek* television series) a few decades ago now seem within reach of a real and useful application.

1.9   Summary

This book focuses on products applied to human health that are regulated by government agencies worldwide, similar in mandate to the US Food and Drug Administration. While the products covered here come from diverse technological backgrounds, all are applied to improve the human health condition by treatment, prophylaxis, or diagnosis. The biomedical (drug, diagnostic, and device) industry is a knowledge-based and science-based industry, needing high levels of risk capital and well-trained personnel and largely relying on differentiation and innovation as a basis for competitive advantage. Firms developing these products therefore cluster in areas where access to these resources is available. The industry value chain components in drugs, devices, and diagnostics have a significant number of elements in common.

Analyzing a value chain can help in strategic competitive planning, identifying robust business models and optimizing investment across the company. An overview of technology trends in each product area shows the increasing trend to integration of software agents and mobile computing technology with medical products in the delivery of care, and a convergence of technologies into combination products (multidisciplinary product development needed), such as drug-diagnostic, device-drug, or device-diagnostic combinations.

Exercises

1.1. Map your technology or product idea on an S-curve (growth curve) for the overall industry and for your own technology platform. What unit will you choose for your y-axis (dollars in gross industry revenues; number of units sold, number of companies in sector, number of products – each choice of unit will give a different insight)? Are you close to maturity in that technology platform? Is the industry segment you are approaching growing rapidly or just taking off? How would you now qualify your product opportunity?

1.2. What new opportunities might arise if you look at the technology trajectories in other industries or technology sectors and their intersection with the biomedical technology industry segment of interest?

1.3. What is the next-generation or emerging technology or regulatory change that might replace or threaten your product?

1.4. Analyze the components of the value chain for your industry in which the highest profitability lies and understand where your organization fits in currently and in the future. Does your business model fit with the industry value chain and your company strategy?

1.5. Do you understand how the primary or secondary functions of the value chain add to improved profit margins or better differentiation in your industry? Doing the analysis, what are the strategies you need to develop in your organization and in what time frame? Are the investments being made now appropriate for maintaining the strategic advantage of the company?

1.6. Given the trends for biological research and technology development identified here, are there specific areas in the ecosystem or technology trends that you could consider investing in to gain competitive advantage in the future value chain?

1.7. Is information technology/software a part of your product or service offering? How could the rapid growth in information technology and in computing power impact the treatment or diagnosis of the disease or disorder your technology addresses? Could information on usage of your device or diagnostic be used to open an additional revenue stream, e.g. offering automated refills with usage monitoring service?

1.8. Is mobile computing impacting the use of your device or service? Do you have a mobile strategy? What approach should be used to safeguard proprietary data? Does it make sense for your product/service to launch a smartphone app

(cont.)

to engage customers or consumers and provide them with information about the industry or products to stay connected with them?

1.9. Artificial intelligence is moving into many applications such as assisting in diagnosis of radiological images, discovering new targets for diseases, identifying patients at risk for mental health disorders, and many other potential uses. Do you have a role for this technology in your business/ research? Could a competitor to your product or service emerge with AI technology as an advantage?

References and additional readings

Burns, L. (editor). (2002). *The Health Care Value Chain*. San Francisco: Jossey-Bass. ISBN: 0787960217

Cheng C-M, Kuan, C-M, and Chen, C-F. (2015). *In-Vitro Diagnostic Devices: Introduction to Current Point-of-Care Diagnostic Devices*. Springer International Publishing. ISBN: 9783319197371

Chiesa, V, and Chiaroni, D. (2004). *Industrial Clusters in Biotechnology*. Imperial College Press. ISBN:1860944981

Collins, SW. (2004). *The Race to Commercialize Biotechnology*. London: Routledge Press. ISBN: 0415283396.

DiMasi, JA, Grabowski, HG, and Hansen, RW. (2016, May). Innovation in the pharmaceutical industry: new estimates of R&D costs, *Journal of Health Economics*, 47, 20–33.

Ecker, DM, Jones, SD, and Levine, HL. (2015). The therapeutic monoclonal antibody market. *mAbs* 7(1), 9–14.

Ernst & Young Annual Biotechnology Industry Reports. (2015–2020). Ernst & Young LLP.

Estrin, NF. (1990). *Medical Device Industry*. Marcel Dekker.: ISBN: 0824782682

Evaluate Biotech and Medtech 2020 in review. (2021). www.evaluate.com/thought-leadership/ vantage/evaluate-vantage-pharma-biotech-medtech-2020-review

Fortune Business Insights Report. (2021). The global monoclonal antibody therapy market 2021–2028. Report ID FBI102734. www.fortunebusinessinsights.com/monoclonal-anti body-therapy-market-102734

Frost and Sullivan. (2022a). Global medical devices outlook. (SKU:MC17024440).

Frost and Sullivan. (2022b). Global in vitro diagnostics market outlook (SKU: HC03451-NA-MT_25793).

Kayser, O, and Müller, R. (editors). (2004). *Pharmaceutical Biotechnology: Drug Discovery and Clinical Applications*. New York: John Wiley & Sons. ISBN: 3527305548

Lu, RM, Hwang, YC, Liu, IJ, et al. (2020). Development of therapeutic antibodies for the treatment of diseases. *Journal of Biomedical Science* 27(1), 1. https://doi.org/10.1186/ s12929-019-0592-z

McKelvey, MD, Rickne, A, and Laage-Hellman, J. (editors). (2004). *The Economic Dynamics of Modern Biotechnology*. Edward Elgar Publishing. ISBN 1843765195

Ono, R. (2013). *Business of Biotechnology: From the Bench to the Street*. Elsevier Science. ISBN: 9781483292236

Porter, M. (1985). *Competitive Advantage: Creating and Sustaining Superior Performance.* New York: The Free Press. ISBN: 0684841460

Roco, MC, and Bainbridge WS. (editors). (2004). *Converging Technologies for Improving Human Performance.* Springer. ISBN: 1402012543

Saltzman, M. (2015). *Biomedical Engineering: Bridging Medicine and Technology.* Cambridge: Cambridge University Press, 2nd ed. ISBN: 9781107037199

Shimasaki, C. (2014). *Biotechnology Entrepreneurship: Starting, Managing, and Leading Biotech Companies.* Elsevier Science. ISBN: 9780124047471.

Springham, DG, Cape, RE, and Moses V. (editors). (2020). *Biotechnology – The Science and the Business.* Boca Raton, FL: CRC Press. ISBN: 9781000159660

Srinivasan, P, Shanmugam, T, Chokkalingam, L, and Bakthavachalam, P. (editors). (2020). *Trends in Development of Medical Devices.* Elsevier Science. ISBN: 0128209615

Websites of interest

www.bio.org – Biotechnology Industry Organization
www.phrma.org – Pharmaceutical Research and Manufacturers Association
www.medicaldevices.org – Medical Device Manufacturers Association
https://dtxalliance.org/ – Digital Therapeutics Alliance
www.advamed.org/ – Advanced Medical Technology Association
www.amdm.org/ – Association of Medical Diagnostic Manufacturers

Visit https://shreefalmehta.com/csbtbook for additional enriching readings around the topics covered in the book, topical updates on the content and for industry viewpoints and news.

Appendix 1.1 Industry Classification System for Government and Other Databases

An industrial classification system is useful for researchers to look into larger government databases and gather industry level information, as companies that work in specific sectors are recognized and classified by standard codes, known as NAICS (North American Industrial Classification System used in Canada, Mexico and USA; the Global Industry Classification Standard – GICS and others such as Industry Classification Benchmark and United Nations Standard Products and Services Code are used in finance and market research). An NAICS four-digit code identifies a high-level industry sector and further digits break down activities within the sector. Companies involved in biotechnology and biopharmaceutical companies are classified as follows (2002 NAICS codes):

Human and animal therapeutics and diagnostics – including biopharmaceutical (drug) companies; also includes tool developers – genomics, bioinformatics, proteomics companies, and companies developing advanced materials for human therapeutics:

NAICS 325411 Medicinal and Botanical Manufacturing
NAICS 325412 Pharmaceutical Preparation Manufacturing
NAICS 325413 In-Vitro Diagnostic Substance Manufacturing
NAICS 325414 Biological Product (except Diagnostic) Manufacturing
NAICS 541710 Research and Development in the Physical, Engineering, and Life
   Sciences

Agriculture, aquaculture, animal health and food – includes seed and livestock development:

NAICS 3253 Pesticide, Fertilizer, and Other Agricultural Chemical Manufacturing
NAICS 32519 Other Basic Organic Chemical Manufacturing
NAICS 11511, 11521 Support Activities for Crop and Animal Production
NAICS 112 Animal Production; raise animals for the sale of animals or animal
   products
NAICS 5417 Research and Development in the Physical, Engineering, and Life
   Sciences

Industrial and agriculture derived processing – including chemical manufacturing companies:

NAICS 32519 Other Basic Organic Chemical Manufacturing
NAICS 32531, 32532 Pesticide, Fertilizer, and Other Agricultural Chemical
   Manufacturing

Environmental remediation – including utilities, petroleum industry:.

NAICS 5629 Waste management and remediation services

Device and diagnostic company industrial classification:
   The federal government NAICS system lists most subsectors of medical devices under the category "Miscellaneous Manufacturing." The following NAICS code groupings are relevant to the industry:

334510 – Electromedical and Electrotherapeutic Apparatus Manufacturing.
   [Includes a variety of powered devices, including pacemakers, patient-monitoring systems, MRI machines, diagnostic imaging equipment (including informatics equipment) and ultrasonic scanning devices.]
334517 – Irradiation Apparatus Manufacturing
   [Includes X-ray devices and other diagnostic imaging as well as computed tomography equipment (CT)]
339112 – Surgical and Medical Instruments Manufacturing
   [This is the largest subgroup category and includes anesthesia apparatus, orthopedic instruments, optical diagnostic apparatus, blood transfusion devices, syringes, hypodermic needle, and catheters]
339113 – Surgical Appliances and Supplies Manufacturing
   [This is the second largest subgroup category and includes artificial joints and limbs, stents, orthopedic appliances, surgical dressings, disposable surgical drapes,

hydrotherapy appliances, surgical kits, rubber medical and surgical gloves, and wheelchairs.]

339114 – Dental Equipment and Supplies Manufacturing
   [Consists of consists of equipment, instruments and supplies used by dentists, dental hygienists, and laboratories. Specific products include dental hand instruments, plaster, drills, amalgams, cements, sterilizers, and dental chairs]

339115 – Ophthalmic Goods Manufacturing
   [includes eyeglass frames, lenses, and related optical and *magnification* products]

Relevant 2002 NAICS codes for the diagnostic (IVD) industry sector are:

325413 In-Vitro Diagnostic Substance Manufacturing [includes chemical, biological or radioactive substances used for diagnostic tests performed in test tubes, Petri dishes, machines, and other diagnostic test-type devices]

621511 Medical Laboratories

325413 – Diagnostic Reagents

While these industrial classifications are of use in collecting data on various trade or labor parameters, the data have to be carefully examined as most companies are self-nominated into these categories. Some companies fall into more than one classification. These classifications are used to compile data for policy direction and impact, for government investments in regions and for economic development analysis.

2 Markets of interest and market research steps

Plan	**Position**	Pitch	Patent	Product	Pass	Production	Profits
Industry context	**Market research**	Start a business venture	Intellectual property rights	New product development (NPD)	Regulatory plan	Manufacture	Reimbursement

---

Learning points

- Goals and methods of market research
- Segment the market and estimate the market size for the new biomedical products
- Assess general drivers and hurdles that help or hinder market growth
- Define the product concept and market positioning
- What is the use of defining an indication early in development?
- Understand what a referral chain is and how to use it to define and evaluate the value proposition of the new product for various stakeholders
- Assess economic impact and adoption hurdles for the new biomedical product in context of the referral chain
- Use market research to help new product development planning and product characteristics

---

A handful of new ventures, out of thousands launched each year, are wildly successful and become the focus for entrepreneurs to emulate; but many ventures built around great technologies fail to make it to market or fail to achieve a profitable return. While there are many reasons for failure, a well-executed product development program can help in increasing the chances of success. The product development process must begin and end with the market need / customer as a clear goal or focus of the effort. The technology becomes a product when you have a clearly identified and confirmed customer and the product becomes real the day the first customer touches it and gives you feedback.

After asking the initial questions about the technology – Can you manufacture it? Is it safe and effective? – the immediate next questions are market focused – **Who Cares?** Why? The Why question should be asked repeatedly until you feel you have gotten to the core point, which is the specific problem and clinical need to be addressed by the new product or service.

Product development activity (see Chapter 5) starts by defining a target product profile (TPP) that identifies specific product characteristics and features and the target market and need, so that the product design and development goals are clear to all. Specific TPP characteristics are derived from understanding the biology and pharmacology or the material's interaction with the body, the specific market, the desirable clinical outcome from addressing this market/problem, competitive positioning, and other internal and external factors (see Section 5.7 for more details on TPP). Additionally, as product development always requires prioritization and optimization of counter-balancing product features or characteristics, detailed market research and subsequent analysis of the needs and economics of the application will help priority rank the product features in design process.

For entrepreneurs, market research helps not only to define the target market segments but also to validate the business plan and overall market potential of their innovation. These market-related points are the linchpin of a business plan. Investors may show interest in a technology breakthrough or innovation out of technical or scientific curiosity, but will rarely invest unless the market needs (demonstrably and uniquely addressable by this innovation) and the path to market are clearly understood. Different investors will require varying levels of market validation. Angel investors with first-hand market knowledge may already understand the needs and fit of the technical idea or breakthrough, venture investors may want early market validation with paying customers, and private equity firms or banks may want to see established and growing multi-year revenues before financing. Investors at all stages will want to see some level of validation of markets in order to assess size and growth prospects.

Market research is used to achieve one or more of the following goals:

- To identify and segment markets for a new technology in building a business plan
- To identify new product and sales opportunities
- To define the required product characteristics from an understanding of market needs and context
- To understand the competition
- To project sales revenues and profits and appropriate pricing for a robust business plan
- To spot current and upcoming problems in the industry
- To identify and evaluate/test hypothesis or assumptions about market need, and better define the target product profile (TPP) – product form, fit, and function

How large is the market for this idea or innovation? This is typically the first market question that new product development projects have to answer, whether in a new venture or a new project within a large company. This is a sweeping question that covers several subtextual questions – What is the market need or pain point identified as the target market for this product and how well does the product meet that segment's need? Is the product differentiated from competition or alternates? What is the market growth potential over time? Dynamic changes with time, unknown adoption cycles, inadequate data availability, and other unknowns can make it difficult to answer the question What is the size of the market? accurately. However, you have to take your best shot; hence specific steps and tools to develop a market size estimate are described in Section 2.3.

The following sections discuss the various methodologies used to address these questions and highlight procedures that can be used to help identify key indication(s) of interest and potential competitors.

2.1   Opportunity recognition

How do you recognize an opportunity to build a business? Most often, this question follows from a technology breakthrough that does one of the following:

- Addresses a specific unmet clinical need or pain point for a caregiver or patient
- Offers new insights into biology or product safety, efficacy
- Incrementally improves an existing therapeutic or diagnostic approach (e.g. complex multiple injections a day replaced by a skin patch with automatic mild, painless, electrical pulses to get the drug into the system)
- Arises in response to changes in regulatory or ecosystem changes in healthcare system
- Some combination of the above

These types of insights and technology inventions sometime occur on a job when a fresh set of eyes sees a way to solve a problem that others are just used to working around, or a laboratory experiment derives an unexpected insight into biology, or a practitioner, who, fed up with available tools, designs a new, more useful tool or technique. In all these cases, there is a desire to bring the technology to market and develop the innovation into a commercial opportunity.

The other way to recognize an opportunity is from involvement in the end market – a specific pain point or need is seen as an opportunity gap from personal experience. The next question that arises is whether this is a viable opportunity for building a business.

There is a strong personal context involved in making a decision to pursue an opportunity to commercialize technology and/or start a company. The passion of the individual founder to see the breakthrough delivered as a product that helps improve human health is usually the biggest factor driving this decision in the biomedical sector. However, as presented here, this passion is best validated with sound market research that can test the beliefs or assumptions underlying the commercial opportunity.

As many inventions take place in a university, it is worth pointing out that the inventor or technologist may not be willing to accept the risk and uncertainty involved in forming a company and building a new business. Many scientists do not have an interest in working in commercialization functions like manufacturing, quality control, or marketing. While many technical founders have made very successful business leaders, the decision to start a company as a technical founder needs to be accompanied with self-awareness and with recognition of the fact that while business knowledge is not technically difficult to learn, people management is not an easy skill to master. A technical founder used to relying on experimental data and facts may not understand why everyone does not automatically fall into agreement with the obvious choice, and in fact may be very uncomfortable in situations where all the data are either not available or cannot be verified before a decision is to be made. Arguably, a businessperson with product commercialization and management experience, with experience in working with non-scientists, would make a great partner to the technologist founder (technopreneur) and provide better chances of success for the new venture. Chapter 3, Table 3.1, and References and Additional Readings in Chapter 3 expand on this discussion on founding team composition.

Fundamentally, the decision on pursuing a potential opportunity can be summarized as: *Given what we know and what we don't know, is the development and financial risk worth taking for the clinical impact and market reward?* In identifying what is known and what is not known in answering this question, one can define the assumptions and beliefs that must be fulfilled for the opportunity to be a commercial product. The following queries are useful in helping to shed light on the underlying assumptions and beliefs:

-   Does the innovation solve a problem for a large enough market? ("Large enough" is defined by the business model and investor context.)
-   What is the potential for the business to make money (profitable) (and when)?
-   What are the assumptions that need to be satisfied in order for the product to be adopted in the market over what competitors are offering or the current alternatives?
-   What are the assumptions that need to be satisfied in order for revenues to grow to reach the market potential? (This important question can help bring an objective viewpoint to the market risk and potential.)
-   Do I understand and know the specific technical hurdles and assumptions for this technology to go from laboratory prototype to reliable device, diagnostic, or drug product?
-   Can we quantitate the impact of introducing the product or service arising from the technology in the existing disease referral chain [Section 2.5]?
-   Do I (scientist, technopreneur) understand and accept the risks that come with a startup company, or do I want to license the technology and stay in my current research position?
-   What unknowns present the greatest potential risk?

Other questions around the business plan and business model (covered in subsequent chapters) are also worth asking in order to make a careful decision and plan around starting a business or licensing the technology (see Box 2.1):

-   Is there a need for a large capital investment (which may be difficult to access) for production?
-   Does the technology need to be demonstrated in full-scale production before a licensee or purchaser will take it over for commercialization? (If the innovation is

---

Box 2.1 Should we start a business or license the patent?

New spinal fusion plate: A biomechanical engineering professor working with an orthopedic surgeon in a collaborative research project realized that a significant percentage of spinal fusion surgeries resulted in failure and required re-surgery due to pseudoarthrosis, the lack of formation of a solid fused bone below the plate that was used to fuse two vertebrae when the disc had to be removed and replaced by a bone graft. The culprit, he felt, was improper design and placement of the spinal fusion metal plate, leading to a combination of either (1) too little or too large a micromovement between the vertebra that disturbed or prevented the bone to grow into the graft interface effectively, or (2) stress-shielding, where loads were not being distributed into the underlying bone due to the spinal fusion plate being too stiff or improperly placed, causing the bone below it to degenerate and not grow across or into the graft material.

An additional need in the surgical process to improve plate placement accuracy would be to design the plate to make it easy for the surgeon to see where the plate was being mounted in the middle of a messy, bloody field of view as the design of all fusion plates, as a flat solid piece of metal often obstructed the ability to see the correct/best mounting location on the vertebrae. A redesigned fusion plate which would flex enough to allow the bone to experience some of the mechanical load and allow just enough micromotion to stimulate the bone to fully grow into the graft, while allowing the surgeon to see more clearly and better place the plate. This redesigned fusion plate was tested in goat vertebrae and then patented by the university. The professor had had some experience in entrepreneurial activities before and was interested in starting a company to bring this to market, but before that was done, the professor needed to decide whether licensing through the university tech transfer office would be the best approach or whether it would be better to build a company to commercialize the product. With over half a million spinal fusion surgeries occurring in the United States alone, a new, easier-to-use fusion plate that promised lower rates of failure/re-surgery seemed to have a good market size (e.g. 1 million surgeries × $1,000/plate) globally. With a few large companies ruling the market for global distribution, would it make sense to license the patent to one of those companies now or would it make sense to start up a company, raise financing, develop the product, get it approved, and then sell the company to one of those larger market leaders?

The professor brought in a local business angel-investor who reviewed the cost of commercialization, the competition, and the size of the opportunity and accompanied the academic's visits to a number of surgical practitioners in order to hear an unbiased "customer" viewpoint. Together, the team realized that the unbiased surgeons clearly found the product attractive and useful in solving existing problems. The team also investigated competition or exit points for license or sale of the technology and found that companies who would have interest in licensing this idea were current competitors whose existing products and revenues would be threatened, and those incumbent companies might be more motivated to shut down the potential new entrant by licensing and putting the technology on the shelf. This business insight about the commercialization pathway, combined with (1) the

---

Box 2.1 (*cont.*)

awarding of a small business innovation research (SBIR) grant from the National Institutes of Health (NIH), (2) an investment commitment from the investor, and (3) first-hand market research showing a need and high potential for adoption by practitioners, convinced the professor to start a company and reduce his position at the university to a part-time role for a few years. The professor is now spending more time at the company as the main technical lead, along with a recent graduate from his lab who has joined full-time, developing the product forward through animal studies. The commercialization path was attractive with a relatively short regulatory approval pathway and reasonable costs to get the product contract manufactured. The goal for the company was to target a sale of the company once it had a market-ready product. The buyer would be one of the large corporate market leaders in surgical implants. Getting the product on the market through a commercial enterprise seemed like the best option to successfully develop and launch this new invention in this particular market.

---

based on manufacturing knowhow or if the production process is complex such as requiring 52 synthesis steps from starting molecule to finished product, the answer to this question helps in laying out a business plan.)
- Is the regulatory path known and investment and risk acceptable given the market size and health benefit?
- What is the competitive landscape and does the Porter's Five Forces (or other) analysis show a path to a successful and superior commercial product position?

2.1.1 How do scientists and businesspeople assess the value in a technology?

Scientists primarily consider the function and rationale of the innovation – *how* or *why* it works – and see much potential in solving problems with a novel approach. A businessperson might take a perspective that they do not really need to know how the technology works, but they focus on answering the questions *What* can it do or *what* can it allow me to do? Is it solving a real market need? *Who cares* to pay for that function?

For example, a scientist who discovers or develops an enzyme to splice and insert genes perfectly into living cells may see it as a platform to solve any genetic disease, and hence see it as a great opportunity with a large market value. The businessperson may be helpful in selecting the specific disease or problem area (diagnosis instead of therapeutics; veterinary medicine vs humans) to which this technology should be applied first, considering how the risks of the new technology balance the benefits of applying this technology. The latter approach lays out a path to market and commercialization of the innovation, and success with a first product launch can help validate the larger platform vision. Picking the right application or target market requires deep knowledge of industry dynamics and technical context, bringing together both technical and business perspectives in a discussion. From the business perspective, a key step in identifying an opportunity and a value proposition is to find an entry applica-tion with a significant pain point for a decision maker, easing the adoption of the new, "risky" technological innovation in the market.

Please read the above points as examples to show different modes of thinking, rather than to box in or stereotype the thinking of a particular group.

2.1.2   Value of technological innovation

The value or commercial opportunity of a new technical innovation that can create a product or service offering is fundamentally defined by the market need it solves and the potential revenue stream from that market.

Many technological breakthroughs occur in the context of a core invention or novel insight, which then becomes the proverbial hammer looking for a nail – for a problem to solve. Identifying that core problem, or market pain point, becomes an iterative process with repeated generation and testing of hypotheses.

At the end of this process, the result will be (i) the identification of a target market segment with a specific unmet need and problem being solved (indication), (ii) a definition of the unique characteristic features that allow the novel product/service to address the problem (target product profile), and, finally, (iii) the target customer among all the stakeholders in the target market (market segment and size).

Two methods that can be used to understand and specifically identify market need and target segment are (1) observation and analysis of stakeholders and referral chains (Section 2.2.3) and (2) hypothesis testing of prototypes or concepts with real custom-ers (Section 2.2.2).

Breakthrough products and new-to-the-world technologies considering commercial-ization require more effort to be placed on repeat observations of customers and stakeholders and garnering more anecdotal learning cycles with prototypes rather than broad market survey and statistics. In some cases, it is a challenge to fully validate your technical and subsequent commercial hypothesis when the technology or approach is new to the world. In those cases, successful experiments carried out sequentially in isolated proteins, cells, animal models, and human subjects will help in building support for the investment thesis. These experiments will also help improve the understanding of the potential impact or outcome and, thus, the commercial viability.

2.2   Methods of market research

Market research information is either primary or secondary information. Primary infor-mation is collected by the researcher through first-hand experience or interaction with the source of information and is usually collected through experiments (experiential), obser-vation, or surveys. Secondary information is collected from data recorded by someone else and is obtained by reviewing and analyzing reports and published information.

2.2.1 Reports, projections, and historical data

Data on past events are used to suggest future trends, but this data must be assessed carefully. Past records of product sales, characteristics of successful products, and specific market segments are useful for technologies/products that are similar to the innovation in consideration. However, if the innovation is radically different in usage, application, or outcome from historical products, this historical secondary data can prove misleading. In these cases, data that could be useful in opportunity validation include historical data on methods of treatment of the disease, growth rates of procedures or costs of existing treatments, or past incidence/prevalence rates of the disease (see Box 2.2).

Box 2.2 Prevalence and incidence of a disease

These are two specific metrics that give you an idea of the current percentage of the overall population that has been diagnosed with this disease (prevalence) and the rate of new cases identified each year (incidence). Knowing the prevalence of the disease allows one to gauge the overall market potential and size (through existing number of potential patients) and knowing the incidence and trends in these two metrics over the past few periods can identify the rate of growth and potential market penetration rates of the new product if prescribed only to new patients.

Example The market size for a newly designed catheter that has an easier hookup into veins for delivery of fluids into the body can be evaluated by looking at sales of intravenous (IV) catheters in general. Assuming usage of this new catheter will be spread evenly over all current patient segments, obtain the total market size/sales for catheters from available historical data (check financial statements of a public company whose main products are catheters; look for market data in company's SEC filings or their presentations, check proceedings of trade conferences; look for data from government economic agencies or trade documents, etc.) and then assume that the rate of market growth follows the trajectory of the past few years. You now know the total size of the market (related to prevalence) and the rate of growth (related to incidence) of this particular market segment. You may also want to check these financial statements to see whether historical data showing increase in sales over time with the launch of a major new catheter brand can be separated out, giving even better comparable data for your new products' projected revenue growth.

For a new drug product, financial data for a drug that is sold to treat the same disease, with a similar mechanism of action (inhibits the same enzyme or blocks the same receptor on the cell), could be used as a benchmark to collect potential market size or growth rate estimates.

2.2.2   Experimental

Testing a prototype is the most reliable way to get assessments from possible customers. For many biotechnology innovations, making a prototype (for early clinical or even animal testing) is a very expensive process and is typically not a commonly used route for conducting market research. For example, assessing potential for inflammatory reaction or toxicity of a device or biological drug (which will better define the market potential) requires that the prototype be made with the same materials and processes to be used for manufacture of the final product. However, some product characteristics that can be evaluated with prototyping methods include selection of inhalation device for inhaled drugs, ergonomic design parameters, or packaging preferences for devices, based on handling of prototypes or other experiments.

2.2.3   Observational

Observational methods could involve visiting and observing physicians and patients in the healthcare setting appropriate to the innovation of interest. Observation methods are useful in collecting data if specific questions and protocols are set up beforehand and comfort and privacy of patients are addressed. Observation of the course of treatment from diagnosis to amelioration of problem ("referral chain") will provide a rich context to gauge the real needs of the market, the actual behavior of patients (e.g. compliance issues, where patients frequently forget or neglect to take medication can impact revenue projections), and the potential for adoption in the application context of interest. Observational methods can also include the collection of data from a focus group. A focus group can be formed from a sample of participants (caregivers) brought together in a space where they are given situations and are closely observed (sometimes by unseen observers) to understand the true context and potential modes of usage of an innovative product. Professional market research firms are usually contracted for this type of research.

This method is best used with a physical product or software that has high interactivity with the stakeholders or where a method of use is important for obtaining benefit of the product – e.g. the surgical technique used to implant an artificial knee joint or a new inhaler design.

Box 2.3 Observational market learning

Observational learning has the benefit of offering new insights that lead to identifying the real problem to be solved. Observational learning also challenges previously held assumptions with first-hand learning in a rich, real-world environment. Noticing behaviors and movement patterns while observing actual procedures being performed can give insights into unvoiced problems (opportunity). In addition, observational market learning can also yield early understanding about the potential for adoption of a new method or device.

Box 2.3 (*cont.*)

Other strategies to identify the real problems use strategic market reviews, or competitor reviews or surveys. However, humans have a large capacity to adapt to circumstances and thus many underlying problems go unnoticed as they are commonplace and accepted as the norm and people have learned to work around or adapt to these problems. Hence, it is very hard for the physician or nurse or patient to articulate a problem that they have not noticed or where their behaviour has adapted to alleviate these problems. Take the following examples below:

- A German company developed a water jet for surgical tissue dissection. Water jets have typically been used for cutting steel, granite, and wood in industry, but this company found that by varying the pressure of the jet between 20 and 40 bar, they could selectively dissect different tissues. This method would have the advantages of reduced bleeding, adjustable dissection of selected tissue layers, and shorter operation times and reduced costs, but the method never got adopted. An observation of the surgeons in the field along with directed questions showed the reasons as follows: (1) there is a long learning curve with this new tool and (2) visibility is affected with the tool, thus hindering the surgeon's work.
- Interdisciplinary students from a university's product design class were sent in to observe various obstetrical procedures in a teaching hospital setting with a view to finding opportunities or specific problems that they could identify in current practice. One group went in to observe the management of maternal hemorrhage postpartum. The Bakri Balloon is a silicone, obstetrical balloon specifically designed to treat postpartum hemorrhage (PPH). The device is used for the temporary control or reduction of postpartum hemorrhage. The students observed the nurse grimacing after pumping the water into the uterine balloon for several minutes. After the procedure, questioning this observation, they discovered that the nurses found it very uncomfortable to grip the device. For added experiential learning, the students carried out the procedure themselves using the Bakri Balloon on dummies and realized that the system was very uncomfortable to hold and use. The students realized the problem could be solved with a better ergonomic design for the handle.
- Another group of students went into a hospital to understand opportunities to apply a new battery technology on medical carts. These new types of batteries could be charged in seconds but only lasted for tens of minutes instead of hours. Earlier surveys had pointed out that the biomedical equipment maintenance department in the hospital had one person in the hospital who had much of their time taken up managing battery-operated carts. The nurses and doctors surveyed by phone said that they had no real problems and that the carts were fine. On spending a few hours observing behaviors, the group of students saw that the nurses, who mostly operated the carts, would wheel the carts into a room and immediately plug it in before attending to the patient. The wire was often unwieldy and sometimes got in the way of movement. The nurses, on being

> **Box 2.3** (*cont.*)
>
> asked about this behavior, said that they had so many carts failing just when needed (as the state of charge on the lead acid batteries is not easy to determine accurately), that they had gotten used to just plugging them in everywhere and not using the battery power at all. Now there was generally no problem. Despite this, someone would forget to plug them in between patients and they would have to get a new cart, as the battery would not recharge quickly (lead acid batteries take hours to recharge). So, the biomedical instrumentation technician would be called up and the nurses learned to always keep two carts in a corridor so that there was always one as backup available. Now that the nurses had learned to manage around the problem and had adapted their behaviors to the circumstances, they did not mention any issue with the carts. A fast-charging medical cart that would charge up in seconds and last for carrying out a few measurements or be reliably quickly charged to be available to monitor a patient for an hour or more while they were being moved would actually solve a lot of the wire and availability problems and reduce the need for a full-time charge maintenance technician. The cart would not need to remain plugged in as it could be recharged when needed in seconds.
>
> Observational visits can reveal new insights and allow one to assess the issues that might hinder adoption of a new product. The observer needs to go in well prepared, taking along note pads, cameras, and audio recording equipment or other recording tools as permitted in the environment. In order to have a useful experience, it is not enough to just observe a surgery – rather, having a specific perspective or a goal creates a fruitful observation experience.

2.2.4   Surveys to collect information

Survey methods involve posing a number of questions to the stakeholders in the markets of interest – these stakeholders could include nurses, patients' family members, doctors, therapists, hospital administrators, insurance companies, pharmacy benefits managers, and other stakeholders along the referral chain. Most typically, survey methodology is used to ask specific questions. An example would be to better identify the need or the price point for a new drug when its advantages are compared to existing therapies in a survey of prescribing physicians. Survey design and appropriate statistical analysis are key to obtaining valid results; for example, an important point is to check the validity of the data received in response to a question by asking the same question(s) in different ways throughout the survey. A challenge is to get responses to a survey from busy professionals who don't gain from this – one could offer to share a summary report of the data collected as a learning point for the responders. Keeping the surveys short is obviously another concern.

Typical response rates for direct mail surveys, telephone surveys, and personal interviews are 1–10%, 20–60%, and 70–100%, respectively. Web and email surveys have a higher response rate of 30–50% and the added benefit of significantly lower cost per response compared to the phone, mail, and personal interview methods. Hence, many times, surveys sent by email or mail require additional follow-up by phone to get adequate responses to make them meaningful.

2.2.5 Potential sources for secondary market information

- Public sources
  - ○ Reports by government agencies, Census.gov, physician associations websites – e.g. American Heart Association has an annual statistics databook on various diseases.
  - ○ The National Center for Health Statistics (www.cdc.gov/nchs) has a wealth of information on disease incidence and prevalence in the United States.
  - ○ The National Hospital Discharge Survey (www.cdc.gov/nchs/dhcs/index.htm) gives detailed numbers of admissions for each disease and numbers of each major surgical procedure performed in the United States.
  - ○ Internet search engines
- Commercial sources
  - ○ Industry and trade associations, equity research firm reports, company financial reports, or securities filings.
- Educational and peer-reviewed publication sources
  - ○ Lexis-Nexis, PubMed; check introductory sections of scientific papers, as they typically review the literature or discuss disease prevalence or incidence.

2.3 Human health-related markets: scope and size

This section will describe market scope and size, useful to get a first estimate of context of application space. The questions to be asked at the early planning stage are of the following nature: Is your technology entering a global $100 million market, a $1 billion market, or $10 million market? And is that market mature or growing or shrinking? Are there specific geographic areas that are growing or shrinking?

A good rule of thumb is to compare the innovative (potential) product in the context of the overall market size and growth rate for a particular application space. For example, the size of the global cancer therapeutics market is known relatively well, but the market for a specific therapy is harder to estimate; however, one can estimate that an effective therapy will grow at least as fast as the overall market or a similar class of chemotherapeutics that are already on the market. Tables 2.1–2.3 give an indication of upper or lower bounds of overall market sizes to provide a quick context for this query.

Table 2.1 Major market segments of worldwide pharmaceutical sales

Therapeutic areas	As percent of $446Bn (Top 10 Therapeutic areas, 2019)
Oncology	33%
Anti-rheumatics	13%
Anti-diabetics	11%
Anti-virals	9%
Vaccines	7%
Bronchodilators	6%
Immunosuppressants	5%
Sensory organs	5%
Top 10 therapeutic areas	US $446 billion (2019)
Total world pharma market	US $871 billion (2019)

[Ref: Evaluate Pharma report, 2020]

Table 2.2 Major therapy segments of worldwide medical device sales

Therapeutic areas	Worldwide market share
In vitro diagnostics	13.0%
Cardiology	11.6%
Diagnostic imaging	9.8%
Orthopedics	9.0%
Ophthalmics	6.8%
General and plastic surgery	5.5%
Endoscopy	4.6%
Worldwide market (for therapeutic areas listed above only)	US $405 billion

[Ref: Evaluate MedTech Report, 2018]

Table 2.3 Major market segments of worldwide in vitro diagnostics (IVD) sales

Application segments	Worldwide sales 2020
Immunohistochemistry	36%
Self-monitoring blood glucose	19%
Point of Care diagnostics	12%
Molecular Diagnostics	11%
Tissue Diagnostics	7%
Hematology	7%
Microbiology	5%
Hemostasis	3%
Blood grouping and typing	2%
Other	16%
TOTAL	US $80.43 billion

[Ref: Fortune Business Insights report, 2020] and [Industry Research Biz report, Dec 2021].

2.3.1 Market size segmented by application

There are three main categories of human medical products regulated by the U.S. Food and Drug Administration, primarily divided by intrinsic differences in methods of use, product characteristics, and distinct development processes:

1. Drugs, including both pharmaceutical drugs (small molecules, chemicals) and biotechnology drugs (biological molecules)
2. Medical devices
3. In vitro diagnostics (IVD)

Within these product categories, the markets can be segmented by the therapeutic application space that they treat.

2.3.2 Market size segmented by geography for drugs, devices, and IVD

The United States is by far the largest single geographic market for most products, somewhat disproportionately to its population rank. The other large geographic medical markets are Western Europe, Japan, Brazil, China, and India, followed by other geographic areas such as South and Central America and Pacific Rim countries (Table 2.4).

2.3.3 How big is the market for my technology/innovation?

Step 1: In the context of the larger market, start by segmenting the market. For example, if your innovation is a new drug that can treat headaches, you can segment the market by several variables, including age, price sensitivity, chronic vs acute, and degree of pain (migraines vs mild headaches). As another example, the heart failure market can be segmented by geography, race, and disease status (Classes I–IV heart failure; see example in Table 2.5 for a drug that is developed specifically for treating patients with severe heart disease). With more insight into the innovative therapy, it may make sense to further segment the market by background medication (e.g. for heart failure market, number of patients on beta blockers at time of eligibility to receive innovative therapy) or other factors such as stage of disease. Several available databases (commercial subscriptions needed) will provide the number of prescriptions per day written for each of these segments, giving the current size of the specific market segments of interest. Pricing the innovation at this stage may not be too reliable, as the final price will depend on many parameters that are not known at the current time, including specific outcome of clinical trials and true benefit vs risk outcomes.

Step 2: How much of the market can I hope to capture and how soon? Predicting the future is difficult, but historical trends and market shares for other products in the same industry segment can give an acceptable benchmark for market share projections. It is best to clearly state all assumptions when presenting these projections. Evaluations of the competition are particularly important in coming up with relevant

Table 2.4 Drugs, devices, and IVD revenues segmented by nation/region

	North America	Europe (all)	Japan	China, India	South/Central America	Australia, NZ, Pacific Rim (Oceana)	Total worldwide US$ millions
Drugs*	52%	25% (top 5 countries only)	17% (incl hospital sales)		4% Latin America	2% (AUS, NZ)	~930 000*
Devices+	40–45%	23%	9%	~9%	~12%	~3%	~370 000
IVD**	41%	31%	11%	NA	NA	NA	~74 000**
Population++	368 million	747 million	126 million	2.8 billion	653 million	42 million	7.8 billion people

* based on Ref: Evaluate Pharma report, Jul 2020 (data for 2017 FY)
** from Ref: BCC research report May 2021
+ from Ref: Frost and Sullivan 2017 reports
++ population from https://www.worldometers.info/world-population/
NA – not available or estimates not reliable

projections. Specifically, issues of market acceptance, reimbursement, and hurdles to growth of the market need to be carefully dissected and analyzed, requiring industry context and experience. This experience can be gained by first-hand observations and interviews.

Table 2.5 Market segmentation example.

Congestive heart failure (CHF), chronic patient population, United States (FY2020)

Category	2020
Total prevalence	6 200 000
Percent diagnosed	60.0%
Total diagnosed population	3 720 000
Percent drug treated	80.5%
Total drug-treated population	2 994 600
Percent compliant	75.3%
Total compliant population	2 254 933
Percent in NYHA Classes III and IV	50%
Total Class III/IV compliant population	1 127 467

The heart failure market for a new heart failure drug that is indicated for treatment of Class III and Class IV patients could be significantly less than the total number of heart failure patients, once the percentage of targeted patients who are actually taking their medication (compliant population) is considered.
Note: New York Heart Association (NYHA) classification of chronic heart failure patients is based on severity of disease and Class III and IV are the two most severe. Data from Virani SS, et al. (2020). Heart disease and stroke statistics—2020 update: a report from the American Heart Association. *Circulation* **141**(9), 139–596.

Box 2.4 Market prediction is not infallible

How inadequacies in market prediction got a biotech success into trouble!!
– *Adapted from* Bloomberg Business Week, *August 12, 2001,*
"Biotech and the Spoils of Success"

Immunex is a former highflier whose good fortune with a blockbuster drug has unraveled into a string of bad breaks. Sales of its rheumatoid arthritis drug, Enbrel, jumped from $367 million in 1999 to $652 million in 2000 – breaking records for a biotech drug in its first 24 months after launching. ... But this year (2001) Immunex can't manufacture enough of the drug, forcing thousands of patients onto a waiting list and leaving the door open for competitors. Only 75,000 of the 1 million patients who might benefit from the drug can get their hands on it, limiting sales to $750 million in 2001 – at least $200 million less than expected, analysts say. ... Much of that lost sales will go to Johnson & Johnson's Remicade, a TNF-blocking drug that was approved a year after Enbrel.

> **Box 2.4** (*cont.*)
>
> Immunex' initial sales estimates focused on the 25% of rheumatoid arthritis patients who failed traditional therapies, so when the FDA approved the drug for children in April 1999, and then as a first-line defense for early-stage patients in June 2000, Immunex was unprepared for the flood of demand.
>
> Immunex is retrofitting a factory to produce more Enbrel. .... But investors are still leery. Immunex shares are down to around $15 each, from a 52-week high of $56. "The pressure is on," says CEO Edward V. Fritzky as he grabs his gut, "and I feel it" (as quoted in *Business Week* article August 13, 2001).

Step 3: Run some benchmark checks on your assumptions as you consider market penetration (adoption) for projected revenue growth and market segment dynamics. List out the assumptions clearly for all stakeholders to review against the projected market size for the innovation. Check the output of the projections in the context of the larger market size and historical growth rate. How have other innovations/product launches fared historically in this market? Run a quick check through the list of drivers and hurdles listed in Section 2.4 to see how the market projections might be helped or hindered by these factors.

2.4   Market drivers and hurdles

A quick list of growth drivers or hurdles that could foster or impede future market growth, development, launch, or acceptance of FDA-regulated products is given here. This is a useful list to run checks and balances for any marketing research output and also for the overall product commercialization plan. Stepping down this list, each topic is used to raise questions against the commercialization plan or projected markets for the product(s).

2.4.1   Drivers

Demographics
For example, an aging population bodes well for continued growth in most disease areas; an increasing urban population in the world also means that diseases linked with urban areas will continue to have a strong market need.

Growth trajectory and history
For example, revenue growth rates in innovative medical devices have historically been higher than in commodity medical technology products; also, emerging areas such as molecular diagnostics have a potentially higher rate of growth than established clinical diagnostics

High levels of investment in biotech – government and venture capital
For example, areas of biodefense research and nanobiotechnology research are receiving high amounts of funding in the United States, whereas certain European and developing nations are attracting investment in biotechnology manufacturing facilities with taxes and grants.

Basic research leading to innovations and new products
In the United States, NIH funding trends often dictate future emergence of new products – e.g. increased funding for cancer research 15–20 years ago has now resulted in a broad and deep pipeline of new anti-cancer therapeutics. The current NIH roadmap is another example of a concerted effort, focused on improving drug development processes. As another example, the recent increase in biodefense funding in the United States portends that newer anti-infectious disease therapies, including vaccines, rapid-response diagnostics and biosensor applications and products will emerge in a few years.

Approvals and streamlined FDA processes
Prioritization of specific disease areas by the FDA shortens product development times, increasing economic attractiveness. For example, many cancer drugs that show good efficacy are given priority review. Typical drug review times vary from 16 months to 24 months.

New laws and regulations (e.g. Orphan Drug Act; Bioterrorism Act, pediatric extension)
Just as regulatory acts can change market attractiveness, the legislative bodies in many countries play a major role in affecting market conditions for specific areas and it is advisable to stay alert to any legislation in medical areas, such as the Medicare Modernization Act of 2005, which extends Medicare coverage to prescription drugs. Also, the Orphan Drug Act of 1983 gave special benefits to drugs for small populations, including grants and market exclusivity on approval. The IVD industry has been affected by changing regulations both in reimbursement (government laboratory reimbursement fee schedules) and regulatory guidelines for new products.

2.4.2   Hurdles

FDA regulatory delays and inconsistency
For example, drug development includes a review period of about 18–24 months between submission of final data from clinical studies and possible approval by the FDA. The FDA has usually always tried to balance the risk/benefit ratio while making approval decisions, but the weights are inconsistent among therapeutic areas and may vary depending on the current leadership of the FDA.

Increasing buyer power
As HMOs, pharmacy benefit providers and other buying collectives increase in scale, their purchasing power and ability to negotiate discounts and enforce price ceilings becomes greater, decreasing the profitability of manufacturers in the medical device and drug industry.

HMOs, Medicare, insurance
The price that an innovative medical device or drug receives in the marketplace is set by a complex process and manufacturers must work closely with insurance companies to make sure they get a fair price.

Pricing pressures
An increasing cost burden of healthcare on companies and individuals, political and social perceptions, inefficient payment and oversight in national healthcare systems, rapidly increasing prices of certain innovative products all contribute to feedback on pricing pressures,

which are exerted at various points in the value chain in a multi-party payer system such as the United States. Awareness to issues surrounding pricing pressures will help a sound product positioning strategy.

Generic competition
Patent lifetimes and patent rights vary in countries and country-specific rules and regulations must be considered when planning global launch and commercialization strategies.

Pipeline of blockbuster products
Companies that aim for blockbuster products could lose opportunities for multiple products in smaller markets that could prove to be more profitable. The scale of large pharmaceutical and device companies forces them to continue to develop products that meet the needs of larger populations.

Development costs and low success rate
In the drug industry, innovative products face a seemingly impossible hurdle – a high cost of development along with a high failure rate – 1 compound out of 10,000 that entire screening makes it through to commercial launch. Sound business models reduce the risk by including commercialization strategies that have multiple product lines or multiple partnerships and diversified early-stage revenue streams. Some companies have a harvest as you go strategy, with early revenues from animal health products or sales in non-regulated markets.

2.5   The referral chain: developing market context and understanding customer needs

Project funding decisions are based on many factors, a key one of which is the size of the market. Once the project is funded to begin development, market research is used to understand the context of product application. With this context, appropriate product characteristics can be selected and weighed against each other. Any successful product fills someone's need – either implicit or explicit. *Understanding the context of that customer need is the key to a developing a good market understanding* and a good product design, development, and sales plan.

The market research methods outlined earlier can be used to answer the following questions:

– What is the referral chain[1] for the disease? Identify steps of diagnosis and treatment until resolution of the disease problem (examples are given later in this chapter and in the book). *This information can help define the value proposition of the new biomedical product.*
– Define the value chains for product development, industry supply chain, and patient care delivery. Who is your customer (purchasing decision makers) and the

---

[1] "Referral chain" is usually used to describe the movement of a patient's care from a primary care physician through referral to a specialist for further diagnosis or treatment. The term is used here to describe the patient's perspective and experience from first realization of problem to referrals to various diagnoses or treatments through to some level of resolution.

stake holders for the product? Who makes the payments back to your company for the product?

- What do patients need? What do caregivers need?
- How does the product meet their needs?
- How does the product get delivered to them?
- How much will they pay for your product based on alternatives or direct competitors?
- Are there issues with patient compliance or other customer acceptance issues (aesthetics or ergonomics of product, price sensitivities, or other impediments in purchasing process)?

2.5.1 Market context: insight into biology or disease pathology

What core element of a disease or medical problem is being addressed by this technology/product and what is the impact on the life of the patient (not just life span, but quality of life)?

Answering this question is a critical step that helps position the innovation in the right context. An exploration of market and technical context can explain the scope and potential of the innovation.

Box 2.5 Addressing disease pathology in the context of specific market application

Case example

The innovative technology is an organic retinal implant that integrates with the neuronal circuitry and provides some visual perception to a blind person while releasing drugs to help retinal cells grow from stem cells and eventually degrades, leaving behind a rejuvenated eye. Where is the market and what key problem is this technology solving? This technology/product seems to address two problems – providing sight for those who are blind and providing a stimulus for regeneration of a degenerated/damaged retina. Keeping a broad perspective while first stating the problem will help to reveal multiple potential markets. The next step is to get as specific as possible to the core benefit – there are different levels of "blindness": Is this going to help one who has been born blind, or one who is legally blind but still has some visual perception? Eventually, it is clear that this technology is intended for a specific type of blind person – one who has become blind due to a retinal problem, not due to a neuronal or cerebral problem. Thus, this product solves the problem of restoring vision to a person whose retinal cells (but not retinal neuronal cells) have been injured or damaged due to trauma or disease. However, for all regulatory product development, identify a specific grade or type of injury and perhaps age of patient based on pathology of the disease or trauma for maximum benefit, depending on more detailed knowledge of the technology and the science behind the product concept. One more step in thinking can help bring more

Box 2.5 (*cont.*)

insight – can the device adjust to any changes in size of the tissues surrounding the eye – if not, then we can only implant it in adult patients. Therefore, the context of usage becomes clearer – the implant device is indicated for adult patients with blindness caused by advanced retinopathy.

Box 2.6 Referral chain
Referral chain case example 1
A wound healing biotherapeutic is made of innovative living cell material that heals quickly without scars or pain. To treat minor skin wounds in daily life a "band-aid" costing $0.50 or less is the common solution. In that context, introducing a new wound healing innovation that heals the wound much quicker but has a limited shelf life (user has to go to pharmacy to buy the product as it cannot be stocked at home) and a cost of $15.00 per wound will demand a shift in economic and usage context that may not be sustained by the market, despite any obvious benefits of the new innovation. "Price sensitivity" of the market is a term used to describe this issue. More market research, perhaps with focus groups, may be necessary to identify better the feasible applications and the segments of wounded people that would benefit from and thus use this innovative living cell skin healing material.

Referral chain case example 2
The innovation is a solution and complex apparatus for perfusion used to incubate an organ before implantation and extends the life of the transplanted organ. In the case of a heart transplant, if the innovative treatment improves long-term survival of the heart but does not change the cost of the procedure itself, the long-term significance may not be as vital to the surgeon doing a short-term procedure. However, the cardiologist who has to care for the patient for years after the surgery will be more interested in bringing this novel treatment in place. Additionally, the cost of treatment over the life of the transplant patient is decreased as the transplanted organ and recipient will be healthier. With $750,000 to $1 million invested into each heart transplant patient, the parties impacted by this new process are the patients themselves, the insurance companies, the cardiologists who care for these patients, and society as a whole. The innovation now must be sold/ marketed to a customer audience that includes not only physicians but also insurance companies that pay for these procedures and the hospitals who likely will pay for the perfusion equipment out of their total reimbursement for the surgery. Thus, understanding the various parties involved in the referral chain as the patient passes from one physician to another gives a better understanding of which physician or other caregiver is affected by the innovative intervention. Marketing the treatment to the consumer may also have some influence on the surgeon's adoption of the apparatus in their practice.

2.5.2   Market context: the referral chain

Evaluate all transactions that start with the end users' realization of a need and end with the fulfillment of that problem or need. This is the referral chain, within which the economic impact can be evaluated. For example, start with the diagnosis of a disease or health problem and step through all the transactions that occur until the end user/ patient sees a final resolution of the problem.

Step 1: determine the condition that drives the first interaction with medical care. Then follow the treatment choices available to the caregiver for that condition. If the treatment branches into various degrees of resolution (early resolution for mild disease), keep to the referral path (diagnosis and treatment path) that is of greatest importance to the innovative intervention.

Step 2: Identify the point of intervention with the innovative product and the cost of the intervention

Step 3: Identify the altered referral path (diagnosis and treatment path) for the patient post-intervention until resolution.

Step 4: Document as far as possible the likely time to healing, savings in materials, other savings in the entire economic system (e.g. reduction in lost time at work due to bedrest).

Step 5: Does the delivery of the innovative product give a specific benefit to any stakeholder to create an incentive for them to adopt or pay for this innovation? Identify and explain this benefit.

These steps will result in an identification of the key value proposition for the new product for multiple stakeholders and a clear economic benefit that can be used to convince payers to cover and pay for the product.

Box 2.7 Market stakeholders and adoption of new technology

Excerpted and adapted from a class report written at the Lally School of Management and Technology, Rensselaer Polytechnic Institute, by Peter Ryan, Brian Monthie, Aaron Germain, and Grant Cochran.

Referral chain example case 3

The innovation is a novel diagnostic technique that uses a novel noninvasive detection mechanism (details are irrelevant in this example) to assess the quality of bone material in a healing fracture. This device has been shown to detect changes in fracture healing with great sensitivity and accuracy. The device is being evaluated for use of early diagnosis of a non-union fracture. Current medical practice and insurance companies regulate the classification of a fracture as non-union – after 3 months of standard fracture care have failed to heal a fracture. Therefore, the referral chain is as shown in Figure 2.1.

Box 2.7 *(cont.)*

NON-UNION FRACTURE REFERRAL CHAIN

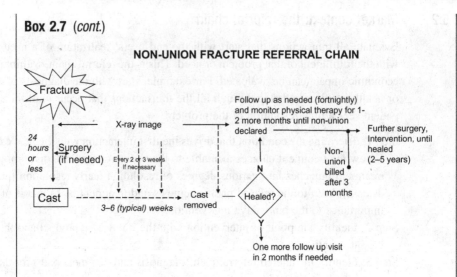

Figure 2.1 Referral chain (diagnosis and treatment pathway) for non-union fracture. The physician has to wait until 3 months post-fracture to be able to establish diagnosis of a non-union fracture.

A common timeline for a simple fracture starts with the patient presenting the broken bone, let's say the wrist, to the emergency department. When a patient comes to the emergency room with wrist pain, and evidence of a possibly broken wrist, the first step is to obtain X-rays of the injured area. If there is a broken wrist, the X-rays will be carefully reviewed to determine if the fracture is in proper position, and to assess the stability of the bone fragments. If the bone fragments are in proper position, then the wrist will normally be placed in a cast. The patient will then see an orthopedic surgeon within the week to discuss the accident and to get an overall health update. In the case of a simple fracture the patient could be seen by their family doctor every 2–3 weeks. X-rays will be taken to monitor fracture healing and the cast may be switched out if desired by the patient, or if the fracture has healed significantly. Around 12 weeks an additional X-ray image will be taken to determine whether proper healing has occurred. At this time the patient may be permitted to ramp back up to normal activity. Non-unions are medically defined as fractures that do not show any clinical progress in healing over three consecutive months. "Clinical progress" is specifically radiographic and a fracture can be classified as a non-union around the three-month mark. The American Academy of Orthopedic Surgeons (AAOS) has this to say regarding non-union diagnosis:

> To diagnose a non-union, the doctor uses imaging studies. Depending upon which bone is involved, these may include X-rays (radiographs), CT scans (computed tomography), and MRIs (magnetic resonance imaging). Imaging studies let the doctor see the broken bone and follow the progress of its healing.

Box 2.7 (*cont.*)

A non-union may be diagnosed if the doctor finds one or more of the following:

- Persistent pain at the fracture site
- A persistent gap with no bone spanning the fracture site
- No progress in bone healing when repeated imaging studies are compared over several months
- Inadequate healing in a time period that is usually enough for normal healing

Blood tests may also be used to investigate the non-union's cause. These could show infection or another medical condition which may slow bone healing, such as anemia or diabetes.

The cost of diagnosing a non-union is substantial. X-ray imagining is relatively inexpensive ($60) but an MRI or CT scan costs roughly $1200. The AAOS recommends multiple MRI and CT scans as well as X-rays for diagnosis. This, it may cost in the upward of $5,000 just to diagnose the non-union. If it takes 6 months just to diagnose and possibly another 6 months to a year to treat this injury, the patient will likely miss work and other important activities. Throughout the diagnosis process there is also a time strain on the physician who is not getting compensated any more than for a simple fracture even though multiple office visits are involved, as each fracture has a flat fee assigned to be paid. Once the case is diagnosed as a non-union this will allow the orthopedist to bill insurance for the visits associated with the condition. Non-unions adversely affect insurance companies also because the treatment of these conditions commonly calls for costly procedures.

The tibia is the most common site of non-unions. Approximately 72,000 tibial non-unions occur every year in the United States. Some tibial cases respond well to non-operative treatments like bone growth stimulators, but most need additional surgery. In surgery internal fixation using plates and screws or intramedullary nails are used to immobilize the fracture site (too much movement is a common cause of non-union and can result in the development of a mid-bone joint called a pseudo-joint), whereas bone autografts, allografts, and void fillers can fill in the gaps. Some surgeons favor the use of bone morphogenic proteins within the tibia to stimulate more growth. Each of these procedures would cost between 10 and 20 thousand dollars and is just a small portion of the total costs – a non-union could take from 24 to 36 months to heal. 36 months of monthly office visits, physical therapy etc, could add up to well above $50,000. Insurance companies should have a significant financial incentive to have a non-union be detected earlier and treated aggressively.

Benefits of intervention: The innovative diagnosis technology will be applied starting at 1 week after the fracture occurs and will be used to track the healing

> **Box 2.7** (cont.)
>
> process of the bone tissue. It will be used to complement X-ray images, but its use should reduce the number of X-ray images that need to be taken. As the technology will be approved for early diagnosis (at 2 months instead of 3 months, for example) of non-unions, the opportunity for earlier surgical or other therapeutic intervention should eventually reduce the healing time for a non-union (due to earlier and more aggressive intervention). The stakeholders that have an incentive to adopt and pay for this treatment are the physician, who can bill for the extra time taken to work with non-unions and the insurance companies (payers), who will see financial benefits from lower costs required due to earlier healing of the non-union. Society benefits generally with earlier healing due to reduced loss of productivity at work and patient is happier to have an early resolution of the problem, which in the past (before this diagnostic innovation), would have taken much longer to resolve.

2.5.3   What competitive or alternate products exist?

An assessment of competitive products must also include radically different technologies that are treating the same problem – for example, while evaluating competition for a new drug for headache, consider all of the following alternatives that a patient might have – an ice bag on the head, rest or sleep, and medicines such as Tylenol and aspirin. For a highly innovative product, consider the biology/pathology of the disease and consider the point of intervention of your product compared to the intervention points of other products.

2.5.4   Defining the end user

The status and role of the end consumer in the value chain (the patient) is very different in medical technology markets as compared to other industries. The end consumer is usually not a fully informed one and does not always have full choice in the marketplace. For example, an average patient scheduled for a hip replacement does not usually know which artificial hips are available or which one is best for them. In other cases, patients with high blood pressure do not know which drug and dosage are best for their particular situation and cannot thus exercise fully informed freedom of choice in the market. Additionally, many insured patients do not know the true cost of the treatments, as they typically make only a co-payment with private insurance, or, in some countries, a national health service program covers all costs (geographic context). There may also be constraints unknown to the patient, such as a limited formulary available due to insurer's decisions. Is a large institution the primary purchaser of the product or are individual users or physician groups directly purchasing from you? Analysis of the market needs for the former will include different

Box 2.8 Competitive product assessment

Competition case example 1

An innovative drug treats the enlargement of the heart in congestive heart failure, by stopping progressive hypertrophy of the heart muscle cells (cardiomyocytes). Its competition will come from devices in development that can prevent the enlargement (i) by providing electrical stimulation to the heart muscle or (ii) by wrapping the heart in a custom-fitted tight glove-like device, or (iii) by providing a left ventricular heart assist device that can be implanted long-term to take the load off the heart and allow it to recover. Other possible competitors also include new and existing drugs that can help improve the heart function allowing it to maintain size and function.

Competition case example 2

Excerpted from a class report written at the Lally School of Management and Technology, Rensselaer Polytechnic Institute by Oscar Perez Prieto, Disha Ahuja, Sam Christy and Linda Yin Chen

Background

The transcranial ultrasonic Head Check is a novel application of low-frequency ultrasonic waves for the detection of brain injury. The underlying concept behind this device is the use of ultrasonic waves to detect disruptions in the right/left symmetry of the normal brain after a contusion has occurred. The novelty of the device is in the application of its method and not in the technology per se. The main benefit of this method is the creation an accurate and portable traumatic brain injury (TBI) detection device that detects trauma rapidly and on-site, leading to rapid triage and more success in treatment.

Medical indications

The transcranial ultrasonic "Head Check" device is a portable device that detects the shift of the midline of the brain or other central structures from the sagittal plane of the head in a human brain after contusion or whiplash effect has occurred.

Markets, competition, and applications

According to the Centers for Disease Control and Prevention, US Dept of Health and Human Services [Ref: report to Congress by CDC, 2015], approximately 4.8 million Americans suffer from traumatic brain injury (TBI) annually. Of this number, 235,000 are hospitalized and in 2010, approximately 52,800 died as a result of their TBI injuries. Some of the common types of brain injuries include contusion and whiplash-type injuries associated with the rapid acceleration and deceleration of the head can result in swelling of the brain. If the pressure within the skull is not relieved through surgery, cooling or medication, the brain will gradually be pushed against the skull, causing the death of the patient. Studies have shown that accurate diagnosis within the "golden hour" after trauma can help in guiding further treatment with successful outcomes. Delayed diagnosis can lead to brain injury, lifetime disabilities, or death. There is a need for early diagnosis in

Box 2.8 (*cont.*)

this indication and current market. Other emerging technologies that claim to offer similar benefits, that is, to provide information about hemorrhages in the brain within the "golden hour," are described below, and the advantage of the "Head Check" device is summarized in Figure 2.2.

Glasgow Coma Scale (GCS)

The GCS is the most widely used scoring system in quantifying the level of consciousness following traumatic brain injury and is based on a subjective assessment of three aspects of the patient: eye opening, verbal response, and motor response.

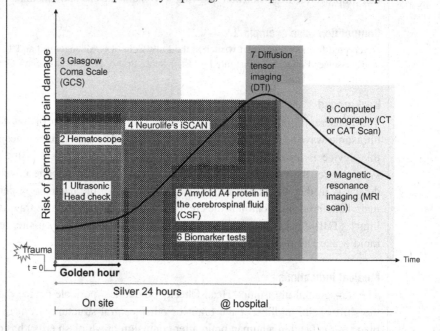

Figure 2.2 Market advantage assessment of Head Check's new product compared to alternate technologies.

NeuroLife's iSCAN

iScan uses fiber optics and proprietary software to measure changes in the eye's blood supply. Small changes in the blood supply to the eye correlate with brain pressure.

Hematoscope

A hematoscope is a hand-held, non-invasive, near-infrared (NIR)–based mobile imaging device to detect brain hematoma at the site of injury

Other technologies such as CT, MRI, biomarker tests, and diffusion tensor imaging (DTI) can be considered complements rather than substitutes to the transcranial ultrasonic Head Check, as they are meant to be used in the hospital setting. These complementary technologies are used after the golden hour as further assessment of the extent of the injury.

considerations, such as portfolio of purchaser or economics of the intermediate supply chain to final consumer.

Therefore, to understand the market for medical products, the researcher must address a broader spectrum of end users - the patient, the caregiver, and the reimbursing institution - in market research.

2.5.5 Defining the indication

An "indication" for a drug or device refers to the use of that drug for treating a particular disease. For example, diabetes is an indication for insulin and a severely arthritic knee joint is an indication for an artificial knee joint. Another way of stating this relationship is that the product (insulin) is indicated for the treatment of the disease (diabetes). *The identification of an indication is a key first step in formal product development of FDA regulated technologies.* Another way to look at an indication is to identify the target market segment.

For example, if a drug is being developed to treat heart failure disease, it could be indicated specifically for congestive heart failure class III, which defines the specific degree of symptoms that the disease has reached when the drug can be prescribed. If indicated for class I heart failure, there will be different criteria for regulatory approval, and the market size will change. As another example, in breast cancer treatments the market is segmented based on the specific indications under which a drug is developed or approved. Breast cancer is the general disease, but specific forms and stages of the disease are indicated for use of approved drugs. Herceptin, as a single-agent drug, is indicated "for the treatment of patients with metastatic breast cancer whose tumors over-express the HER2 protein and who have received one or more chemotherapy regimens for their metastatic disease" (https://www.accessdata .fda.gov/drugsatfda_docs/label/2010/103792s5250lbl.pdf). Or: "Arimidex is indicated for the treatment of advanced breast cancer in post-menopausal women with disease progression following tamoxifen therapy" (https://www.accessdata.fda.gov/ drugsatfda_docs/label/2011/020541s026lbl.pdf_/). Or: "Xeloda is indicated for the treatment of patients with metastatic breast cancer resistant to both paclitaxel and an anthracycline-containing chemotherapy regimen or resistant to paclitaxel and for whom further anthracycline therapy is not indicated" (https://www.accessdata.fda .gov/drugsatfda_docs/label/2015/020896s037lbl.pdf). All three are breast cancer drugs but address slightly different segments of the breast cancer market.

A device is also approved or cleared for a specific indication. Take for example, this approval for a balloon catheter:

This device is indicated for the dilatation of stenoses in coronary arteries for the purpose of improving myocardial perfusion in those circumstances where a high-pressure balloon resistant lesion is encountered. In addition, the target lesion should possess the following characteristics: discrete (<=15 mm in length) or tubular (10–20 mm in length) with a reference vessel diameter ranging from 2. 0 mm to 4. 0 mm; readily accessible to the device; light to moderate tortuosity of proximal vessel segment, non-angulated lesion segment (<45 degrees), smooth angiographic contour; and absence of angiographically-visible thrombus and/or calcification. (www.accessdata.fda.gov/scripts/cdrh/cfdocs/cfPMA/PMA.cfm?ID=2595)

Another balloon catheter was approved for the following:

... indicated for balloon dilatation for the stenotic portion of a coronary artery or bypass graft stenosis for the purpose of improving myocardial perfusion. The balloon dilatation catheter (balloon models 2. 5 mm - 4. 0 mm) is also indicated for the post-delivery expansion of balloon expandable stents. (http://www.accessdata.fda.gov/scripts/cdrh/cfdocs/cfPMA/PMASimpleSearch.cfm?db=PMA&id=8034)

The indication for the product, as worded by the FDA in their approval letter, restricts the marketing, prescription, and sale of the product only to patients with that particular indication. The wording of the indication is derived primarily from the results of the studies that product developers design to test the efficacy and safety of their products. In the above examples of cancer therapies, the drug developers cannot claim in any marketing message that their product can be useful for any other state of breast cancer other than that indicated in the approval letter.

Hence, it is important to define the population and specific benefit of the technology/innovation as best as possible in the early marketing research (Boxes 2.9 and 2.10). Marketing research can be used to identify the most beneficial potential application of the product and the medical need it fills. A manufacturer must define this indication as early as possible in the development process in order to better understand the market and to guide product development with greatest efficiency.

2.6   Market research in the context of medical device design and development

Output from market research feeds into the medical device design and development processes in specific design parameters, where there is a constant balancing act between features and cost of the product (Figure 2.3). Quality has to be maintained throughout this process, but data from market research can be used to assign weights to parameters for decision-making between cost and features of a device. This is done formally through user or patient interviews, surveys, and internal engineering planning (see Chapter 5 for more details and examples of the use of market research in product development).

2.7   Summary

A major challenge in the biomedical industry is to identify the actual customer for the product or service. This chapter has offered perspectives on how to go through a referral chain pathway to identify the customer, most-affected stakeholders, and the actual economic impact of a new technology or intervention. Taking your own project

Market research information

Figure 2.3 Market research data impacts new device development by influencing choices between cost of product and features. In new drug development, market research information can steer the product toward more profitable indications or can suggest a market expansion strategy, or help make choice of product characteristics like oral or infused delivery.

Box 2.9 An insulin pump – referral chain, indication, and basis for competition
Case study

As an example, consider the *referral chain and specific indication* for application of an insulin pump. A company decides it has some innovation in insulin pump design and fabrication and wishes to enter the market. A full featured insulin pump that is technologically superior to its competitors seems like the best way to enter and capture the huge diabetes market. But a little research into the application *referral chain* will reveal a different picture, also indicating the segments in the market. Most diabetic patients do not immediately go to an insulin pump prescription after being diagnosed with diabetes. Most of the patients will be treated by their physicians first with oral medication or combination of various orals. Failing to respond to that, they will have insulin added to the oral medications and then multiple daily insulin injections will be tried to control their disease before the doctor suggests an insulin pump. Finally, an insulin pump has few differentiators for increasing competition for this small group of patients. Even though the overall diabetic patient population is increasing, better oral medications will mean that fewer patients will reach a stage that is indicated for insulin pumps. Lastly, the fact that competition will increase in the insulin pump business means that eventually the competition will turn into a cost-based competition with commoditization of the insulin pumps. Therefore, a new entrant will be better suited to give more weight to cost than to features during product

Box 2.9 (*cont.*)

design, giving the product a better chance of success by the time it gets to market (Figure 2.4).

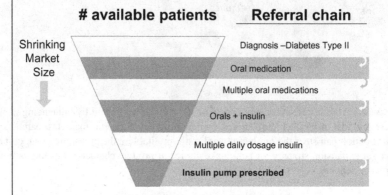

Context – diabetic patient

 # available patients   **Referral chain**

Shrinking Market Size

Diagnosis –Diabetes Type II

Oral medication

Multiple oral medications

Orals + insulin

Multiple daily dosage insulin

Insulin pump prescribed

Final indication – diabetes Type II patients who are refractory to insulin or other medications

Maintenance/management of the disease
Competition = commoditization

Figure 2.4 As the referral chain for the disease is identified, the market segments get defined and the true available market becomes clearer. Additionally, this market segmentation can help guide clinical trial design. For example, after initial approval in the last segment, a market expansion strategy might be planned to move up to the next largest market segment.

Box 2.10 Percardia Inc. learns about the market

Contributed by Laurence Roth, Vice President, Business Development and Operations, Percardia

Case study

Percardia Inc was developing a unique system, VPASS™, for treating advanced coronary artery disease (CAD). In the early stages of CAD, patients are typically treated with drugs to improve heart function, reduce the risk of acute closure of a vessel due to thrombus formation (resulting in a myocardial infarction or "heart attack") and to mitigate the pain (angina) associated with reduced blood flow. If medication is unable to provide relief of symptoms, physicians will attempt to restore blood flow to the heart by either catheter-based techniques (inserting a stent to prop open the obstruction) or surgery (creating a bypass around the obstruction

Box 2.10 (*cont.*)

with a vessel graft). VPASS is a catheter-based procedure that supplies blood to the heart by connecting the left ventricle (the heart's main pumping chamber) to the coronary vasculature with an implant placed directly in the heart muscle. It is intended to allow the interventional cardiologist to treat patients who are otherwise not amenable to current therapy; these patients are commonly referred to as having "refractory angina."

Although physicians were universal in agreeing there is a strong unmet need for a therapy (such as VPASS) to treat refractory angina, literature-based estimates of the number of refractory angina patients varied from a low of 75,000 to a high of 800,000 patients annually. In addition, initial clinical evaluation of the VPASS system was proceeding much slower than anticipated, creating questions about the potential market. An accurate understanding of the market size and patient referral process was required in order to develop an effective commercialization plan.

Finding that there were no commercially available research reports that analyzed the refractory angina patient population, Percardia contracted a firm to perform a semi-quantitative market research study. The firm combined primary research through direct physician interviews with secondary research through literature reviews to characterize the refractory angina patient population and to provide an understanding of the referral chain in the management of this target patient. This research led to several key findings:

- The refractory angina market in the United States consists of ~400,000 patients with an additional 45,000 patients diagnosed annually.
- Only those patients with the most debilitating symptoms (40% of the total or 160,000 patients) would be considered candidates for VPASS.
- Initial clinical studies should target physically active patients considered most likely to benefit (40% of candidates or ~64,000 patients).
- Significant long-term data would be required to expand the use of VPASS in patients with less debilitating symptoms.
- Additional indications were identified providing a long-term pool of up 750,000 potential patients.
- Primary care of refractory angina lies with either the clinical cardiologist or primary care physician while the cardiac surgeon and interventional cardiologist perform treatment.
- Clinical trials and initial marketing efforts should focus on the clinical cardiologist as the clinician who will recommend VPASS.

Understanding your market is critical to the development of a viable business plan. By identifying patient flow in the referral chain and market segmentation, Percardia was better prepared to design effective clinical studies and market launch plans.

through the market segmentation exercises and asking the questions "So what?" and "Who cares?" will help crystallize a value proposition for the product. The importance of identifying an indication is also clearly explained and tools to achieve this are presented. Market research drives business planning and is invaluable if done thoroughly not only for successful launch of a product but also for a product's lifecycle management.

Exercises

2.1. In simple terms, briefly describe the invention and summarize the significance of the underlying technology innovation.

2.2. Define clearly the specific market need or core problem that your technology addresses.

2.3. Identify the specific indication for your product and identify other potential indications for future development.

2.4. Also brainstorm to identify non-medical or unregulated applications (animal health, other industries) for your technology (if so needed in the business and financial model, the product can be commercialized in a market where time to market may be shorter and revenues might help fund medical product development).

2.5. Describe the referral chain for the indication and the economic context (costs).

2.6. Describe the patient population and identify the payers (also see Chapter 8).

2.7. Identify the purchaser, user, prescriber, and other stakeholders who are affected in a value chain for your product (e.g. other stakeholders could be nurses who will see the patient each week for 4-hour infusion of the drug).

2.8. Describe the benefit from the new product to each stakeholder or to the primary purchasing decision maker.

2.9. Describe the savings in the entire referral chain from the intervention made by the new product. This will also help define the actual economic value proposition for the insurers/payers/purchasers.

2.10. Write, in one sentence, why your technology will be purchased over others and over current treatment options – in other words, define the value proposition for your technology/product for the chosen indication in one sentence (limit the punctuation marks).

2.11. Identify competitors based on the specific problem the new product/ technology is addressing. Describe very briefly why a certain competing solution is a competing technology/product.

2.12. Classify alternatives and competition in a comparison table highlighting advantage of your product considering appropriate parameters – efficacy, safety, price, etc.

2.13. Collect specific value propositions that the stakeholders will value and translate those into product characteristics that would be appreciated by the stakeholders. Give more weight to the primary purchaser's perspectives.

(cont.)

This list of preferred product characteristics will serve as input to a product design.

2.14. For a market research project follow these steps:
   a. Define the specific goals of the market research program – Examples include: (i) Define and segment markets in order to clarify reimbursement strategy; or (ii) Define potential market opportunity in order to drive investment decision; or (iii) Provide input on specific market needs for better defining desirable product characteristics.
   b. Decide on the extent of primary vs. secondary research to be carried out and make sure adequate resources are available
   c. Collect data and tabulate results without any bias entering the summary tabulation
   d. Analyze the data and then present the specific output required – based on the defined goals of the market research project.

References and additional readings

BCC Research Report. (2021) In vitro diagnostics: technologies and global markets. #HLC186C. BCC Publishing. www.bccresearch.com/market-research/healthcare/in-vitro-diagnostics-technologies-and-global-markets-report.html

Bloomberg Business Week. (2001, August 12). Biotech and the spoils of success.

Centers for Disease Control and Prevention, National Institutes of Health, Department of Defense, and Veterans Administration. (2013). Report to Congress on traumatic brain injury in the United States: understanding the public health problem among current and former military personnel. Atlanta: Centers for Disease Control and Prevention. (www.cdc.gov/traumaticbraininjury/pubs/congress_military.html)

Denault, J-F. (2017). *The Handbook for Market Research for Life Sciences Companies: Finding the Answers You Need to Understand Your Market*. Boca Raton, FL: CRC Press. ISBN: 9781351773492

Evaluate Medical Technology report, World Preview 2018, Outlook to 2024. (2018). Evaluate Ltd.

EvaluatePharma report: World Preview 2020, Outlook to 2026. (2020), pages 26–27. Evaluate Ltd. www.evaluate.com/thought-leadership/pharma/evaluatepharma-world-preview-2020-outlook-2026

Fortune Business Insights. (2021). In vitro diagnostics market size, share, and COVID-19 impact analysis with regional forecasts 2021–2028. Report ID FBI101443. www.fortunebusinessinsights.com/industry-reports/in-vitro-diagnostics-ivd-market-101443

Frost and Sullivan. (2017). Global medical device industry snapshot, 2017: new era of healthcare transformation set to disrupt medical device industry. Report Code: MC3C-01-00-00-00. https://store.frost.com/global-medical-device-industry-snapshot-2017.html

Industry research Biz. (2021). 2022–2029 global in-vitro diagnostics (IVD) professional market research report, analysis from perspective of segmentation (competitor landscape, type, application, and geography).

Osterwalder, A, Pigneur, Y, Bernarda, G, Smith, A, and Papadakos, T. (2014). *Value Proposition Design: How to Create Products and Services Customers Want*. New York: Wiley. ISBN: 9781118968055

Ulrich, KT, and Eppinger, SD. (2003). *Product Design and Development*, 3rd ed., New York: McGraw-Hill. ISBN: 0072471468 (Note – this book also has a practical section on working out client/customer needs through various steps of market research.)

Yock P, Zenios, SA, and Brinton TJ (eds). (2015). *Biodesign: The Process of Innovating Medical Technologies*. Cambridge: Cambridge University Press. ISBN: 9781107087354

Visit https://shreefalmehta.com/csbtbook for additional enriching readings around the topics covered in the book, topical updates on the content and for industry viewpoints and news.

3 Starting up your company

Plan	Position	**Pitch**	Patent	Product	Pass	Production	Profits
Industry context	Market research	Start a business venture	Intellectual property rights	New product development (NPD)	Regulatory plan	Manufacture	Reimbursement

---

Learning points

- How do you build a team for your new business?
- What is in a business plan?
- What are investors looking for in a company business plan?
- What key steps should you consider when starting a company?
- How to structure equity sharing amongst founders?
- What is the use of a board of directors?
- Where to raise financing at different stages?
- What is a business model and how many flavors does it come in?
- How to manage personnel issues as you grow?

Many excellent books and online guides cover the writing of business plans, creation of business models, preparation of investor pitches, and consideration of the legal aspects of incorporating a new business (see list of References and Additional Readings). This section, however, focuses on the key practical aspects and issues to consider when starting a new business. Reading this chapter will give the entrepreneur some thinking points and lead to asking the right questions for achieving greater success in structuring and growing their company.

The first step to starting a company, or starting to write a business plan, is to bring together a team. With a good team you can make magic happen even if your idea turns out to be a pumpkin. If the innovation is excellent, like a top-of-the-line, superbly balanced canoe, but the team is not capable and cannot move in one direction together, the canoe has a poor chance to survive any perturbation, let alone make it to the finish line. So, let's discuss the team formation and composition first before we look at starting a business.

3.1   Building the team

3.1.1   How do you start attracting and building a team in a startup company?

The very simple way is to start talking about the concept or innovation with people around, especially people that have some experience in starting companies. The excitement about the concept will be contagious and these people or others will connect you to others. In this way, the founders typically meet or are introduced to people who voice their interest in helping join to further develop the concept.

Selection at the management level can be based on gut feeling or chemistry with the people, as well as logical fit within the gaps in the founders' skills or a fit for the company's future needs. The founder has to be a credible and passionate salesperson from day one forward (and continuously thereafter). People will naturally be attracted to exciting technology or visions of change espoused by the company mission and prospects. In my experience, hiring people and assigning roles on the basis of aptitude, not experience or degrees alone, pays off in a growing company. As the company grows very quickly in the first few years, functions and roles are likely to change; therefore, if people have the aptitude to continue learning and can retain a strong sense of self-awareness, they will continue to grow with the company and add value.

A very important marker for successful hiring is the fit of that person with the company culture (and its founder[s]'s values). As startups, many companies wonder how to define their culture as it may be too early to spell it out or codify it, but most founders will be able to voice their personal value systems and goals as a starting point. Culture is not created by inserting a few points into a poster or presentation, but is built every day in how the leaders and others act and react. Learning what that means and how the company sees its path to excellence and success allows one to identify quickly those that do not seem to have a fit with that culture. For a young company, ensuring that recruits fit with its values is a key marker for long-term success.

Another way that works to recruit people or co-founders is to pitch the concept in business plan competitions or pitch days at your local incubator or startup forum. Local business schools usually have entrepreneurship pitch competitions or programs where business students pick university technology to develop business plans. These programs are useful to develop ideas and find a team or co-founders that bring in a different perspective or some business-world experience. Here is an example from personal experience: A medical doctor who had an idea to solve a serious unmet clinical problem but did not know quite how to proceed, met the local university incubator director and was introduced to an MBA student with biomedical research experience. The two decided to work together, attracted by the possibility of solving a real problem that would save lives and founded a successful biopharma company. Their efforts were fruitful, and they were able to survive several "near-death" situations for the company, because they adhered to the joint vision and mission that had founded the company, and developed and demonstrated mutual respect for each other's skills and perspectives. They found they had enough of a cultural fit with each other's value systems when it came to decision making in the company, even though they both came from very different ethnic and social backgrounds. Thus, when they contributed diverse ideas each time when it seemed they had hit a dead end, there was a willingness to listen, evaluate and adapt.

3.1.2　Who should be on the team?

If the founders are from technical backgrounds (technopreneur), it makes more sense to try and attract a co-founder from a different background, especially if the other person brings experience in building companies, or is thinking predominantly from the business or the commercial perspective (market perceiver). The value of having a different perspective brought in by people from different backgrounds (economic, cultural, education degree) cannot be overstated and should be a part of recruitment strategy at all levels, from board of directors' members to management and staff.

A technopreneur and market perceiver make an excellent team, as these two distinct roles effectively represent very different, but complementary, functions and perspectives at the various stages. An invention or discovery is often the starting point for the biomedical technology–based company, and in many cases, the invention is like a new hammer, looking for a nail. The market perceiver can help identify the entry markets and narrow down the product direction, while the technopreneur can describe the potential for the technology to meet various needs. At each stage of development of the company and product, these two roles have something specific they add with their different perspectives, as seen in Table 3.1. Balanced by these two complementary and differing viewpoints, the company has a better chance for successful outcomes.

3.1.3　Hiring a CEO

In many cases, the founders manage the early days of the company formation themselves without adding outside executives. Sometimes, the founding CEO can grow and learn as they manage the growing enterprise and people, and a good board can help that transition. However, after the company achieves a major milestone and raises larger financing from institutional investors, often, a growth-experienced CEO or other management is brought in to help ensure sound execution of the company plans and vision at a larger scale. For the founding CEO, it is helpful to discuss this possibility at an early stage, and it is a sign of maturity and self-awareness to prepare by defining a possible new role to continue in the company.

Some points to consider in the recruitment process and effort:

– Use an experienced advisor, investor, board member to help with interviewing and selecting a CEO.
– Hire a recruiter who has specific experience in your industry sector and ask them for references. Talk to the references about the recruiter's working style and process. Note carefully the terms of the agreement with the recruiting agency,

Table 3.1 Different roles of technical and business founders

Value-added perspectives from two complementary viewpoints

Venture stage	Market perceiver	Technopreneur
Recognize opportunity	**HAS:** A comprehensive knowledge of markets. Clear definition of product characteristics that are needed. **NEEDS:** To find appropriate technology to meet the market's specific needs.	**HAS:** A solid understanding of and expertise in specific, well-characterized technology. Established credibility with peers, investors, and customers (academic researchers, biotechs, or big pharma). **NEEDS:** To confirm that there is a market for the product. To define product characteristics.
Secure Intellectual Property (IP) rights	**HAS:** A clear understanding of the market application, so IP claims can be formulated easily. **NEEDS:** To comprehensively harvest IP portfolio.	**HAS:** A strong position to easily license their own invention from the university into the startup. **NEEDS:** To erase any perceived conflict of interest that may arise by being on both sides of license negotiation, as academic inventor and employee and company executive. An understanding of future IP needs; the inventor often overestimates the novelty of the invention and breadth of patent protection.
Fund team and build company	**HAS:** Business credibility in creating sound commercialization and business plan with clear market needs. **NEEDS:** To carefully evaluate timeline for technology development, balancing the attraction of large markets with a plan for growth as technology matures. Strong history of experience to overcome credibility gap if technology is licensed in without inventor participation.	**HAS:** Strength in early phases of company, where main efforts are focused on research and most of the personnel are technically oriented. Credibility with investors due to technical expertise. **NEEDS:** To manage investors' questions in business and commercial areas. To learn how to manage nonscientists.
Develop technology to product	**HAS:** An external perspective that brings a strong product focus to the development process. Focus on scaling up production to commercial manufacturing levels, market acceptance, and regulatory acceptance. **NEEDS:** Understanding of the transition from R&D to commercial manufacture – to manage the technopreneur's expectations.	**NEEDS:** Experience of commercial product development, particularly issues in scaling up. Unbiased perspective to evaluate the technology's realistic potential versus its elegance.
Survive	**HAS:** Sensitivity to the needs of the business and finances so as to strategize and manage IP assets astutely.	**NEEDS:** To understand that their appropriate position within the growing company may not be at the helm, but rather in a specific technical leadership position, such as CTO, CSO, or on the Scientific Advisory Board.
Market	**HAS:** Ability to deliver the market potential message to multiple stakeholders on their terms, put a commercial team in place, and strike appropriate business partnerships.	**NEEDS:** To shift focus from developing technology to building a strong commercial team speedily and efficiently.

Source: Table reproduced from Mehta S (2004). Paths to entrepreneurship in the life science. *Nature Biotechnology* **22**(12), 1609–1612, with permission.

especially clauses about guaranteed performance and no-cost replacement of hired candidates for a period of time after hire.

– Board members, advisors, startup company lawyers, and investors are all likely connectors to potential CEO candidates.

The background and experience of the CEO must be appropriately considered before hiring, depending on the type of product, stage of the company's growth, and level of financing raised to date. For example, if the medical device product is in final stages of clinical testing, it is important to hire a CEO who has had commercialization experience and has previously launched at least one related medical product before. As another example, if the company has preclinical drugs in development, it may be important to hire a CEO who has recently done partnerships with other biopharma companies and/or has experience with raising large amounts of capital.

A common mistake made by many startups is to bring in a mid-manager or director-level person from a larger company, thinking that their industry experience would be perfect for the company. The mindset of a mid-level or senior manager from a large company frequently needs to be readjusted for the startup environment, a transition and gestalt that sometimes do not work out. The dependence on a robust office infrastructure support for all minor or major matters and the bureaucratic processes that were part of their behavior/planning in the large company can make these senior executives from large companies a poor fit with the culture and environment of a smaller startup company. Not intended to be a broad indictment against this type of potential candidate, a thorough evaluation of the applicant, their aptitude to adjust, and their sensitivity to the difference in environments must be considered.

3.1.4   Why do we need a board of directors? when? and who should be on it?

The board of directors usually include (1) the representatives of the largest shareholders (usually investors or founders as non-executive directors), (2) the CEO and possibly other management (executive directors), and (3) independent, unaffiliated persons (independent directors). The directors primarily have a fiduciary duty to *all* the shareholders of the company and act on the behalf of shareholders/investors to oversee management. The board provides oversight for major strategic decisions made by the management team. Final approval for major changes in the corporate structure usually requires the board's approval – such as issuing new stock or options, accepting a new equity or debt financing, hiring or firing the CEO, and giving input to the management team on the company's stategic choices. Hence, it is important to choose board members prudently so they can guide and contribute to the growth of the company.

For an early-stage company when the main activity is R&D, there is little need to build out a corporate board, with one- or two-member boards being adequate to carry out formal functions of the corporation. At this early stage, an advisory board (without the fiduciary responsibility to shareholders) may be a useful construct to affiliate important people with the young company, such that their network or their fame can be tapped into to help the company's efforts.

The board can play a significant role in guiding and growing the business. The first-time CEO should recruit investors and independent directors who can give the energy and time to help the startup and who bring hands-on experience in building companies. Many board directors play an active role in helping with financings as needed or by opening doors using their extensive networks. Others with more hands-on experience in managing teams can provide guidance and emotional support for a young CEO and help resolve conflicts in a management team.

When a company brings in larger amounts of outside financing (usually by selling shares in the company), the new shareholders will usually appoint a person of their choice to the board as a condition of the financing. Thus, appointments to a board will naturally grow with the company. One way to set terms on board positions is to base board director membership on percentage of shareholding. If not managed well, the board could quickly grow and become cumbersome to manage for a small startup. Having an odd number of directors is helpful to ensure that there is no impasse in voting for controversial resolutions.

Some practical dos and don'ts about working with boards:

- Discuss and confirm the time commitment and availability of a director for a startup company during the initial interview and discuss expectations on both sides.
- For a startup with no revenues, the remuneration of the director should only be in equity (stock options with vesting schedules). An experienced company lawyer can advise management on current best practices.
- Do assign roles with the board such as compensation committee, personnel committee (e.g. to provide input on senior management hires) and the CEO should also feel free to request board members to take on appropriate strategic tasks to help with major initiatives (like M&A discussions etc).
- Ad hoc calls of the CEO with individual board members between meetings is helpful in order to understand each other's thought processes better, build trust, and follow up on agreed-on tasks.
- Prepare board members with sufficient level of detail before the board meetings and structure the agenda so that most important strategic issues are presented at the beginning of the meeting, giving enough time for discussion.
- Board meetings should promote two-way dialogue – reporting from management team to board and gathering feedback from board members.
- Diversity of thought and background is important while building a board.
- The CEO, representing management, should drive the board's tasks, manage the interactions, and create a culture of transparency and honest dialogue to get the best out of the expertise on the board.
- It is important to have structured meetings with agendas.
- Useful sometimes to have a corporate counsel attend the meetings as the funding rounds increase, most law firms will do this at very low or no cost (as the startup company will rarely have a corporate counsel in house).

3.1.5   Advisory board members

Placing an academic professor on the board of directors of a newly formed company is usually not a fruitful use of that professor's capabilities, especially if they have not been in that role before. Research scientists or medical physicians are best added to the scientific advisory board, where they do not have fiduciary responsibility to the shareholders and can freely offer guidance or oversight to the technology development or help with writing grants or making contacts. The advisory scientific board may also help the CEO by offering independent feedback on the CTO's or CSO's technical judgment. It may also seem very attractive to have a large corporation's senior executive on the board of directors, but most executives in large companies do not always understand a small company's culture and constraints or do not have the time needed to help the young, inexperienced entrepreneurs. A board of advisors can be formed to show the support of important names in the field and add credibility or bring specific relationships to the young company.

3.1.6   Roles in the startup company

In the early stages of starting the company, one executive will likely wear several hats. While people may feel capable of understanding and doing many diverse tasks, there must be a clear designation of the duties and responsibilities so that there is accountability for goals and objectives that need to be met. Given that in the frenzy and excitement of the starting days, the designated management roles will need to be flexible and adapt to the tasks at hand, including dealing with more routine tasks that may normally not be a part of the assigned role of the executive role. As the leader, make it a point to hire people smarter than yourself and demand accountability, but give these smart people resources and freedom (from micromanaging) to achieve the mutually agreed-on company goals and vision. Playing to the strengths of individuals in various roles will grow the organization's capabilities and allow it to meet challenging milestones.

However, sometimes adjustments are needed if the aptitude of an individual turns out to be a poor fit with their roles or with the culture of the startup company that requires collaboration. The main element for success in a startup is the team, and maintaining trust and mutual respect must be a key part of the cultural values of the company and management team.

One role that is needed from the beginning is that of a leader (CEO, or in some countries Managing Director) who is the voice and face of the company to current and potential stakeholders (outside and inside the company, e.g. interface with potential investors, current staff, new hires, board members). Very few companies can continue working through a team consensus mode with multiple people in the leadership role. While this may be possible in an academic setting, it is rarely a good fit for the functioning of a high-tech company. The other role that is needed from the beginning is the role of head of science (CSO) or product development, or (CTO) technology (or a VP of Engineering if the project is far enough along in the process of its development) and the rest of the roles are dependent on the circumstances. For example, many life sciences companies have just two management-level roles – CEO+CSO – for many years (sometimes occupied by the same person) while the company is carrying out prolonged R&D.

Some dos and don'ts with respect to roles:

- Do specifically assign roles and responsibilities, even though they may seem too formal in the beginning (even in a three-person company).

- Do follow through on accountability – even though many people may contribute and share in a specific task the one person functionally responsible for the task should lead, be making the major decisions, and be responsible for the outcome.
- Decision making in a small group of founders is usually accomplished by establishing consensus, and while it is important to have open dialogue and challenge decisions, decisions need to be made by the person assigned to that functional area. When a decision is made after due consideration for inputs, the decision needs to respected by the other management team members.
- Do build a culture of trust so that the management team comes together as a united group. Note that discussions and arguments can happen in private management team meetings; however, once completed and a decision made, the message is unified and followed through by all leaders in larger staff meetings. Disunity can hurt working toward goals, and the culture slowly may become one of mistrust and back-biting.
- The best laid plans of mice and men . . . when the company is struggling, do keep the roles and their responsibilities steady unless it is clear that someone is overwhelmed or incapable or has consistently shown poor judgement – in which case, the CEO or appointed leader can give direction or reassign responsibilities . . .

3.2   What's in a business plan?

A business plan lays out the vision for the company, with a set of signposts and milestones that describe the way that the company is going to be able to grow and become profitable. The business plan lays out the vision of why the world is better with the launch of that new venture and describes why the company is being formed or why it exists, and how it plans to grow and become profitable.

Today, most investors will review business plans presented as a collection of the following parts: (1) a one- or two-page executive summary document; (2) a slide deck that covers the business plan; (3) pro forma financial statements for a few years into the future; (4) biodata of the key-founders/management; (5) a second slide deck with technical data for secondary level deeper dive or diligence review; and (6) technical documents or patents reviewed in due diligence. Not many investors require a full written multipage business plan. However, the issues that must be worked through, planned for, and explained to stakeholders are the same as seen in an extensive written business plan, and the process of planning has great value for any startup team. Writing a business plan or preparing the above elements helps to explicitly identify assumptions and clarifies the vision of the founding team, for themselves first and then for others.

A business plan made with these components coupled together can be changed quickly and extensively as circumstances change, as new facts are learned, or as the company matures. A saying often ascribed to President Dwight D. Eisenhower makes the point: "In preparing for battle, I have always found that plans are useless, but planning is indispensable." Most experienced private investors or venture capital investors understand that in an early-stage startup company, the business plan is rarely executed exactly as written, but it does help them understand the maturity of thinking and degree of planning. In fact, a majority of the successful companies that have raised financing from venture capital pivoted to a different business model from the one that was funded originally. The success of a startup company hinges on the founding management team, and experienced investors make a decision to invest in a startup largely based on their appraisal of the founders/management team.

Once a market need is targeted and the innovative solution described, the following two parts serve as the base for a business plan: (1) a commercialization plan, which defines the market need, the development pathway to build and launch the product/ service, and an associated budget and resources; and (2) a financial plan that lays out a strategy to raise financing and meet budgetary needs to execute on the commercialization plan. Thinking through these two preparatory plans can help create a sound and credible business concept. The following important questions that the business plan must answer are discussed further below: What is the market and the need that is being addressed in that market? What is the competitive landscape and the competitive advantage? Who are the people leading or managing this business? What is the business model?

3.2.1   Key contents of a business plan

People: A business plan must include a short biography of each key management team member (and the founders). The plan should focus on showing why each person is the best suited for the role they play in the company. The overall purpose of this section is to establish the credibility of the startup management team in being able to execute the business. Be sure to include the board of directors or advisory board members, showing how they can contribute to the company's growth. For more detailed discussion on people, roles and team see Section 3.1.

Need and market: The specific market segment (Chapter 2) and market need should identify the specific compelling need or problem being addressed by the product or service and, in so doing, establish a compelling investment case for an investor. The evidence you provide must also describe a credible market size and demonstrate why this product/service will be adopted in this market and how it uniquely serves this market need. The market size is usually best described from a bottoms-up perspective (see Chapter 2). Note: it is imperative that the market need for the product/service proposed, is demonstrable as a 'must have,' not a "nice to have."

Competition: Include current state of medical practice and care/treatments available on market as current alternatives to be replaced with the new innovation. Includes specific competitors in development or on the market and other viable alternatives. Highlight the clear competitive advantage of the product/service offering and the basis of the advantage as perceived by the customer.

Identifying the customer: This is a fairly complex quest in biomedical products/services: the product may be purchased by a hospital but prescribed by a doctor or preferred by a patient. The actual user may never make a purchase directly from the manufacturer and may be dependent on the local formulary pharmacy benefit plan decisions, for example. The purchaser may not be the decision maker. The insurance company may restrict availability. See Chapter 8 on reimbursement and Chapter 2, Section 2.5.4 for more details.

Path to market: This is a slightly different approach than the business model, which is a more generalized view of the revenue model. This topic is all about establishing credibility around the first sale and the next steps to grow revenues. How are you going to access the customer? Can you show specific partnerships that you plan to initiate, or have you already obtained a letter of intent/interest from key distributors? What else needs to be done to reach the purchaser of the product/service? For example, in a company with a licensing business model where the first revenues are from licensing the product/service/technology, the path to market would highlight the potential licensees and why they are/would be interested and showing how many nondisclosure or confidentiality agreements (NDAs/CDAs) have been signed to date with such licensees. The path to market shows a skeptical investor that the management team understands how to reach its first revenues (a very important milestone for a young company).

Product/service: The product or service must be described in relation to how it uniquely addresses the market needs and must also be compared to alternate solutions along the parameters that matter to the customer. For example, if the product is a new drug-delivery method or formulation, explain clearly the problems encountered with dosing and administration of the current drug and the impact of those problems. That will frame the problem with the alternate solution. Showing results from the first prototypes or lab or animal tests, if available, would be greatly helpful.

Business model: The business model clearly establishes how the company will generate income in the future and should contain such elements as sales revenues, licensing income, and service contracts. This aspect is further discussed in Section 3.3.

Valuation/liquidity: This is a controversial subject. Valuations are highly subjective and yet important to communicate some threshold for the investors who may be lay investors in early stages of the company's fundraising. The most objective way to do this is to select comparable companies and state their valuation (at IPO, at acquisition, or current market value). Giving the investor some sense of current or future valuation is also accompanied by giving them some idea of when to expect a cash return (liquidity event) and at what valuation that could happen. A liquidity event or financial exit for the early shareholders is either an acquisition by a larger player who pays mostly cash, or a public listing of the company. Since investors are focused on the financial outcome, this is something they will all be looking for.

The business plan can be presented as written paragraphs in a document with references and data, or in slide format, or in some combination of the two. Many elements of the business plan are useful in presenting the company vision and status in a fundraising document called a private placement memorandum (PPM), used to convey these plans and bring additional investors into a financing round (see Section 3.6, Box 3.6).

3.2.2   The commercialization plan

The commercialization plan lays out the pathway and key milestones to achieve in order to bring a new product through development to a new or existing market. The output of the commercialization plan is (1) a recognition of key milestones for the company and (2) a critical path outline with the time and resources (budget, staffing functions, headcount, etc.) required to bring the product/service to revenue. Planning the manufacturing requirements and scale-up stages and processes, planning the clinical trial protocol and costs, understanding the regulatory path and gateway for access to the specified market, and going over all the other steps covered in the various chapters ahead will thus be a useful exercise. From this timeline of cash and resource needs, a financial plan can be built up. The commercialization plan and financials create, in this way, a bottom-up plan. A bottom-up plan is so called because each milestone is practically reviewed from the basic steps of how to carry out the plan, rather than taking a generalized view from the top down. This bottom-up planning as a basis for the business plan is much more credible to investors and gives the founders a sound base from which to address questions from all company stakeholders, including staff and advisors.

3.2.3   The financial plan

The financial plan gives investors information about the cash investment: how much is needed and when. The plan correlates that funding schedule and budget to key milestones in a credible way and describes quantitatively the business model by which the company will earn revenues and reach profitability. The financial plan may also identify specific sources of funding or revenues that could likely be available as the company develops (see Section 3.5 for details). The amount of capital needed over time to get the company to the point of break-even or early profitability is calculated from the bottom up and thus is more realistic.

3.2.4   Identify the assumptions

An important part of going through the above planning steps is to identify clearly the assumptions that need to be addressed for each milestone to be met. The major assumptions that have to be made true in order for the plan to succeed are the key risk factors for the management team to focus on. Identifying these assumptions and risks will point to the credibility of the plan and gives a higher probability for successful execution and for closing on financing rounds with investors who can get comfortable with the outlined risks. As a simple example, in creating a therapeutics company around a new biological target molecule or pathway, a major assumption is that the new molecule has a meaningful outcome in a specific disease. In this case, the team may plan to conduct an in vivo experiment with that target /pathway knocked out in a relevant disease model, helping to reduce the risk perception associated with that assumption as early as possible.

3.2.5   Investor Perspective Summary

In summary, an investor will be looking to answer the following questions when reviewing the business plan:

1. What is the specific market need that is being addressed by the company? Is it a big enough market, and one that is growing?
2. How much further investment is needed and when could the investor get their capital and profits back (also known as an exit point or liquidity point)? What kind of valuation is to be expected at that stage, assuming the company has achieved its

milestones and the market size potential is validated? The potential return to the investor is calculated based on calculations that consider the total capital needed to be raised by the company, valuation at each such raise, and the valuation at IPO or sale, which in turn is based on the assessed market size and growth opportunity.

3. How is the proposed solution unique and competitive in the market compared to others?

4. What is the defensible moat created by the company's patents, trade secrets, or business model that will allow the company to maintain healthy profit margins and keep competitors at bay or make the company an attractive acquisition target?

5. What is the experience and background of the founders or management team, board directors, and advisors that makes them capable to serve in their roles successfully? Has the team worked together long? What is their dynamic? How invested and passionate are they? What is their aptitude to weather future difficulties?

6. Does the company have multiple chances (e.g. products in different therapeutic areas) to get revenues from a product or service to the market, given the unknowns in human biology and the new science in application?

7. Does the product development or business model strategy adequately mitigate the perceived risks?

8. Is the financial plan and next proposed milestone compatible and achievable?

9. What are the underlying key assumptions that are needed to be fulfilled for this business to be successful? Has the business model and plan addressed or considered those assumptions ?

3.3   What is a business model?

A business model describes the ways and means by which a company makes money. For example, one business model is to carry out R&D, file patents, and then license them out to earn revenues. Another model may be to develop and provide an R&D service for product testing and development or as engineering design services. A common business model is to develop a product, take it through the approval stage, and market and sell it for a profit such as a fully integrated pharmaceutical company that carries out its own R&D, manufacturing, clinical trials, marketing, and sales. As another example, the business model for an early-stage R&D-focused company could be to offer its unique skills/technology as a service and earn revenues, or for it to license or sell some of its patents. A single company may change business models as it evolves.

A functional approach to defining your company's business model is presented in the value chain schematic in Figure 3.1, where the current company status can be clearly depicted in terms of key value-creation functions. This same chart can be used to define future activities and engage in strategic business planning. For example, working with the chart in Figure 3.1, place an X in a box in the first row (build internally) where the company's current strengths and expertise lies and another X in the box it would like to grow internally (in the same row). Now the other functions are sourced from outside contractors and a mark can be put in those remaining boxes in

the second row. This is a simple but effective tool to visualize the company's business
model and strategy. The following include some of the questions that frequently arise:

- Are we focusing the company on R&D – services or products?
- Are we going to manufacture or partner or outsource?
- Are we going to build sales and marketing or go to resellers and commercial
  partners?

	R&D	FDA Process & Clinical Trials	Manufacturing	Marketing & Sales	
				US	International
BUILD Grow Internally					
BUY Strategic Partners/ Contractors					

Figure 3.1 Build or buy business model planning using strategic functional categories to grow
internally or contract out

Box 3.1 Business model changes from R&D to product development

As a company develops and its technology advances, it may adopt different
business models. Millennium Pharmaceuticals started out in 1993 by building a
strong R&D platform technology in the newly emerging field of genomics. Their
early business model was to make money by licensing and partnering their R&D
capacity with diverse companies in human and animal health and with agriculture
applications. Several years later, having collected a large amount of cash from
these partnerships and from investors who saw these partnerships as validation of
the technology platform, Millennium transitioned to a product development and
marketing business model. Not finding any successful human health products from
their own discovery platform, they took on the business model of buying or in-
licensing products at a clinical stage of development and furthered them through
regulatory approval to marketing and sales for growing revenues. The change in
business model was also accompanied by a change in the company's operations
and personnel – from genomic scientists doing a lot of innovative, exploratory
research and generating patents, the company had to shift to manufacturing and
quality control, managing clinical operations, regulatory development, marketing,
patient advocacy, reimbursement strategy, and sales and marketing and distribution
of product. The change to a sustainable product model with revenues paid off for
the company's investors, with the sale of the company to Takeda Pharmaceuticals
for $8.8 billion in 2008.

Box 3.2 Changing business models

Ophthalmology app startup pivots business from AI diagnostic to ophthalmology portal

Written with assistance from Dr. Ankur Gupta

As a fourth-year medical student, Ankur Gupta got excited by an idea for a mobile app that would diagnose ophthalmic disease. The app would run an AI-based program, trained on datasets of images and diagnoses, to provide accurate diagnoses from images taken by a smartphone. Through widespread use of this app, blindness could be prevented by catching the onset of retinal disease early, making this app easy and cheap compared to finding and seeing a trained ophthalmologist for diagnosis.

A company was formed to develop and patent a software process for capturing the image data from any smartphone with variable lighting. By raising seed money from family and friends, Ankur, who also had an engineering background, was able to hire some talented engineers and build up a team. Soon, they had a pilot app working.

On successfully reaching this first milestone, the company was funded by Y Combinator and the StartX Fund (Stanford), both major U.S. West Coast incubators that had a special mentoring and training program to help first-time entrepreneurs. These high-profile incubators connected the company to name-brand venture investors who looked at these incubators for a steady pipeline of quality startups. These investors initially insisted that the company raise a significant amount of money to run a rigorous clinical trial program for a few years before seeking FDA approval for their software and then commercialize this platform technology for many applications. However, Ankur and the team realized that the business model for these early-stage venture investors was to fund a lot of companies that seemed positioned to make it to a billion-dollar valuation company, and hope that one out of many would actually make it there. They were actually betting (expecting) that 90+% of the companies would fail and hence would only invest in companies that had a clear path to revenues in large markets very fast, and thus reach billion-dollar valuations in a few years, ideally before the end of the typical 5-year life of the fund. Additionally, the model for most of these venture investors was software as a service (SAAS) companies that had highly scalable, large industry applications. The VCs were comparing Ankur's ophthalmic app opportunity to other nonmedical software companies (the majority of the startups going through the incubator were "software as a service" projects) that were quicker to revenues and could scale with lower costs and risks. It also meant that these investors would likely pull the plug quickly if it seemed that the company was not making progress along a hockey stick–like trajectory.

The pressure from this incubator and investors to broaden the platform and "think big" forced the company to revisit its original business plan. In order to capture a large valuation and market, they would need multiple indications

Box 3.2 (*cont.*)

approved to be diagnosed by the AI system and it would clearly take a few years and cost tens of millions of dollars to get through large clinical trials with multiple diseases and diagnoses. Even if the trials were successful, the adoption and acceptance of this diagnostics-assist tool into the workflow of the primary care practices was uncertain. In addition, the reimbursement level for this type of software diagnostic device was also highly uncertain. In addition, the significant amount of funding would need to be raised at a low valuation (the company would be seen as early stage with no revenues for a long time, and product market fit still needed to be determined), which would dilute other shareholders and the founders significantly.

Faced with this feedback, the CEO and his team re-evaluated their product roadmap. They were able to refactor their software platform to make it one that allowed actual ophthalmologists to make the diagnosis. They would continue to use these inputs to train their AI software in the background from these images and diagnoses made by actual ophthalmologists. This, in effect, allowed the company to run a prospective clinical trial while generating revenue, instead of raising a lot of money to run prospective trials. The company changed their software into a portal to allow consultant ophthalmologists from around the world to address and help patients who would upload high-quality, clear images of their eyes using the patented methods in the app. The company would charge a fee for the service and pay the consultants from those fees, making the company potentially profitable very early on.

This revision of their roadmap had several near-term benefits – they did not need to rely on venture financing and did not need to force-fit their markets and business into a billion-dollar growth-company mold. They put the AI-driven automated image diagnostic program on the shelf for a Phase II growth of the company. The attractive certainty of early go-to-market feedback allowed for quick product optimization, as opposed to waiting several years to assess for "product-market fit."

Lessons learned in this pivot

The feedback from markets and investors is valuable and can bring about important reassessment of the business plan, but the company and founders also need to pursue their vision instead of force-fitting it into something that they think others want. Going down the nonregulated path and still providing healthcare to patients, cheaper and faster, allowed the team to achieve their founding vision of improving patients' access to care. The other program on AI-assisted diagnosis would continue to develop and grow from all the learning data gained while treating actual patients.

A business model can be quantitatively defined by looking at all possible income streams for a company over time (one-time, ongoing sales of disposables, recurring maintenance contract licensing royalties or milestone payments, revenue through distributors, revenue from retail sales, software annual license or upgrade fees). The amounts invested to get to the revenue streams can be used to calculate a break-even point, where the sum of money invested is equal to the sum of revenues generated. The other critical calculation is to identify the time point when the company will become profitable and when it will become cash-flow positive (these are two different calculations). The key difference between cash-flow balance and profitability is that cash flow represents actual In/Out funds in a given period, whereas profit usually looks at booked and planned income and expenditure in a given period. A company can be cash-flow positive and not be profitable and vice versa. The cash-flow positive prediction is usually an important point for investors to review. These above calculations are obviously less meaningful for investors in a biopharma company that has 10 years of development before its product reaches market revenues. However, all startups need to tightly monitor and manage their cash position in order to make sure they can execute their plan and reach the next milestones; hence, these cash-flow calculations and projections are critical.

What are the key drivers for reaching this point of positive cash-flow? If the break-even point or profitability is dependent on reaching X customers, you would need to identify how many of those customers are currently in hand and how many of the X are unknown, even in the profile of a likely customer. This exercise gives the management and investors an idea of the probability of successfully reaching the milestone in the given time – which in turn impacts the financial planning. These assumptions in the financial and business plan have cascading effects that create even greater levels of uncertainty for the future of the business, which the management team must continuously find ways to reduce.

3.3.1   Business models for drug, device, and cell and gene therapy innovators

In the case of drug innovators, early drug discovery is an expensive and challenging area, full of early failures. It is common for most early-stage drug discovery companies to aspire to forward integrate along the value chain, but increasing levels of investments and continued high risk make it a reality that they will have to license their innovative product idea to the large pharma companies that have clinical trial infrastructure and marketing and investment power to build portfolios that mitigate risk.

Device innovator companies might find it easier to go to market as each device can find a niche market, but again, the economics of venture capital type investors typically focus a company toward large markets, which are likely best addressed by licensing to one of the top device companies that dominate the markets.

Cell and gene therapy tissue engineering companies (regenerative medicine companies) have typically vertical, fully integrated models, since their unique tissue constructs or engineered cells often require novel manufacturing and delivery mechanisms.

Innovator diagnostic companies have a particular challenge in getting their test to market. The test is usually performed on a technology platform that is accepted in the market and has an established base of labs that are familiar with that platform. This is an established market base that is captured by the technology platform manufacturer/ larger diagnostic company. If the new test is not accepted rapidly and launched across established platforms, the innovator diagnostic company runs the risk of being beaten by a fast-follower who copies their test on another established technical platform. Thus, the innovator diagnostic company has to build an acceptance base and reach a threshold market share to have their test carried by the major centralized diagnostic labs (high volume). In order to capture various technology platforms and rapidly gain market acceptance and usage, innovator diagnostic companies can nonexclusively license their technology to several large diagnostic companies. They can continue to develop and sell the test themselves, while establishing the test as a "standard of care" through the increased market share gained by allowing larger partners to market the test. An additional and significant consideration is the opportunity to gain early revenues by launching the test as a laboratory developed test (LDT) and work hard to establish a level of reimbursement from payers (Chapter 8, Box 8.8). A good model for in vitro diagnostic companies is to target a niche market with a unique biomarker or test format and build sizeable revenues for the company while establishing its new platform; and then channel other tests through the same platform to grow its market.

3.3.2 Dominant business models among life sciences and biotechnology/ biopharma companies

This section refers predominantly to the evolution of business models in the drugs and research tools segment of the biomedical technology companies discussed in this book. The phrase "biotechnology companies" is used in this section to refer to drug companies that are making either small molecule chemical drugs or biological drugs. Within each type of business model are notes on licensing or on intellectual property management and financial management strategy. These should serve as broad guidelines. The list of business models is ordered by the appearance of these dominant strategies in the industry over time, but is not meant to imply that the vertical or horizontal strategies of others are not relevant today.

The dominant business models that emerged in the biotechnology industry are:

1. Vertical (product)
2. Horizontal (tools and services)
3. Hybrid I – Discovery tools or platform forward integrating to product development
4. Hybrid II – Services back-integrating to discovery
5. Venture capital–led, virtual incubated companies

1. *Vertical model:* A company with a vertical business model has a product focus with vertical integration over the whole value chain to discover, develop, and market a single technology or set of technologies (usually therapeutic drugs) that are end products sold to the consumers. The first wave of "vertical model" biotechnology companies were founded in the late 1970s and early 1980s, based on breakthrough innovations that enabled increased efficiency and scale of biopharmaceutical manufacturing, making protein therapeutics a reality. Traditional large pharmaceutical

companies have a vertical business model. Many smaller biotechnology and device companies focus on research, and work in a "short vertical" model, where the product prototype (having achieved proof of concept in human) is sold or partnered out to a commercial partner.

Intellectual property strategies in this vertical model would typically include building and owning patents and in-licensing patents to maintain or gain market access. Patents typically focus on novel compounds or blocking patents on processes or formulations. Process trade secrets would be an important strategic advantage for certain companies. Forming brand recognition for technology platforms can also build advantage.

Financial strategy – Large up-front investments with long-term wait for returns. High-risk models as resources are focused on a few products that take a long time to get to market. Balanced by high return on investment. This business model captures maximal value of investment.

The successful ones have emerged as the giants of the small but fast-growing biotechnology industry. Examples of companies with varying degrees of success with the vertical business model are Biogen-Idec, Immunex, Amgen, Genentech, Medtronic, Boston Scientific.

2. *Horizontal platform model:* Companies following a horizontal model operate at a specific location in the value chain and sell a broadly applicable product/service across various sectors or industries such as human, animal, agricultural or industrial biotechnology verticals. Radical innovations that industrialized biological research and created new insights with "big data" like genomics, were sources of many new companies which adopted horizontal business models. These companies captured intellectual property, proprietary information or new processes (high throughput gene sequencing, expression arrays, etc.) and sold or licensed them as services, products (e.g. array chips) and intellectual assets (licensing use of platforms, partnerships to sell output of platforms). Drug delivery technology companies also fall into the horizontal business model.

Intellectual property management strategies would typically include owning patents around a core technology area and would include a strategy to out-license them to gain licensing fees.

Financial strategy – Companies need lower up-front investment, with technology platforms generating revenues through services or licensing or through sales of non-regulated research tools relatively early. Companies may start with venture financing or governmental financing but have a potential to be profitable with a positive cash flow within a few years.

This business model is represented in the second wave of new company formation in the biotechnology industry in the 1990s. Of these companies, a few successful ones built up rapidly on partnership revenues or made products that were widely licensed and used in all fields of biological research – academic and industrial. Examples of this horizontal business model include Affymetrix, Lion Biosciences, and Tripos. Also existing within this type of business model are service companies that obtain a comprehensive fee for their service and have no plans to develop an FDA-regulated product on their own. Examples of this type of service include the custom manufacturing houses and the clinical research organizations (CROs) such as Parexel, Lonza, and Quintiles.

3. *Hybrid model I – tools to product development:* This business model occupies a horizontal and vertical position over the value chain. The company services various industries but then aggregates its resources in one industry or sector. The company usually starts with a horizontal business model to make money quickly and then builds resources to vertically integrate for higher value creation. The popularity of this business model, which appeared after the genomics stock market bubble collapse (September 2000), reflected two shifts in the environment (i) reduced early-stage venture capital investing and (ii) diminished value on tools and IT for a few years following the collapse meant companies had to generate revenues with services to get the attention of investors.

Intellectual property management strategies for these companies would tend to have a mix of internally generated patents and a strong drive toward in-licensing IP as product portfolio and vertical integration strategies emerge.

Financial strategy: Early revenues derived from platform tools and services sold across the value chain are then invested in integrating forward into drug development over a therapeutic product development value chain, merging the two business models described earlier.

Examples of hybrid I companies include Curagen, Millenium, Incyte, and Celera.

4. *Hybrid model II – services backward-integrating to discovery:* This business model, similar to hybrid model I, predominantly occupies a horizontal position over the value chain. Shared risk models, where the services company takes on a performance risk and gets incentives based on success, have been increasing in recent years. Some companies have also leveraged the existing R&D service functions and revenues to carry out internal R&D on drug discovery, adopting a model similar to the early-stage biotech companies. These hybrid business models have been used with caution in the past, as there is a real risk in alienating existing customers who may feel the service business is now becoming a competitor. Some examples of service companies that developed their own internal R&D and product lines include Albany Molecular Research (Allegra), Jubilant Biosys, Structural Genomix, and Accelrys.

5. *Venture capital–led, incubated virtual company:* This model is seen largely in drug therapeutic or device companies. A venture capital firm will partner with inventors and scientists to incubate projects that are too nascent to become companies and need to have one or two critical experiments conducted either to validate or to "kill" the idea. The venture capital groups can hire experienced pharma mid-career executives who can have the comfort of knowing that of the multiple projects at least one may succeed to give them a further path, and the venture capital company invests a small amount of money to get the experiment done in exchange for equity alongside the founders. Usually, these companies are run virtually or with a small, shared lab footprint and majority outside contracted work (partial virtual model), leveraging the growing number of contract research organizations (CROs) worldwide that have made this model a reality since about 2006. The goal of the optimized business model is to find out at low cost whether the new target or drug platform can be validated before funding the setup of a fully staffed company and laboratory. These models have

proved to be financially successful for the venture investors and the founders. In addition, the company can get access to expertise and support at a very early stage, something that would not have been as easy to access as a single-product funded company. For more information, check the References and Additional Readings section at the end of the chapter.

3.4   Some practical tips on presenting the business plan

- Preparing a business plan or pitch deck is an exercise for the entrepreneurial team to think through and review in detail the assumptions and hypotheses underlying their projections.
- Typically an executive summary, slide deck/presentation, and separate financials in pro forma statements are used to make the business case. Technical presentations are prepared separately for in-person confidential discussions and follow-on meetings.
- Different versions of the business plan presentation are likely to be needed, and modifications for varied audiences are common.
- Financial statements (budgets) for the first 18 to 24 months should be presented in monthly or quarterly layout. If further future financials are required, they can be presented as quarterly or annual statements. Some investors want to see projections out to initial revenues on a monthly basis. Preferences vary and companies should be prepared.
- The exercise of creating those long-term financial projections can help the founders to specifically identify assumptions behind the projections and flesh out the potential for scaling the business. These assumptions can be formulated into hypotheses that can be tested. In fact, iterating through ideas early in the venture formation process can be codified as a "learning plan," with a series of hypotheses and specific tests, market research studies, prototype demos, etc., to validate or eliminate such hypotheses.
- The business plan focuses on the benefits of the product or invention (not on the" how," but on the "what" it does) thus not needing to disclose confidential information in the first meetings and discussions.
- A go-to-market strategy is critical to add credibility to the business plan pitch and should include any possible validation from prospective customers.
- Competition must be addressed credibly – for example, it is helpful to present alternatives and competitors and highlight the reasons why customers will switch to the new product/service from current gold standard medical practice and adopt a new product from a startup company.
- A key issue for startup companies is management experience (or lack thereof) in the eyes of investors, and the business plan should present a team and not just an individual founder. Having an experienced board of directors or advisors can help mitigate the perception of inexperience.

- Usually, business plans do not propose a valuation for the company unless you already have part of the round closed with a lead investor who typically sets the valuation for the investment round.
- Typically, companies that are raising early rounds of financing will bring in a lead angel-investor and then circulate a private placement memorandum (PPM) to bring in other investors. This is typically useful when reaching out to many diverse angel-investors, as the PPM answers key investor questions without needing to do a presentation to each one. When bringing in a venture capital round, typically, once a lead venture firm puts across a term sheet and it is accepted by the company, that term sheet is circulated to its known syndicate partners or other venture firms as the terms for the financing round, unless the company can bring in a competing lead investor with better terms.
- Hire a lawyer to review the term sheet from the investors before agreeing to terms. It is also advisable to have a securities lawyer review the PPM before circulating to individual investors.

3.5   Starting a company

When you incorporate a company, you are birthing a new entity into which you are breathing life. The entity now requires ongoing maintenance activities such as accounting, purchase orders tracking, invoicing, payroll management, financial controls, and management approval processes, which all take effort and time to organize and implement. The business formation process is usually driven by retaining a legal firm that helps identify the type of legal entity that would be best for the future development planned. The following points are shared with a view to identifying a few key issues not always discussed in business books.

3.5.1   Distributing equity

Founders should have an open discussion on how the founding equity is to be distributed. Many founders decide to share equity evenly, although some split the founding equity based on perceived contributions to the company (e.g. author of a key patent) or based on efforts and time to date of founding. All these (and other) ways of sharing the founding equity are acceptable and largely depend on the context, but it is most important to put all members on a vesting schedule so that if someone leaves in a few weeks or months, they keep only that amount of equity they have earned over the time they worked. Vesting equity can be just time based or a combination of time based and milestone based. Milestone-based vesting may not work very well for a university spin-out where technology is at a very early stage, as business models and plans can change rapidly and earlier milestones may be discarded along with the change in strategy. Box 3.3 contains an example case and links for methodologies and tools for distributing founders' equity.

Early sale of equity to founders should be undertaken with guidance from legal counsel and business advisors, so that appropriate board and shareholder approvals can be correctly put in place in the company records.

Box 3.3 Equity distribution among founders

There are many ways to divide equity in a startup among founders, the most common of which is to equally assign equity to all founders on day one. Some well-publicized stories of high-tech startups such as Facebook and Zipcar show that roles and goals change; some founders choose other life paths while others continue to build billion-dollar companies. It seems inherently unfair to have given 50% of the startup to someone who did not stay and build value in the company or treated the startup as a hobby for a little while and then left the bulk of the heavy lifting to the other founders. The easiest way to balance out committed effort and equity is to create a time-based vesting schedule in which founders' shares vest in a monthly or quarterly vesting schedule over four or five years. Keeping a one-year cliff for vesting is also a frequent mechanism to ensure that founders stay committed for the first year. Many founders prefer this method – everyone starts with an equivalent amount of founding shares, and those who stay longer and commit to executing the business plan earn in all their assigned equity. Those founders who leave in a short period of time, say, 18 months, leave with some portion of equity, but the others who have stayed now increase their percentage of ownership of the company by the proportion of the unvested equity.

Founders who contribute more resources such as cash or equipment and feel that that they should fairly be assigned more equity in exchange for their contributions will want to have a look at some internet articles and blog resources at the end of this box. Creating a table in which points are assigned to various contributions, risks, and commitments gives founders a framework with which to engage in detailed conversation about contributions and efforts, goals and commitments. Mostly, these different methods encourage discussion among founders by creating a framework to have what is typically a difficult conversation. Typically, these equity distribution frameworks assign weights to (1) the significance of the idea or the patent; (2) business knowledge using key functional attributes such as who is typically drawing up the business plan, bringing in investors, etc.; (3) domain expertise such as a chief architect of the product – e.g. a chemist or biotechnologist or mechanical engineer – (4) commitment and risk such as who is keeping their full-time job or leaving other job opportunities aside to work on the company; (5) responsibilities such as taking leadership in most items and staying up late worrying about payroll or delivering on planned milestones etc.; and (6) additional contributions that may not be returned to the founder (such as equipment brought in or cash invested for legal expenses or salaries and not written as a loan to the company).

Each element through discussion among founders to clarify the points can be assigned a weight from 1 to 10. Then a set number of points (say, 10) are assigned for each such functional point and need to be distributed among each of the founders, giving more points to the founder(s) who represent or contribute to that area most. Summarizing the points for each founder gives a percent ratio of allocation of founding equity. This method (over the equal assignment to each

Box 3.3 (*cont.*)

founder) has the main benefit of promoting discussion and improving founders understanding of each other's motivations.

Case example

Contributed by a StartX Stanford University technology founder

A businessperson (Founder A) with a PhD in molecular biology, two decades of corporate leadership, and startup founder experience joined a volunteer program at Stanford University looking for the next big idea that he could help develop. Teams of such advisors/volunteers were formed by the technology licensing office to take its technologies and develop a commercialization plan over a 12-week period. A team of four people came together: Founder A, Founder B (a post doc completing his postdoc work in a lab unrelated to the technology they worked on in the project), Founder C (a clinical faculty member with some name recognition but relatively junior, who had had some experience with starting up companies), and Founder D (another business-experienced Stanford alumnus who was giving back to the community). The team liked the technology and they decided to continue past the completion of the program. Founder A was probably the most excited and pushed for the team to form a company and raise financing, as he was willing to put full-time effort into this project going forward. Founder B, the postdoc, was similarly excited and also ready to join full-time once he finished his postdoc position in a few weeks' time. Founders C and D were glad to put in time and effort but decided they would be continuing only in a consulting advisory role. They had a discussion on equity split when they decided to found the company in 2015 and ended up deciding to allocate equity to the founders A:B:C:D at 35:35:15:15 percent, respectively. They had a 4-year vesting schedule with a 1-year cliff. They decided to backdate the vesting schedule, as some time had passed since they had started the Stanford program together. When they founded the company, therefore, some partial vesting had already taken place. Founders A and B put some cash in to get the company started and with largely the efforts and contacts of Founder A, they brought together a seed round of family and friends, enough to get a lab rented and experiments started to validate the technology. Only one class of shares was authorized as common, and options were put aside for employees and advisors.

They rented the lab in June 2016, and it became clear within a couple of months that Founder B was ill suited to the specific lab work and technology development needed. He soon agreed/proposed to move to a consulting position and gave up some shares in that transition so that he was vesting at the same rate as Founders C and D. By November that year, Founder A fired Founder B, since it was clear that he was not much help as an advisor as well. By then, the 1-year cliff vesting had occurred. A few months later, in March of the next year, Founder C passed away unexpectedly. Founder D continued to support the company development as needed

Box 3.3 (*cont.*)

and continued vesting. Two years after starting the company, while progress was slowly being made on the technology development and prove out, two of the founders had not really contributed much time or value to the company but they (or their estate, in the case of Founder C) were holding about 15% each. Founder A and Founder D tried to buy back some shares and Founder B was cooperative in the process, selling back some of his shares but the estate of Founder C was not interested in selling any back. Founder D joined a large pharma company as program head full-time and did not really have much time to offer for the startup but was still supportive of the efforts. At this point, 5 years after the founding, Founder A was the lone founder in the company full-time and holding 50% of the equity, Founder D had about 20%, Founder B had 10%, and Founder C's estate had 10%.

Does this seem like a fair distribution given the efforts, commitment, and risk taking by Founder A, who really had been a solo founder in retrospect?

The issue was not just where the equity distribution ended up, but where it would need to go in order for the company to successfully motivate the next executives who would be hired and continue to keep Founder A motivated to continue his multi-year commitment. Founder A needs to decide whether to re-capitalize the company in the next round of financing in order to free up more equity for new executives and employees without facing too much dilution himself. This is not an easy decision, as he ended up in this position through a series of unpredictable events and circumstances.

However, this situation is all too familiar for those who have gone through starting up companies with co-founders. What would you do going forward?

Among many available resources on the Internet, the reader can visit the following posts and resources for ideas and suggestions on how to distribute founder's equity:

- https://cofounders.gust.com/ (an interactive equity allocation calculator from the world's largest online platform connecting founders and angel-investors)
- www.andrew.cmu.edu/user/fd0n/24%20Founders'%20Pie.htm (a framework and article by Dr. Frank Demmler at Carnegie Mellon University)
- https://fi.co/insight/how-to-split-equity-with-cofounders (an Excel spreadsheet from the Founders Institute)
- https://slicingpie.com/ (a book and pay-for-use tool by Professor Moyer that helps dynamically change equity allocation over time)
- https://feld.com/archives/2011/07/finance-fridays-getting-started-allocating-equity-and-founders-investment.html (case example with suggestions)

Note: there are many articles and resources available on the Internet on how to allocate equity to other founding team members –board directors, advisors, investors, and employees. A few more resources are also mentioned in the References and Additional Readings list at the end of the chapter.

3.5.2   Doing science for product development vs academic science

The difference between doing academic research and carrying out research to develop a product is subtle and yet significant. In a science-based enterprise, the scientific methods or tools may be the same as academia and the discussion may be based on academic research findings or publications, but the questions asked and the decisions made on what experiments to pursue may be vastly different. Even though the inputs to the decision seem similar, i.e. reproducible results and hypothesis-driven experimental methods, additional considerations such as commercial fit with the market needs, regulatory strategy, and milestone-driven financing strategy all play an equally important role in decision making on projects.

The decision making in commercial product development is driven by a desire to identify technology proof-points (or failure-points) as early as possible so as to know whether the technology works or to be able to change directions before running out of money to reach a significant milestone. In academic projects, exploration of all possibilities and following outliers for academic discovery are suitable actions, and the academic scientists may explore many ways to prove their ideas, either to gain a publication or to gather data for the next idea for a grant application. While companies can and do engage in scientific discovery, the process ultimately will differ from the academic environment by virtue of the input considerations. The thinking process of a researcher, who may be trained to search for and follow outliers in biology or to ask questions with the goal of getting grants or publishing, has to shift to a different gestalt in a company setting. This is one fundamental shift in moving from scientist to CEO.

3.5.3   Setting up a lab

Infrastructure in larger organizations such as academic institutions and companies is taken for granted, and entrepreneurs who come from that background are sometimes surprised by the amount of time it takes to replicate those functions for their fledgling startup. It may not be easy to set up credit accounts with larger suppliers of bulk materials who also may not be willing to sell quantities smaller than their minimum (kilograms or tons when the need may not be more than a few grams). Have a credit card ready. An incubator can be helpful in extending the relationships or infrastructure of a larger institution to the member startups. For example, this can help in ordering specialty gases or chemicals where it can be costly to meet the regulatory materials handling or disposal requirements. Setting up a biological laboratory or animal testing facility requires significant compliance-related activities that may strain the limited resources of a young company with a handful of employees. Some universities have formulated guidelines to allow young spin-out companies to contract work back to the lab where the original discovery or intellectual property was developed. Others have built incubator facilities with wet labs.

Box 3.4 Working with academic labs – lessons learned by a startup CEO
Based on interviews with Andrew Radin, CEO of twoXAR

A startup company spun out by a graduate student at a top U.S. university had developed a software-based approach with advanced data-mining to identify new approaches to treating diseases that had not been recognized before. Their software also allowed them to discover new drug candidates with novel modes of action in treatment of the disease. Their most immediate and complex next step was to synthesize the molecules and test them in appropriately developed preclinical models of the diseases (in isolated cells in the lab and in animal studies) to verify that the new approaches they discovered. The company hired, early on after their first round of funding, a Director of Preclinical Studies with significant industry experience in running these scientific studies.

For a young startup company, it was convenient to reach out to scientific and medical experts in specific diseases, from their own university campus and to contract with them to run studies. These experts were world-renowned researchers in each disease area.

The first such collaboration contract was put in place through the Office of Technology Transfer/Licensing (OTT) with Dr. A's large laboratory group that had expertise in the animal model of the disease of interest. Dr. A had a very disciplined postdoctoral staff member as head manager of the lab, and this lab manager had established a culture of sound laboratory practices with consistent record-keeping and reporting that were followed by the students and postdoctoral fellows in the lab. The company's project manager was the liaison to this lab manager and was regularly a part of lab meetings with the group so there was a significant amount of oversight by the company. This project ended up with a robust set of data generated and several presentations at conferences that got attention from potential licensees / partnering companies. However, the CEO found that the licensing partners all wanted the study repeated in an outside commercial laboratory in order to accept the data as verified.

Another leading research scientist-clinician fellow, Dr. B, at the same university, had an interest in modeling a different disease with the tools developed by the startup company. The company was willing to work with Dr. B to generate a new disease dataset from clinical samples that Dr. B could collect from his patients. They agreed to co-write a grant to a private philanthropy supporting research in this disease. When the private philanthropy indicated that they typically gave the grant in exchange for some (equity) share in the specific revenues from the funded project, the discussion with Dr. B also turned to his sharing in the equity of the project. The idea had been generated by the company, the tools and IP belonged to the company, but Dr. B felt that he should own 50% of the project returns for providing access to his patients and their biopsy samples. When told by the company that it was not a reasonable request given his minimal input into the

Box 3.4 (*cont.*)

project and the significant amount of work still to be done, without his future involvement to develop a therapeutic, he retorted with examples of his clinician colleagues who had told him they had much larger stakes in the startups they worked with. The company's founder failed to convince Dr. B that those colleagues had different contexts as co-founders or had added much more value to those companies' development, and Dr. B felt he was getting cheated by the startup company. This project never went ahead, and they abandoned the conditionally approved grant.

In a third example, the startup company contracted to do an animal study with a clinician-professor in a major cancer clinical research center in another city. The professor had two medical students running the lab. The clinician-professor altered the experimental setup, using far fewer animals per study group, claiming that she knew what she was doing in this disease model and that the drug, if good enough, would show differences with just a couple of animals per group. The students were caught between conflicting demands of the company project manager who tried to guide the experiment long distance, and their boss's changes in the protocol. These infrequent and inconsistent communications and reduced statistical strength protocols (only two animals per group) led to a technical failure in which the positive control and negative control groups gave no reliable reading to interpret the study.

The following lessons were learned from these experiences by the startup company CEO:

- Many academics feel that their contributions are outsized, as their fame or reputation itself is transformational to the young company (in most cases it is not) and, thus, negotiating with them on a viable, equitable distribution of equity or gains becomes a very emotional discussion. Giving them large chunks of the company may make it impossible to raise financing from savvy investors. Sometimes their understanding of the experiments and motivators (publication) varies significantly from the company's needs to show commercial viability (inadequate controls or statistical strength, or lack of comparison to competitor products in the study).
- Students run the labs and thus the quality of the labor available for running experiments is highly variable, lacking professional discipline on timings and consistency. In another case with a fourth academic project, a contract was signed but the postdoc who was supposed to join the lab that semester did not, leaving no one to run the study.
- Most painful and important learning for the startup CEO from these experiences was the feedback from the pharmaceutical or biotech companies that were approached as potential partners for licensing out these new products. These companies typically gave very little credence to academically generated data and

> **Box 3.4** (*cont.*)
>
> trusted the data only if replicated in a professional, commercial laboratory. So, the academically generated data may be useful for brand building (prestigious academic institute and researcher) but have little commercial value except perhaps in some cases where they have developed rare or complex disease models that are not available in commercial labs.

3.5.4   Setting company culture

Setting a culture (see also 3.1.1) for accountability and decision-making processes in a young company is very important. A self-aware executive team is sensitive to the collaborative leadership styles that are often correlated with successful science-based companies. When the company is small, the culture is that of the individuals' own behaviors, but as the company grows, the way people communicate and behave toward one another, their integrity, and approach to problem solving all depend on the culture, which can be codified and explicitly acted out. Entrepreneurs are well advised to make conscious decisions about the culture they want to see in the company and embody those characteristics in their own behavior from day one. See Chapter 7 for how one biotech drug manufacturing executive established a culture that led to efficient manufacturing operations and successful technology transfer to manufacturing.

3.5.5   Role of the CEO

A company needs a figurehead at the helm, even for a "flat" organization, and that person is the talking head for the company and the rest of the team. The investors place their trust in that person to deliver the milestones/results to budget and continually strategize on how to make the company successful. The CEO is answerable to the board and ultimately to the investors. In the early stages of the startup, the CEO communicates the vision of the company, lays out a path to the next milestone, sets up the strategy, organizes responsibilities, and assigns tasks. They hold people accountable and course-correct behaviors and tactics with the team regularly as needed. The team is mission driven, and it is primarily the role of the CEO to bolster that mission and vision and build a cohesive culture in the company.

As the company grows, the CEO has to hire strong second-in-command people with specific expertise and skills in functional areas such as business development, HR, finance, technology, quality control, and regulatory affairs. The strength of the company lies in the CEO identifying people much smarter than themselves in those specific functional areas and then letting the smart people they have hired do their jobs without looking over their shoulder, thus building trust while holding them accountable to meet the agreed-on deadlines. The CEO needs to continue to build and reinforce the culture of the company, coordinate goals between group and functional leaders, and communicate the vision of the company to all stakeholders.

3.5.6   Milestone-based planning

The business plan must lay out specific, significant milestones on the path to commercialization, and the financing strategy of the company is usually focused on these milestones, with each milestone usually representing a significant increase in value (reduction in risk). The budgeting and fundraising planning must support the achievement of major milestones within planned timelines. Assessing the resources needed to achieve the milestone and executing to plan is a major challenge for the early-stage management team. Many professional investors will look for an 12- to 24-month history of successful delivery on promises by the executive team before investing. Some examples of key milestones for the different types of companies are given in Table 3.2.

3.5.7   Customer and product development feedback

Bringing a new product to market in a startup company has many challenges, the most significant of which can actually be the initial faith of the founding technopreneurs in their conceptualization of a product to meet a market's needs. This faith, while serving as the impetus for the company, can blind the team to the actual market or customer needs and drive them to make a product that they (the founders) think the market needs rather than a product that fills a real need and that the market will actually buy and use. Hence, getting a first customer who is ready to buy gives an invaluable focus to the product development efforts. The sooner the company can reach the milestone of putting a product or concepts to trial with a "customer," the better chances for success for the startup. These concepts are discussed in more detail in Chapter 5. Customer in this case can also be a strategic corporate partner who is a customer as licensee or acquirer of the technology/product.

Early feedback from strategic partners helped one drug delivery company shift its product focus and helped it succeed. This drug delivery company designed a catheter to deliver a drug for atrial fibrillation in a pericardial space, solving a problem with current delivery methods. In an early meeting with a large cardiovascular medical device company as potential strategic partner (potential licensee and investor), the feedback received was to change the target indication to post-operative atrial fibrillation rather than the broader market of arrythmia atrial fibrillation they were aiming at. The company team took the feedback seriously, changed direction, using the same drug with a redesigned catheter, and found quicker adoption and success in their first market application.

Table 3.2 Typical milestones that are relevant to an investor

Stage	Biopharma Life Sciences company	Medical device company	Diagnostics company
Early and preclinical stage	- Patent applications filed / patents issued / owned by company or license agreement signed with university (or other licensor) - Proof of concept demonstration data: e.g. in vitro data in cell culture, enzymatic data, mechanical simulation and design, binding assay in diagnostic testing, in vitro cell culture, first proof of concept prototype - Key technical thought leaders in the field validating the new technology by joining advisory board / corporate board / management team		
Prototyping stage	- In vivo efficacy data in at least one disease model - Manufacturing to support clinical trials - Toxicology testing done - Regulatory clearance to start human trials (IND)	- In vivo data in relevant animal model - Manufacturing process qualified - Safety data in relevant context - Regulatory clearance for human trials (IDE) or approval to market (510(k))	- Data collected with human disease samples showing accuracy of diagnostic and false-positive rate - Regulatory pathway and path to first revenues validated
Clinical stage	- Human safety data (Phase I study) - Randomized blinded human clinical trial proving efficacy (Phase II)	- Randomized human clinical trial proving efficacy	- Large human clinical trials validating diagnostic specificity and accuracy
ANY STAGE	Strategic partnership or licensing deals with large brand in that sector brings strong validation for investors and can spike valuation		

NOTE:
IDE – Investigational Device Exemption
510(k) – procedural pathway for marketing approval based on a predicate device

3.6   Financing a company

"If an investor is offering you money – take it!" is the general advice to a biotech entrepreneur. Given long development times, variable human biology, regulatory gateways for most marketed products, and complex reimbursement payment structures, it is prudent to have more cash than you think you need. The rule of thumb for all science-based startups may well be to double your time estimates for product development and triple your cost estimates.

However, the above statements are generalizations and the financing needs and situations for each startup will vary. Human drug therapeutics usually have the longest gestation time and are the most expensive to commercialize among biomedical products. Medical device, diagnostics, and software-based companies may need less time and money to get to market. For most startups, first-time entrepreneurs are well advised to overestimate the amounts needed to fund the development and commercialization of their products/services.

How do you estimate the amount of capital needed to get the product to market? A few tips: The budget has to be built from the ground up. For example, it may seem like a herculean task to consider raising the tens to hundreds of millions estimated in publications as the average cost to bring a drug to market, until one realizes that estimate includes all the failed attempts and parallel developments with multiple drug candidates carried out by a large pharmaceutical company. A small biotech startup is more likely to focus on one candidate with much more outsourced work, and the actual cost of the development for one new drug may range from tens of millions to a few hundred million dollars, depending on the disease area and type of therapeutic. On the downside, this focused approach also increases the chances of failure of the enterprise, making the outcome of the one product a binary event for the project and the company. Raising investment for a single drug product company is much more difficult, as an investor sees little chance to rescue their capital on the event of a negative outcome at any point in development. Most biopharma companies thus have a platform technology with multi-product pipeline strategy planned out. Some license or buy in drugs to show rapid progress to a clinical stage product in order to make themselves attractive to investors. Hiring experienced executives needed in internal positions to guide clinical development and other functions, is very challenging in a single-product company. There is not enough work to justify hiring someone full-time, and so the company may need to reconsider their business model.

3.6.1 Sourcing of financing to early milestones

Early development activities up to the first milestone (a proof-of-concept test of the technology or service model) are typically funded through the following sources: existing research grants (by staying within the university environment before formation of a company); money from founder, friends, and family, angel-investors (also known as business angels); early commercialization grants from governments (US-SBIR, EU-EASME, India-SBIRI, etc.), philanthropic foundations, or venture capital. Occassionally through strategic partnerships. The References and Additional Readings at the end of the chapter contain links to recommended templates for financing agreements. The financing need for this first phase (proof of concept) is relatively small.

Further development of the product with animal data or testing of the prototypes or service in a system environment (as opposed to an isolated cell study) is the next key milestone that also gives early validation of a specific market or application. Safety and product reliability data must be submitted to regulatory bodies, and while some of it is collected in this product validation stage, usually the tests involved continue into the next development stage. Funding for this stage is typically obtained from the following sources: venture capital, advanced government grants, ultra-high high-net-worth angel-investors (business angels), potential corporate partners, or licensing or sale of intellectual property.

Completion of product safety, reliability, and manufacturing product design processes is a key gate/milestone in order to launch into human clinical testing. Significant funding is required to test drugs in humans, and sources for the tens of millions of dollars of risk capital are few – venture capital (financial investors) or strategic corporate partners. In some rare diseases, philanthropic organizations or rare disease associations may direct their funds to support this clinical stage of development for specific therapies.

Once there is sufficient clinical proof of concept data in the relevant disease population, the company's next financings for the larger clinical studies, marketing, and manufacturing are usually sourced from public markets, but funding sources can also be private equity, venture capital, or corporate partner financing.

The above scenarios are generalizations for biopharma therapeutic products or services. Diagnostic devices or healthcare IT software or services will have different milestones and funding needs. For medical device companies, milestones that get investors more interested include mechanical testing (as applicable), toxicology/safety/immunogenicity testing, animal efficacy data, and completion of first meeting with FDA to classify the device and get approval for a human clinical testing program.

For diagnostics companies, key milestones for investors include product design and commercial feasibility (which include confirming identifying the specific need and confirming commercial opportunity through multiple diverse stakeholders) and validation in a human clinical setting.

For a medical device or diagnostics startup, early funding for developing the design concepts or proof of concept in vitro can be sourced from government grants, self-funded by entrepreneurs, or sourced from seed funds from universities or local angels. The next round of funding is usually from venture funds or angel-investor groups. Many companies are boot-strapped (sweat equity from founders, credit cards, loans or investments from friends and family, service contracts, business competition awards, etc.) and it may take a few million dollars to get to animal studies proof of concept with a relatively small number of samples needed to show efficacy of a device or therapeutic. The further funding to manufacture for human trials and to obtain regulatory approval to do human testing can then be gotten from strategic corporate partners or from venture capital.

3.6.2   Valuations

A major step of completing a financing is getting agreement on company pre-money valuation, which is a highly subjective measure by any count. Common methods to support valuation positions include market comparables for the company, a product valuation taken from licensing or financing transactions, and/or the current value of discounted cash flow calculations from projections of future revenues. Practically, the best method to support a targeted valuation is to bring in competing offers (term sheets) from multiple investors or corporate partners.

Usually, a professional investor or a seasoned angel-investor sets the valuation by first offering a term sheet to the company (see Box 3.5 for typical investment terms), but sometimes the company may offer the term sheet or may already have a private placement memorandum circulating with terms (see Box 3.6), depending on the circumstances.

It is important to note that the term sheet valuation is not final, as completion of further due diligence before signing the investment agreements may change risk perceptions. However, once negotiated and signed by both parties, the valuation and terms are by and large maintained in the final closing documents barring any material findings in post-term sheet diligence. Depending on the legal jurisdiction in which the term sheet agreement is signed, even if the term "non-binding" is used on the document, both parties may be obligated to negotiate in good faith based on the agreed-on terms in the signed term sheet.

The management team is well advised to review the term sheet, and later the closing documents, with an experienced securities/venture financing lawyer. Take time to understand the implications of each clause and key terms of the term sheet or investment agreement before signing. Most terms in an agreement are negotiable, no matter how "standard" or "boilerplate" they claim to be, but the company must choose its negotiating points and be willing to give up some positions in order to get what is important. A main point to note is that the investment terms are not only to be negotiated on the company pre-money valuation price/share, but also on other terms such as "preferred participating" or "liquidation preference," which can be used by investors to gain more value on the back end at a sale or public listing of a company. Hence, for the entrepreneur, it is important not to get too stuck on a specific valuation figure, but to consider other terms in the investment offer along with the nonquanti-tative benefits that the particular investor may bring in. Taking a lower valuation for the seed round than may have been desired may actually help with long-term success, including successful next round financing.

Metrics provided by Y Combinator [https://blog.ycombinator.com/how-to-raise-a-seed-round/] indicate that a company may give away up to 20% of the company equity in the first seed round and about 15–30% in the next round of funding. However, these are generalizations based on software companies, and metrics for biomedical device companies will vary. Biopharma companies may give up a lot of equity at early stages to raise the large amounts of cash needed, but valuations rise rapidly so that at the time of IPO or sale of the company a very significant cash return to the founders is still available despite lower equity holdings. A common question for founders is: "Would you rather have 50% of a $1 million value company or 10% of a $1 billion company?" The main challenge for biomedical companies is that the risk versus value equation is quite high until completion of human clinical testing, hence there is usually a strong drive to source nondilutive financing in early rounds.

Box 3.5 Key clauses in a typical investment term sheet

Key terms	Notes to consider
Pre-money valuation	This clause lays out the current (pre-investment) price per share and determines the percentage of the company that will be owned by the investor. The best way to increase the price/share is to get multiple term sheets – simple supply and demand. Typically, investors buy preferred shares and founders and management all hold common shares with none of the protections or preferences described below. Valuation is usually the most hotly debated term by entrepreneurs, but there are other terms that can more significantly influence the take-home returns to entrepreneurs.
Dividends	A dividend may or may not accrue on the preferred shares the investors typically are buying. This money is to be set aside when the company is sold as additional payment on the capital or each share purchased by the investor.
Liquidation preference	This clause dictates how much investors and preferred stockholders are paid before everyone else when the company is sold. A *1x liquidation preference* means they get 100% of their money back (plus accrued dividends if any) before anyone else guaranteeing a return to them in the downside case that the company is sold for much less than expected. Any more than 1x preference means they get paid more than what they invested. But the payout to preferred shareholders does not end there. *Participating preferred* shares means they get paid first, then convert all preferred shares into common shares and again share proportionately in the distribution to all common shareholders of the remaining moneys left from the sale of the company. *Non-participating preferred* means that they take their preferred (1x or more) payment and do not get to participate further in the distribution of the remaining money paid to common shareholders.
Protective provisions	The most significant clause in this section is typically an anti-dilution clause that needs to be carefully understood.
Option pool	Typically, the investors will require a certain percentage of the common shares set aside as an incentive option pool for future hires or investors before the financing (pre-money). This will increase the dilution of the founders but does not affect the investors as this option pool is included in the denominator of issued shares to calculate the fully diluted pre-money value per share. Allocating options toward each planned future hire until the next financing, is a good way to argue for limiting the option pool size.
Investor rights provisions	An important clause in this section is the right to one or more seats on the board of directors for the new investors. The board sets the direction of the company and is the decision maker for most major decisions such as firing of key management, issuance of new shares, taking on debt, and any winding down, sale, or IPO of the company. Thus, the founders have to consider how much voting power they wish to give up to their investors while encouraging maximum investment and attracting the best participants.

Reference the Term Sheet template available online from National Venture Capital Association website of model legal documents – https://nvca.org/model-legal-documents/

Review the References and Additional Readings at end of chapter for sources for financing agreements.

Box 3.6 Difference between a business plan and an offering memorandum or private placement memorandum (PPM)

After hearing a presentation/pitch and on getting interested in the opportunity, investors' next step is to carry out a detailed due diligence on the technology, targeted markets, prospective customers (if any), intellectual property status, past financials, and future projections. The company can address these due diligence questions by loading individual documents in an online repository with secured access (a data room).

Once an investor is ready to invest and the key terms are negotiated with the first (lead) investor, the company will usually need to bring more investors into the round. In many cases, the company may issue an offering memorandum, also called a private placement memorandum (PPM), and circulate it among interested parties to collect additional investors. Investors typically will read this to get familiar with the company and the business opportunity and see details on the terms of the investment offering – debt or equity, price per share, current share structure of the company, and other details. This document is written with the intention of fundraising efficiency so that investors may make their decisions without needing the management team to do detailed presentation pitches to each investor. Companies often also develop and add a follow-on FAQ from the feedback questions received after circulating the PPM.

A PPM contains many of the elements of a written business plan and specifically includes content about the current securities offered for sale (valuation and price per share being offered, capitalization table), risk factors, state and global securities rules, historical financial statements, etc. The narrative created in the content of the PPM is written so a general investor can follow, but it gives sufficient information to satisfy the investor who is knowledgeable in the field. The narrative is written in simple language so any reader can get a very clear sense of the company mission and investment thesis (opportunity).

The inclusion of a 'disclaimer' section detailing potential risk factors is not typical in a business plan and is drawn from the practice of prospectuses in public markets where the regulators require management to clearly state the risks for the general public.

A table of contents from such a PPM for a biotech point-of-service diagnostics company is presented here. The page numbers give a general idea as to the depth of the narrative.

Executive Summary 3
Risk Factors 6
The Need for Rapid Medical Tests 13
CE Mark and FDA Application Status 14
Description of NewCo's Rapid Test Products 15
Competitive Advantages of NewCo's Technology 15
Competitive Advantages of NewCo's Products 16
The [Disease] Market and Targeted Marketing Strategy 19
NewCo's Other Products 22

Box 3.6 (*cont.*)

Market Penetration Strategy 23
Future Product Development 25
Production and Manufacturing Scale-Up 27
Pilot Production 29
Bio Data of Management Team 30
Use of Proceeds 32
Intellectual Property 32
Description of Capital Stock and Capitalization Table 33
Legal Representation 34
Financial Statements (historical and projections with notes) 34

3.6.3   Preparing for due diligence

Basic materials to prepare in advance for investors requests for due diligence are the following:

- Technical presentation (confidentiality agreement could be requested to share this if really needed)
- Bios of key management and founders
- Capitalization table showing number of shares issued and authorized, price per share sold, and top shareholders of note
- Financial pro forma statements (projections) as requested by investors (24–36 months is the minimum)
- Letters or emails from potential customers or corporate partners; any memoranda of understanding or agreement signed
- List of intellectual property and any published patents (no pre-publication materials should be shared without confidential materials)
- Competition evaluation

Second level of due diligence could include any or all of the following:

- Discussions with the main technical lead/founder/inventor and existing investors
- Calls with current or prospective customers
- Detailed technical evaluation by third parties
- Intellectual property (IP), any licensing agreements, review of patents for freedom to operate, review of IP by third party may be requested
- Background checks and references for each of the key management/founders
- Incorporation documents, bylaws, or operating agreement of the company
- Copy of the stock option plan, warrants given out, option grant agreements
- List of debts and convertible note holders (if any) and past signed agreements
- Current (past 12 months) profit and loss (P&L) statements, balance sheet
- Employment agreements, consultant agreements, nondisclosure agreements signed
- Lease agreement, any loan agreements signed by company

3.6.4 Selecting your investors

Multiple investors should be approached in parallel rather than working sequentially with one investor at a time. The company must focus on investors who clarify that they independently make a decision (take the lead) in making the investment, as many other investor groups (followers) will not make a decision without having such a lead investor complete the due diligence and give their decision on the investment. Investors are also looking at multiple companies and opportunities in parallel and hence may take time initially to process due diligence. Typical process time from initial meeting to investment ranges from 3 to 24 weeks. If investors feel that they will be left out of a big opportunity, they will move more quickly than normal. Hence, the best way to move an investor and get closure on investment is to get other investors to the same decision point in parallel. A company should also carry out due diligence on the investors by speaking to CEOs of other companies in the investor's portfolio and identifying and getting to know the working style of the person who will potentially be appointed to the board of the company from the investor group. Additionally, the strategy of the venture fund is important to align with the company's stage. For example, a fund in its fourth or fifth year will typically not invest in an early preclinical stage company as the potential exit event for the investors will exceed the life of the venture fund.

3.7 Nondilutive funding sources

Startups in biomedical technology (including biopharma) face unique financing challenges given that the question on risk cannot be truly addressed until regulatory safety gateways are passed and testing in human patient populations can be carried out. Thus, business models have developed to address the high cash requirements through nondilutive funding:

1. *Service revenues:* Develop the innovation and use the new technology to offer specialized R&D activities and earn profits by offering services; or license the platform or specific parts of it to others in areas that are not conflicting with long-term plans of the company. Use the profits to develop your own products to a major milestone. This is difficult to do, as profits do not usually cover all costs of development, but this positive cash flow can reduce significantly the financing needs. Thus, a revenue-generating company may be more attractive for some investors.

2. *Grants:* Get government grant funding through NIH/NSF Small Business Innovation Research (SBIR) grants or similar programs to support development to clinical trials. This takes more time as the agencies have their own granting cycles and review processes, making this nondilutive path more tortuous than getting investor funding, but for those successful in getting the larger multimillion-dollar grants, it offers a level of technical validation due to the intense peer reviews and

also allows the founders to retain more of the company equity until major milestones and higher valuations are reached.

3. *Philanthropic institutes or individual grants, loans, or other creative financing structures:* Raise investment from philanthropic disease foundations that have an interest in developing new products for patients with these diseases.

4. *Work with a strategic partner* who will take a share of later revenues in exchange for offering in-kind services or materials that reduce fundraising needs significantly – e.g. a new diagnostics company that needs next generation gene sequencing equipment to run clinical trials may get this equipment from a manufacturer in exchange for a license to the manufacturer for their product in a specific area or for a share of a future revenue stream. Another example is to give some equity or a royalty in specific products to an R&D services company that offers cash burn reduction by carrying out required tests at cost or cheaper in exchange for future gains if that product/project is successful. The R&D services company has the benefit of covering their operating expenses and building a portfolio of potential future revenue streams.

5. A type of *debt financing*, called venture debt, is typically offered by specialized financing companies right after a venture financing event (from an equity investment round) in the form of a (relatively) low-interest loan securitized by the new capital infusion or other security. This debt financing is helpful in extending the venture financing timeline without giving up more equity. Standard business loan financing from banks is typically not available for biotech companies, as revenues streams are too far away and risky for most banks.

3.8   Operations in a growing company

The following sections contain tips on organizing the company operations in a startup mode when everything seems urgent, and everyone is overburdened with multiple tasks.

3.8.1   Meetings

The frequency and method of meetings set the pulse of an organization and help organize the flow of information and decision making. According to the Gazelle methods (see https://scalingup.com/), the following types of meetings are relevant to have regularly:

- A daily operations coordination meeting, which is a standing check-in-only, 5- to 15-minute meeting held with each team or team leader every morning without long discussions, but during which people report in on activities for the day and the previous day. This meeting can be used by team leaders to set the tone for the day . . .

- A weekly management review meeting with only team leaders or upper management to discuss and resolve tactical issues and lay out priorities for the next week
- A monthly strategy review meeting with management and may include advisors and board members as pertinent
- A quarterly or monthly all-hands-on-deck meeting with review of deliverables, overview of the progress toward quarterly or annual goals of the company

3.8.2   Conflict resolution among staff and management team

- The CEO has to deal with any behavior issues that do not seem resolved after one interaction.
- Hire a part-time HR manager from a local firm so you can have a regular HR person who can guide the first-time management leadership, help resolve conflicts between staff, help with recruiting efforts and mediate in personnel differences, offer training for state-mandated sexual harassment education, etc. for all levels of the organization.
- Interpersonal conflicts, egos, misunderstandings, and downright bad behavior are all best addressed head-on and in person rather than with written emails.
- The person with poor behavior must be brought in by a team leader or with the CEO for a one-to-one discussion in private as soon as the incident is reported or known, and an explanation of why this does not suit the culture of the company must be conveyed clearly.
- Clear corrective actions and follow-up must be laid out in writing; in smaller companies, the best approach is usually to recommend that the person find another place of employment as they would not be / are not a good fit with the culture of the company and would not be happy or productive in the long run. As hard as it is to find and recruit good talent, it is much harder, in the long run, to root out bad habits and productivity-sapping behavior from the company culture once people see leaders ignoring or condoning poor behavior. The tough decision to let a problem person go is usually seen as the best decision that a CEO or team leader has made once they have taken that step. Many other staffers who have been exposed to this person often ask what took the manager so long to take the step to terminate the person.

Box 3.7 The technical genius who is too important to fire

Many executives from larger companies want to join startup companies because they want to make a greater impact from their efforts by accomplishing a significant outcome or by sharing their gathered wisdom and experience with everyone around. They feel that they can bring in practices and processes from their experience that make workspaces more productive and efficient. While startup

Box 3.7 (*cont.*)

companies can benefit from some discipline and processes, all practices of the highly hierarchical large company do not always work in an environment where everyone is empowered and contributing in a very "flat" organizational structure. This case study highlights challenges in establishing startup company culture and also the topic of management of too-important-to-fire personnel.

A science-driven, small startup healthcare products company (five people) hired a new CTO who was previously head of engineering in a large company. There, she had been in charge of several large production facilities. During the interviewing process, this person came across as highly versatile and adaptable while being very technically deep. Subsequently, she was selected over other candidates because of her "hands-on" attitude and being capable of "getting it done." She turned out to be technically proficient, reinventing the technology platform successfully within 3–4 months, against other industry expert naysayers who said it would take more than 1–2 years, and got the technology stabilized for pilot production to start.

The next level of junior younger engineers was brought in to support her plans for scaling up production from prototype to first pilot production line to support animal studies. The CTO set up a reporting process similar to the large companies she had formerly been a part of. These experienced engineers were joining a small startup to have more autonomy and collaborative work environments, which was not the approach the CTO was taking. The hierarchical do-as-you're-told approach may have kept things humming in a larger production facility that was already established, but it caused friction in the developmental stage of the technology that was invented as new production equipment was being designed simultaneously. The less-experienced staff needed to follow her directives. The engineers did not get enough feedback when they were overruled, as the CTO felt that her extensive experience gave her insight that did not need explanations for her decisions.

The staff meetings she led seemed to become a tongue-lashing and personal-bashing session because of her use of sarcasm and taunting language to bring people in line, rather than a collaborative learning or sharing session. Discussions in technical group meetings, which were initially more engaging, gradually almost ground to a halt. In this small company, the hierarchical approach and reporting processes ended up creating silos. She did not realize that her behavior patterns, which may have worked very well to protect her team and job function in a larger company, did not work very well for the fluid environment of the smaller company. A key production engineer whom she had hired left after 18 months, citing a toxic work environment, despite attempts by the CEO to intervene several times between the CTO and the engineer.

Despite her clear technical prowess, and potential for future guidance in the growth of the company's product and production line, the CTO's behavior was (although not purposefully) undermining the company's attempts to create an

Box 3.7 (*cont.*)

inclusive and collaborative culture and killing their motivation, all in name of bringing in discipline, process, and hierarchical approaches, implemented too early in the small company. The CEO and board debated several times whether to let her go. However, during this period the company was also facing serious financing headwinds and running out of funds to carry out another search and retain enough runway to meet next milestones. The CEO could not see how she could let the CTO go in this situation. Who else would be credible and could pick up the technical slack when making presentations to potential investors for their desperately needed next financing? It might take another 2–3 months to find and hire a replacement CTO, and given the small commercial field of 3D printing, it was questionable whether they would ever find a replacement with a similar depth of technical knowledge and experience who would also be willing to join a small startup. This was the quandary – the person who was not a good cultural fit with the company was also the most important technical person. Was there a better solution or should the board and CEO let the CTO go?

The CEO wondered:

Should I have let her go early on when it became clear that her actions had created significant discord between the team members?

Were the other staff members just not capable enough (as the CTO expressed) or skilled enough to appreciate the technical challenges and understand the complexity of the technology and production platform?

What else could I have done to ensure success in this situation with the CTO?

It is always a difficult decision to fire a bad actor who is also a technical genius and can make miracles happen for a company. The only way a small company can survive such a person is if that person can be put to work in isolation on specific projects; however, that is uncommon and further, cannot be sustained. For example, Steve Jobs (founder of Apple Computers), in his (first and only) job as an employee at Atari, used to come in late evenings and work through the night when very few staff were around.

After attempts at reconciliation or correction and warnings of behavior problems, if there is no improvement, then it is better for the survival of the company to take the decision to let the technical superstar go. The problems created by people who cannot adapt to working with others, who cannot adhere to company culture, or who seek to prove their superiority by disparaging others are only exacerbated when the company goes through a hard time with funding or market growth. Growth salves many personnel ills in a startup company. With growth in revenue, products, and personnel, many interpersonal issues are ironed out as more people come into the organization, functional groups and a hierarchy of reporting build up, and people are focused on reaching the next positive growth milestone.

3.9   Summary

This chapter illustrates the nuances involved in building and managing a team, a board of directors, and investors in a young company. The transition from scientist to CEO will be easier for readers who have absorbed the points and examples in this chapter. After building a team, the art of pitching investors, constructing a credible business plan, and milestone-based financial planning are all covered in this chapter. This chapter also provides practical, operational tips toward successfully pitching investors, writing business plans, preparing for due diligence, and managing a team of executives.

Exercises

3.1 What is the value proposition of the innovation/company? Write it out in one sentence (only punctuation marks allowed are commas). This sentence should be constructed as the answer to a cocktail party question from a potential investor: What does your company do?" This sentence should clearly state the market need and the product innovation value proposition and should only serve to interest the investor to ask to meet later or ask further clarifications – the sentence should not aim to give an explanation of the how, just state the impact and create interest to ask follow-up questions or meet. An example of such a sentence is:

"My company, [name], . . ."
". . . [status] . . ." e.g. is developing/is testing/has launched
". . . a [product category] that . . ." e.g. service/product that
". . . helps [target group] / addresses [market need] . . ." e.g. indications/ unmet need/ . . .
". . . [use case/benefit for target group] . . ." e.g. save money/improve health outcomes by XX%
". . . [USP/special sauce that makes you different from the competition]" e.g. "in seconds / with a patented new drug aiming at a new target / improves diagnostic accuracy by 50% / at a third of the price."

3.2 Prepare a slide deck for your business plan and read it aloud with the founding team as practice and ask them to put themselves in an investor's seat and ask questions. Make two versions of the deck based on (a) a lay audience and (b) an audience who is well versed in life sciences investment.

3.3 With your team, write out a sequence of simple practical actions that you need to take in the first few days and weeks and then after the first 6 months of starting the company. OR Take the random list of actions steps below and put it in sequence with the group discussing the order:

Feasibility determination; Marketing of technology/product; Product development; Build out patent portfolio; Organize Team; Plan financing

(*cont.*)

strategy; Raise money; Write business plan and presentation; Form Commercial Partnerships; Incorporate business; Talk to a lawyer; Lease office/lab; Buy equipment and start R&D

3.4 Describe the value or utility of your patent (application or issued patent) as though to a business investor to help them recognize the value of what market it permits your company to defend. Keep it simple without describing all of the technical details (especially if it is a pending patent application or [United States] provisional patent).

3.5 Explain your company's business model(s) as though to a 5th grader. Use that simple explanation and then build on that to present to an investor. Simplicity brings clarity.

3.6 Have a discussion among your founding team about equity sharing, expectations and write out the meeting minutes afterward. Also write out a forward-looking personnel plan for the company management that you have all agreed on. Use that personnel plan to quantify how much to put aside into a stock option plan for those future hires.

3.7 Write out all the milestones for your company for the next 12 months; assign responsibilities for these milestones; discuss with the assignees the resources they need to get there. If warranted, break down the milestones into specific tasks that are on the critical path to reaching that milestone. Describe what will define success in reaching that milestone, keeping it as quantitative as possible. Make sure you achieve team agreement.

3.8 From Exercise 3.7 output, create a personnel time or effort (percentage) allocation by tasks (existing and projected new hires) for each set of tasks leading to successful milestones.

3.9 Create a budget from the above exercise completion on a monthly basis, including supplies, equipment, overheads (e.g. rent, utilities), and assuming a 30–40% increase in time to complete the task.

3.10 Do a discounted cash flow–based net present value calculation for your company and write out the key assumptions on the market revenue levels and then outline what the product performance needs to deliver in order to address those assumptions. (See various online resources, including www .investopedia.com/ask/answers/021115/what-formula-calculating-net-present-value-npv-excel.asp.)

References and additional readings

Blank, S, and Dorf, B. (2012). *The Startup Owner's Manual: The Step-By-Step Guide for Building a Great Company*. K&S Ranch Publishing. ISBN-13: 978-0984999309

Feld, B. (2019). *Venture Deals: Be Smarter than Your Lawyer and Venture Capitalist*. 4th ed. New York: Wiley. ISBN-1: 978-1119594826

Fled, B, and Cohen, DG. (2019). *Do More Faster: Techstars Lessons to Accelerate Your Startup*. 2nd ed. New York: Wiley. ISBN-13: 978-1119583288

Harnish, V. (2012). *Scaling Up: How a Few Companies Make It ... and Why the Rest Don't (Rockefeller Habits 2.0)*. Gazelles Inc Publishing. ISBN: 978-0986019524

Mehta, S. (2004). Paths to entrepreneurship in the life sciences. *Nature Biotechnology* 22(12), 1609–1612.

Mehta, S. (2019). Chapter 2: From scientist to CEO, in *Bio and MedTech Entrepreneurship: from Startup to Exit*, edited by H Flaadt and J Dogwiler. 2nd ed., Stampfli Verlag. (Note: accompanying text for the BioBusiness course taught by the author, with details at http://www.biobusiness.usi.ch/).

Shimasaki, C. (2009). *The Business of Bioscience: What Goes into Making a Biotechnology Product*. New York: Springer. ISBN-13: 978-1441900647

Shimasaki, C. (2014). *Biotechnology Entrepreneurship: Starting, Managing, and Leading Biotech Companies*. Elsevier Science. ISBN-13: 978-0124047471

Websites of interest

A good read on an investor's viewpoint is found in various blog posts at https://lifescivc.com, and a post that discusses business models on how to start up a biotech from an investor's perspective is at https://lifescivc.com/2019/08/the-creation-of-bio tech-startups-evolution-not-revolution/ [last accessed Dec 2020].

Other Venture Capital blogs that provide general information about startup investing and issues (not biotech specific) are [last accessed Dec 2020]:

- http://avc.com/ Union Square Ventures' Fred Wilson provides life advice, commentary, and practical know-how for growing a tech startup.
- www.bothsidesofthetable.com/ This blog is written by Mark Suster, a successful entrepreneur who became a VC, blogging about his experiences and offering advice to aspiring CEOs on topics ranging from leadership advice to thoughts on how to build a thriving company.
- www.davidgcohen.com/ David Cohen, the founder and Managing Partner of Techstars, the number 1 ranked internet startup accelerator, provides actionable advice for dealing with the day-to-day activities of growing a startup.

Visit https://shreefalmehta.com/csbtbook for additional enriching readings around the topics covered in the book, topical updates on the content and for industry viewpoints and news.

Links for additional readings on financing documents

- National Venture Capital Association website had many model legal documents that are used by both investors and founders – https://nvca.org/model-legal-documents/
- Y Combinator website – has many useful tips and blogs (however, mostly focused on software companies) and has templates for SAFE (Simple Agreement for Future

Equity) financing documents (a type of debt/equity financing instrument) – www
.ycombinator.com/documents/ – and the blog with notes on financing details is at
www.ycombinator.com/library/6m-understanding-safes-and-priced-equity-rounds
(last accessed Dec 2020)
- Some law firms make basic incorporation and related documents available as a
package – and the Cooley law firm website can generate relevant agreements and
documents at no charge — www.cooleygo.com/documents/ (last accessed
Dec 2020)

4 Intellectual property and licensing

Plan	Position	Pitch	Patent	Product	Pass	Production	Profits
Industry context	Market research	Start a business venture	Intellectual property rights	New product development (NPD)	Regulatory plan	Manufacture	Reimbursement

The Patent System added the fuel of *interest* to the fire of genius.

Abraham Lincoln

> Learning points
> - What are the different types of intellectual property and what is their purpose?
> - What is a patent and how does one interpret a patent?
> - What is patentable material?
> - What is the process of filing and obtaining a patent in my country and globally?
> - When does an invention or an idea become patentable?
> - What are the different models used to manage or make money from patents by licensing?
> - What are the key terms to understand in a licensing or a technology transfer agreement?
> - What is the value of a patent?
> - What is a reasonable "royalty rate" and other financial terms for licensing my patent?
> - What are the damages I can claim if someone infringes my patent?

4.1 Types of intellectual property

"Intellectual property" is the term used to describe a set of commercially valuable rights that result from formally codified inventions, literary or artistic works, or other representations of creative thought or particular symbols of commerce. Intellectual property includes patents, trademarks, copyrights, and trade secrets. Patents are concerned with discoveries, inventions, methods, data, designs, or practical implementations of algorithms. Trademarks are concerned with symbols of good will in a business, such as brand names and logos. Copyrights are concerned with the unique artistic expression of an author or artist. A trade secret is "know-how" that is protected only so long as it actually remains a secret.

Intellectual property is increasingly viewed as an asset, a tradable piece of property. There is clearly some value attached to intellectual property, just as there is value attached to a physical asset such as land. However, just as land by itself usually has little economic value unless there is development, intellectual property or "IP" has value only when the underlying work is being developed and commercialized. In this book, we are primarily concerned with patents, which are comprise a set of laws and legal documents that protect new and useful discoveries and inventions in most countries of the world. Specifically, each country has its own patent laws, which allows for the development and commercialization of innovative products. The perceived "market" value of a patent is based on the status of development of products, size of market, strength of particular intellectual property to block market access, and many other parameters. Thus, there is intrinsic value in registering and holding intellectual property.

The origins of modern patent law are in Venice, Italy, where a decree was issued in 1474, by which new and inventive devices had to be communicated to the Republic in order to obtain legal protection against potential infringers. The origin of the word "patent" is the Latin *patere*, which means "to lay open" (for public viewing). During the medieval period in Britain, "letters patent" (today the short form "patent" refers to this Letters Patent) were documents issued by monarchs granting exclusive rights to a person to practice their instrument or equipment and thus creating monopolies. In 1624, British Parliament restricted the term of the exclusivity for which the monarch could grant the Letters Patent and made these "patents" applicable only to new inventions. The Industrial Revolution in the West flourished due to this legal protection that also stimulated further invention.

What are the different types of registered intellectual property?

Patents
Trademarks
Copyright
Trade secrets

4.2   Patents

Patents are a widely used form of intellectual property registration to capture the value of a discovery, invention, or innovation and are particularly important in the commercialization of biomedical technologies. In exchange for disclosing an invention to the public, so that it can be made, used and improved upon, the inventor or patent holder receives protection for the invention, for a limited period of time. For the life of the patent (usually 20 years), the patent holder can exclusively block others from making, using or selling the invention.

4.2.1   Patent Rights

In the United States, a patent is a property right granted by the government to an inventor "to exclude others from making, using, offering for sale, or selling the invention throughout the United States or importing the invention into the United States" (as described in United States Code Title 35, Part II, Section 154) for a limited time in exchange for public disclosure of the invention when the patent is granted. Issuance of patent rights is in effect the granting of a limited monopoly on the market by the government in exchange for full disclosure of the detailed invention and subsequent innovation (reduction to practice of the invention and its application). The advantage to society is the dissemination to the public of the detailed methods by which the invention works and this stimulates the further development of new technology based on this invention. Thus, societies that protect inventors with patents have an opportunity to advance rapidly in science and technology applications.

What is equally important to note is that this granting of a patent is NOT an exclusive right to make, use, and sell the invention, but only a right to block others. There may be other existing patents that may block aspects of this invention from being made, used or sold. For example, if Company A has the patent to a new salt of a useful compound that makes it easy to deliver in oral formulation, but Company B has already patented the use of the compound for treating a particular disease, then Company A has the right to block others from making and selling the compound, but cannot itself sell the compound for treating the disease as it will infringe the patent held by Company B. More about this in Section 4.3.

4.2.2   Types of patents

Utility patents

- This is the most common type of patent filed to protect the way something is made, how a tool or device operates, or how a process can accomplish some useful purpose. Improvements or any combination of these are also filed as utility patents. In the broadest sense, a discovery or invention is patentable when it is (i) new, (ii) useful, and (iii) non-obvious. Among other things, a patent includes a specification, which describes the invention, and a set of claims, which defines the legal boundaries of what the patent protects.

Typically, biomedical utility patents are of four types:

1. *Composition of matter* – e.g. a newly synthesized pharmaceutical drug molecule composition. A composition of matter is patented for all of its uses, whether now known or developed later – so long as the composition has one practical use when the patent is applied for.
2. *Process or method* – e.g. methods of treating patients with a given disease through the use of a particular gene or protein or through targeting a gene or protein with a drug. Even if a patent on the gene or protein may be granted, if a new use is discovered, a new patent can be filed on that later invention.
3. *Device or machine* – where a new machine or medical device is developed to treat a disease, including how the device is engineered and how it is used.

4. *Article of manufacture* – according to the U.S. Supreme Court, the production of articles for use from raw or prepared materials by giving to these materials new forms, qualities, properties, or combinations, whether by hand-labor or by machinery, e.g. ceramics, cast-metal articles, mousepads, shoes. However, a natural article, say, a rind of orange impregnated with borax to prevent decay, is not patentable as an article of manufacture.

For example, pharmaceutical product patents could include a composition of matter, describing the possible combinations of new chemical compounds that could be used to make the invented drug for the purpose of treating a disease; device patents could include construction of a device, composition of matter of new materials, and specific applications of the device; biological drug patents would typically include the process of manufacture of a protein, the engineered protein-producing organism, and any final delivery modifications to the protein as the functional aspect of the protein.

Practically, utility patents are just referred to as a "utility patent" rather than classifying them into these subgroups.

Design patents

A design patent is a new, original, and ornamental design for an article of manufacture. These are more applicable in medical device field; for example, the handle of a surgical tool could be designed with a certain texture or contour for better grip, to increase comfort, or to simplify operation. Buttons may be grouped by frequency of use or in functional groups to make it more intuitive for the user. These features may have been developed after extensive feedback from users and redesign processes. The device company should file a design patent to prevent competitors from copying the look and feel of the handle into a competing tool and thus block others from benefiting commercially from the extensive design efforts of the company.

Plant patents

A plant patent is an asexually reproduced plant variety by design. This patent is granted for the invention, discovery, or asexual reproduction of a distinct and new variety of plant. The plant cannot be a tuber propagated plant or a plant found in an uncultivated state. A link for information on plant patents is provided in the reference section at the end of the chapter.

Note: We will not be discussing design and plant patents further in this chapter.

4.2.3 What cannot be patented (from the United States Patent and Trademark Office [USPTO] website)

- Laws of nature; physical phenomena; abstract ideas; literary, dramatic, musical, and artistic works (these can be copyright protected)
- Inventions which are:
  - Not useful (such as perpetual motion machines)
  - Offensive to public morality

Can living things be patented?

The following subsection is reproduced by permission from the Biotechnology Industry Organization

Some living things can be patented, but not all. Like any invention, a living thing must be "new" in order to be patented. More importantly, living organisms under consideration for patenting cannot be those that occur or exist in nature. Thus, one cannot obtain a patent on just any living creature, such as a mouse, because mice have been around for a long time. If someone makes a kind of mouse that never existed before, however, then that kind of mouse might be patented.

Microbes: As long ago as 1873, Louis Pasteur received a U.S. patent for yeast "free from organic germs or disease." With the growth of genetic engineering in the late 1970s, the patentability of living organisms was reexamined, and confirmed. A landmark case involved Dr. Ananda Chakrabarty's invention of a new bacterium genetically engineered to degrade crude oil. In 1980, the U.S. Supreme Court clearly stated that new microorganisms not found in nature, such as Dr. Chakrabarty's bacterium, were patentable. Dr. Chakrabarty received a patent in 1981 (U.S. Pat. No. 4,259,444). In the Chakrabarty decision, the Supreme Court stated that "anything under the sun that is made by the hand of man" is patentable subject matter. Therefore, if a product of nature is new, useful and nonobvious, it can be patented if it has been fashioned by humans.

Plants: In 1930, the U.S. Congress passed the Plant Patent Act, which specifically provided patent protection for newly invented plants that are asexually reproduced. In 1970, Congress provided similar protection for newly invented sexually reproduced plants.

Animals: In the 1980s, the question of whether multicellular animals could be patented was examined. The key case involved a new kind of "polyploid" oyster that had an extra set of chromosomes. This new, sterile oyster was edible all year-round because it did not devote body weight to reproduction during the breeding season. The USPTO found that such organisms were in fact new and therefore eligible for patenting. It found this particular type of oyster to be obvious, however, and thus did not allow a patent for it. Nonetheless, the polyploid oyster paved the way for the patenting of other non-naturally occurring animals. In 1988, Philip Leder and Timothy Stewart were granted a patent on transgenic non-human mammals (U.S. Pat. No. 4,736,866) that covered the so-called Harvard mouse, which was genetically engineered to be a model for the study of cancer.

Natural compounds, proteins, and nucleic acids: Natural compounds, such as a human protein or the chemical that gives strawberries their distinctive flavor, are not themselves living, but occur in nature. A compound that is purified away from a strawberry or a protein or gene that is purified away from the human body may be patented in its purified state in some countries and not in others. In the United States (as a result of the 2013 Supreme Court Decision in *Association for*

Molecular Pathology v. *Myriad Genetics Inc*), nature-identical proteins, genes, and bacteria are not considered patentable. The act of extraction or isolation of the protein is not deemed sufficient to allow patentability of that protein sequence or gene. However, naturally occurring nucleic acids and proteins can be patented in Europe provided that at least one use (utility) for the nucleic or protein is known, and that the use is written into the patent application when it is first filed. Such a patent would not cover the strawberry or the person. For example, a newly discovered protein, where the function of the protein is not known, would be a discovery as opposed to a patentable invention. However, if the protein is shown to be involved in arrythmia, and at least laboratory data are available, making it plausible that modulation of the expression level of the protein can alter the occurrence or severity of arrythmia in cells, this would be considered an invention.

In the United States, naturally occurring biological subject matter, including bacteria, viruses, and human or animal stem cells and cell lines, are considered to be patent ineligible. However, stem cells or cell lines that have been genetically engineered are considered to be patent eligible as they are not products of nature. In Europe, bacteria, viruses, cell lines, and cell models of disease can be patented, provided that they have a substantial, specific, and credible use. Similarly, in India and China, the isolation of a protein or nucleic acid sequence, without indication of an industrial application or useful function, is not by itself a patentable invention.

For biomarkers, the *Myriad* case cited above makes it very difficult to get a patent in the United States for a new protein or microRNA that is used in a method to detect a particular pathology or condition (biomarker or end use claims not permitted). In the United States, to have some chance of success in patenting biomarkers, claims need to be directed to new detection methodologies, detection reagents, or equipment. However, in Europe, it is possible to claim the biomarker itself or a method of detecting a condition using the biomarker.

The human body, at the various stages of its formation and development, and the simple discovery of one of its elements *cannot* constitute patentable inventions. *The USPTO (and other national agencies) does not allow anyone to patent a human being under any circumstances.* An element *isolated* from the human body or otherwise produced by means of a technical process, including the sequence or partial sequence of a gene, may constitute a patentable invention, even if the structure of that element is identical to that of a natural element. Criteria of novelty, non-obviousness, etc., still need to be satisfied. As techniques to isolate and sequence DNA and deduce function based on similarity to other genes are becoming more and more commonplace, the USPTO looks at whether the invention is novel and includes an inventive step (non-obvious), plus other criteria to determine patentability. Among these important criteria are that the patent disclosure provides a written description and an enabling disclosure, sufficient for persons of ordinary skill in the relevant field to make and use the invention. The disclosure must also include the "best mode" for practicing the invention that is known to the inventors at the time they apply for the patent, as described further in the next section.

4.2.4   What type of invention or discovery is patentable?

Patents are filed with government agencies, and their patent examiners evaluate, issue, and register patent rights. These agencies evaluate the validity of an application based on the following criteria (from the USPTO criteria; similar criteria are used by patent offices in most other nations). Any invention or discovery that is the subject of the patent application therefore must have the following characteristics covered in the application:

- Novel (not previously known, used, sold, on sale, marketed, publicized)
- Utility (useful task, some use for invention)
- Non-obvious to a person with knowledge in the field
- Adequately described to the public at time of filing
- Enable a person with knowledge in the field to make and use it
- Best mode or effective mode must be disclosed
- Described in clear, unambiguous, and definite terms

Box 4.1 Flowchart for determining patentability of subject matter

The flowchart in Figure 4.1 is taken from the U.S. Patent and Trademark Office (USPTO) website www.uspto.gov/web/offices/pac/mpep/s2106.html. In this flowchart, each numbered claim of a patent application is evaluated.

Step A allows a claim to be eligible patentable material if it is self-evident and does not seek to tie up or include a natural material or a similar exception. As an example, a claim directed to an artificial hip prosthesis coated with a naturally occurring mineral is not an attempt to tie up the mineral and thus would be an allowable claim at this step. Also, if the claim is a clear improvement to a technology or to computer functionality, then it also qualifies as eligible subject matter for the patent.

However, allowable subject matter can be quite different among various countries. For example, the discovery of a protein (biomarker) that is correlated with a disease detection or outcome is not patentable in the United States (without some added subject matter on applications), as it is deemed a naturally occurring substance, but would be allowable as subject matter for a patent in Europe. As another example, until 2001 in India, composition of matter claims on medical drug compounds were not allowable subject matter and only the process of manufacture was deemed to be allowable (mainly to encourage low-cost

Box 4.1 (*cont.*)

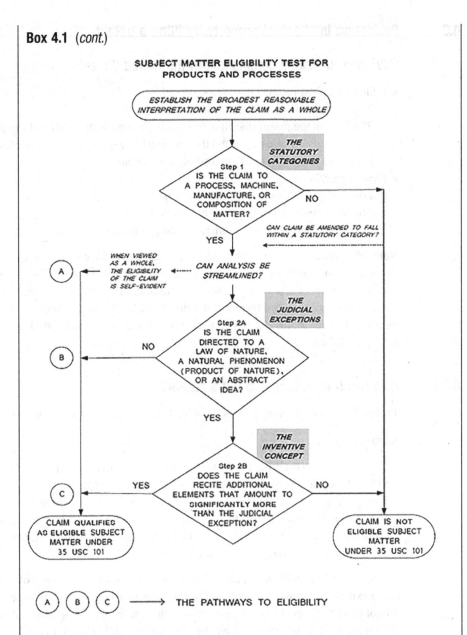

SUBJECT MATTER ELIGIBILITY TEST FOR
PRODUCTS AND PROCESSES

ESTABLISH THE BROADEST REASONABLE
INTERPRETATION OF THE CLAIM AS A WHOLE

THE STATUTORY CATEGORIES

Step 1
IS THE CLAIM TO A PROCESS, MACHINE, MANUFACTURE, OR COMPOSITION OF MATTER?

NO

YES

CAN CLAIM BE AMENDED TO FALL WITHIN A STATUTORY CATEGORY?

WHEN VIEWED AS A WHOLE, THE ELIGIBILITY OF THE CLAIM IS SELF-EVIDENT

CAN ANALYSIS BE STREAMLINED?

A

THE JUDICIAL EXCEPTIONS

Step 2A
IS THE CLAIM DIRECTED TO A LAW OF NATURE, A NATURAL PHENOMENON (PRODUCT OF NATURE), OR AN ABSTRACT IDEA?

NO

B

YES

THE INVENTIVE CONCEPT

Step 2B
DOES THE CLAIM RECITE ADDITIONAL ELEMENTS THAT AMOUNT TO SIGNIFICANTLY MORE THAN THE JUDICIAL EXCEPTION?

YES

NO

C

CLAIM QUALIFIES AS ELIGIBLE SUBJECT MATTER UNDER 35 USC 101

CLAIM IS NOT ELIGIBLE SUBJECT MATTER UNDER 35 USC 101

A B C ⟶ THE PATHWAYS TO ELIGIBILITY

Figure 4.1 Flowchart to determine eligibility of subject matter for patenting

production of needed drugs by local manufacturers who became adept as process modifiers).

4.3   Protecting intellectual property by filing a patent

4.3.1   How long do issued patents last in the United States?

- Utility and plant patents:
  - Twenty years from the application date
  - The term of the patent (the right to enforce) begins with the date of the grant and usually ends 20 years from the date at which the patent was first applied for
  - subject to the payment of appropriate maintenance fees.
- Design patents:
  - Fourteen years from the application date
  - The term begins from the date the patent is granted
  - No maintenance fees are required for design patents

Note: Pharmaceutical and some medical device patents have mechanisms by which the patent term can be extended beyond the standard 20-year term. The period is dependent on specific regulations and circumstances that vary with context and over time, and the specifics are therefore not discussed further in this book. For further information refer to the Hatch-Waxman Act and a legal counsel versed in regulatory and patent affairs.

4.3.2   How much does it cost to get a patent?

Table 4.1 contains the various fees involved in obtaining a patent in selected countries.

Additional costs:

- Lawyers' fees to draft the first patent application could range from $3,000 to $15,000, depending on the complexity of the topic and patent.
- Foreign filing fees, adding up various nations' patent offices where you choose to file, can range from $30,000 to $50,000 and to that you can add the local lawyer's translation and representation fees, which can range from $40,000 to $80,000 depending on how many countries are selected.

These numbers will vary widely based on factors such as the type of patent, the complexity of the subject matter, and the number of countries. Also, these estimates do not include the cost of continued prosecution steps where examiner's denials and objections need to be addressed by the lawyers in order to successfully get a patent issued. Various filing strategies can be used to delay the payments (in consultation with your lawyer), giving time to determine commercial value of the invention, but the amounts due will have to be paid eventually.

4.3.3   Considerations before filing a patent

What are the steps involved in capturing an idea and defining it as new intellectual property? Table 4.2 summarizes the process for three different entities, including individual inventor(s).

Table 4.1 Patent fees in selected countries

	DE	GB	CN	JP	US
Application filing fee	40 €	30 £	950 ¥	15 000 ¥	280 $
Search report	300 €	130 £			600 $
Request for examination	350 €	100 £	2 500 ¥	118 000 ¥	720 $
Granting fee	– €	– £	255 ¥	– €	960 $
Total	**690 €**	**260 £**	**3 705 ¥**	**133 000 ¥**	**2 560 $**

DE= Germany, GB=Great Britain, CN=China (Chinese Yuan), JP= Japan (Japanese Yen) , US=United States)
Note: In the United States, a reduction of 50% or 75% of total fees is available for small entities/individual inventors.
Note: Additional fees are charged for patents with more claims (e.g. in Japan add 4000¥ per claim; others charge per claim over a threshold).
(Table adapted from www.patent-pilot.com/en/obtaining-a-patent/costs-of-obtaining-a-patent/)

Table 4.2 Steps to consider before filing a patent for academics, corporate scientists, or independent inventors

Process steps	Independent inventor	Academic investigator (PI)	Corporate scientist/ inventor
Confirm novelty of idea	Search USPTO and European Patent Office (EPO) websites and conference proceedings, papers, etc.	Search USPTO and EPO websites and conference proceedings, papers, etc.	Search USPTO and EPO websites and conference proceedings, papers, etc.
File disclosure before publishing anywhere	Find a good lawyer and discuss process with them – read other similar patents	With technology transfer office (OTT) of university	Through supervisor, to Chief Scientist/ Technology Officer or other designated person
Confirm business case – decision to file patent (this Section 4.3.3 has more details)	Market research activities	PI should write up commercial significance in disclosure to guide OTT personnel in their market research	Depends on company process, but often, the scientist has to write up preliminary business potential (impact) in disclosure. A committee of commercial and scientific executives will make the decision to invest in patent
Patent filing	Typically a provisional patent is filed to allow time for further investigation of commercial potential or feasibility	Typically a provisional patent is filed to allow time for further investigation of commercial potential or feasibility	Typically a provisional patent is filed to allow time for further investigation of commercial potential or feasibility

Considerations before filing patents

- Cost:
  - A decision to file a patent is an investment and business decision.
- Ability to obtain and enforce:
  - How clear is the patent space for your invention (are there many people patenting similar inventions)?
  - Are you likely to be able to detect infringement of the patent in the future?
- Business need:
  - This is a key point in filing decisions: Will your business be helped with this patent. Can you really protect the market exclusivity that a patent provides? Is there likely to be much competition and can your product be easily reengineered, going around your patent claims ? Will your patent claims, when finally issued (after prosecution through patent agency) be broad enough to be of value in blocking others?
- Return on investment:
  - Finally, when can you gain a return that justifies the cost and time to file? An economic evaluation or estimation must be made at each stage in the prosecution.
- One more technical assessment that is important in this context is a *freedom to operate (FTO) assessment* (see Section 4.3.13), which essentially attempts to answer the questions: Is it possible to practice this invention/technology without infringing the patent of a third party? Does someone else have a broad patent claim on a particular use case area that would prevent you from practicing your invention? These questions typically arise once the invention is developed further as a product with defined use cases.

  The FTO search can be expensive and technically challenging to do in a thorough fashion. This search is quite different from searching for prior art (see Section 4.3.6), which is done to determine the **novelty** and hence the patentability of your invention.

4.3.4   Is the idea or invention ready to patent?

The idea or invention is considered ready to patent when it has been conceptualized, which is a two-step process:

Step 1: Conception exists when there is a formation in the mind of the inventor of a definite and permanent idea of the complete and operative invention as it is to be used.

Step 2: Conception is completed when someone ordinarily skilled in the field could perform the process or make the composition, when the concept is conveyed to them, without unduly extensive research or experimentation.

"Actual" reduction to practice of the invention by the inventors is *not* required,. meaning that the inventor(s) can file a patent without physically building or actually demonstrating the invention working. Note, however, that actual data or experimental evidence always bolsters the case for a patentee in the process of review and rebuttal with the patent examiner.

Once conceptualized, it is important to file a patent as quickly as possible to get the earliest filing (priority) date. However, in the case of biomedical inventions, it is equally important to have an adequate amount of data in the patent (e.g. to show therapeutic effect or beneficial diagnosis of conditions with a new protein or molecule) and a balance must be struck between collecting sufficient data and filing as soon as possible. Drug companies will wait until clinical trials start to file a patent, preferring to maximize the commercially useful patent life. In academia, there is an additional pressure to be the first to publish, but in general the inventor should file before publishing in order to best capture the broadest global rights, as certain countries will treat any publication before to the filing date as "prior art" and reject the patent application. (See Sections 4.3.8 and 4.3.9, and Box 4.4.]

4.3.5   Determine the inventor(s)

This step should precede your preparation to file the patent application, as inventor-ship determines ownership. In general, the inventors are the (joint) owners of the patent unless their work contracts or other obligations requires them to assign the invention to their employer or other parties.

How do you identify who is an inventor? Each inventor, to be considered an inventor, *must* have made a contribution, individually or jointly, to the subject matter of at least one *claim* of the application.

Box 4.2 Who is not an inventor?

Thanks to Karl Hermann of Seed IP, USA for his input on this and other topics

Assisting in reduction to practice, is not enough to establish inventorship, regard-less of how arduous or time consuming the effort was. However, during the process, if an assistant (i) encounters a problem that requires more than ordinary skill to overcome, (ii) comes up with a defined way to resolve that problem, and (iii) that defined resolution is reflected in a claim, then that assistant could be an inventor.

So, who is not an inventor? Someone who:

Reduces another person's invention to practice
Contributes an obvious element to the invention
Merely suggests a result without a means to obtain it
Only follows instructions from others
Only explains how or why the invention works
Adopted derived information
Became involved after conception occurred
Contributes an aspect to a journal article or patent application, but this aspect is
   not claimed in the application

> **Box 4.2** (*cont.*)
>
> Provides information of a well-known or general nature
> Publishes information relied on for an invention
>
> Improper inventorship can be grounds to invalidate a patent, but inventorship can be corrected during prosecution and even after issuance of a patent.

4.3.6   Steps to prepare a patent filing

1. Search for keywords describing your idea among existing patents and published patent applications at the USPTO or the European Patent Office (EPO) websites (www.uspto.gov or www.european-patent-office.org) or in your local public library. Once you have determined your idea has not been covered by other patents, you can do the next search for prior art. The process of comparing your invention to prior patents or to the literature can be complex, and includes factual and legal considerations. For example, the legal meaning of terms in the claims of a patent have to be determined before a proper comparison can be made. Note that patents are classified by technical area into over 70,000 codes (International Patent Classification, IPC) making searching easier once you identify the specific class and code for your technology.
2. Find out if any prior art exists – do at least a preliminary search on conference abstracts or scientific or technical publications where your idea might have been published by someone else. If you do not find any, you can move ahead to prepare and file an application.
3. Choose an attorney (unless you are doing it yourself) and file a patent application (see later sections for more details on the process of filing patents) after consultation with them. Get the attorney to draft the claims. The first to file a patent for an invention gets the rights (the United States shifted its practice to this worldwide system in 2013).

4.3.7   What is in a patent? How to read an issued patent

Patents are composed of four main sections:

Face page: This normally contains the patent number, date of issue and application, inventors, assignees (owners of the patent rights at time of issuance), abstract and title of patent, and other miscellaneous information. This important information includes the application date and the issue date (from which one can determine remaining term of the patent), the inventor(s), and most importantly, the assignee (the owner of the patent rights).

Claims: This section has numbered statements. It tells you specifically and pointedly what invention is made and claimed for protection under the patent laws. The claims consist of independent claims (these have no reference to any other claim in that patent and can stand alone) and dependent claims. Even if the independent claim may be found to be invalid, the dependent claims may still be allowed, as they may include additional features that allow the prior art. A dependent claim adds one or more features to the independent claim to which it refers. For example, if a chair per se were patentable, an independent claim might be: "(1) A chair comprising at least three legs." A dependent claim might be: "(2) A chair as in claim 1, having four legs and further comprising a cushion and a seat back." Thus, if an independent claim is patentable (new and non-obvious), its dependent claims will also be patentable (because adding features cannot take away the novelty or make the invention obvious). The language in the patent claims is critical to legal definition and defense of the patent holders' rights and the use of single word can change the legal rights of the patent holder. The use and interpretation of even a single word can change the legal rights of the patent holder as seen in the example in Box 4.3. This is one important reason to have a lawyer experienced in patent prosecution construct the final patent and claims. Claims define the scope of the invention. Much like a fence around a property, the claims of a patent define the limits of protection, or rather the ability to block others from practicing the claimed aspects of the invention.

Description: This section further adds detail to the claims by virtue of examples and background. This section is very important as it is often used by the examiner to determine patentability and the meaning and scope of the claims. Also, the claims may be modified after the patent applied for, if supported by content in the description, but the description cannot be modified once it is filed. (In special circumstances the claims may be amended after the patent is granted, in what is called a "reissue" proceeding.) The *description and specifications* must be carefully drafted and content must support the claims and scope of the invention.

Drawings: These could be images, photos of gels, schematics, process flow diagrams, or graphs of data. They help to convey more specifics on the invention and help to substantiate the application and reduction to practice of the invention.

Box 4.3 Why you need legal counsel in writing a patent

An inadvertent choice of words that may seem innocuous in regular spoken or written language can have different meanings and significance in a legal context and may end up reducing the scope of the claimed invention and its uses and thus also reduce the subsequent commercial value of a patent. The following example serves to highlight the reason why a well-written patent is key to commercialization.

An inventor in a university isolated and discovered a novel protein that seemed to have significant application in curing a major cardiovascular disease. Patents to the invention of isolating and preparing the specific protein were filed by the university, claiming the specific sequence, subsets of the protein (active sites)

Box 4.3 (*cont.*)

and DNA, antibodies, etc., resulting in a patent family of over five patents. Company A licensed in the patents that were filed by the university, gaining exclusive rights to the patented technologies, thus ensuring that no one else could use that isolated protein to develop drugs (synthetic small molecules or antibodies) against that target. Drugs targeting that protein were then developed in the company A over many years, requiring significant investment (millions of dollars), on the presumed strength of the patent around the target and on subsequent patents on the composition of matter of new small molecules developed by the company. The market was predicted to be greater than $800 million/year for a successful drug against that target. Six years later, Company A discovered that Company B was also developing drugs against that same target, obviously making and using the target protein to identify the drug and potentially infringing the exclusively-licensed patent on the target protein. Company A wrote a letter alerting company B of its patents and asking them for an explanation. Company B wrote back stating that they were not infringing on the patent , because as they were making and using a fragment or a subset of the protein (an active site peptide). A closer review by company A's attorney (different from the one who filed the patent) revealed the surprising news that in fact, company B might be right. The patent claims used the wording "consisting of" when claiming and identifying specific subset sequences of the protein. E.g. One of the key claims read: "The protein claimed in claim 1, which *consists of* sequence 1 or sequence 2, etc. . . ." Company B was making a peptide of the active site of the protein (relevant for analyzing drug targeting activity) that had a few more amino acids than those listed in the subset sequence 1 of the claimed protein.

Anyone who made a peptide that had one or more amino acids than the specified sequence 1 would literally not be infringing on the patent (see Section 4.3.12 for more details on infringement). Company A would then have to argue in court, if possible, that Company B's peptide is a legal "equivalent" to what was claimed – an exercise with an uncertain outcome. It would have been better to have sought broader patent claims in the first place. In this example, company A learned that it would have been far better to use the phrase "comprising" rather than the restrictive phrase "consisting of." The broader language could have included peptides that were longer than the reported Company A sequence 1, such as Company B's peptides. In other words, the patent claims in this simple example turned out to be too easy to design around. This subtle difference in claim drafting language, that was not a significant difference in normal conversational usage or to an untrained person, had significant implications for Company A. Company A potentially lost the opportunity (at least the inexpensive negotiated option, rather than the expensive and uncertain option of going to court) to clearly and unambiguously block a competitor drug developer who was using the patented technology (in spirit, if not in the letter).

> **Box 4.3** (*cont.*)
>
> In general, the term "consisting of" is more restrictive on the scope of the patent claims than the term "consisting essentially of," which in turn is more restrictive than the term "comprising." The goal of the inventor is to claim the broadest possible scope of their invention and it is the job of the patent office to in turn try to use the most precise language to describe the invention. This example highlights the need to use trained patent counsel in the writing and prosecution of a patent.

4.3.8 Provisional patent application process in United States

The U.S. patent system is now a first-to-file (as in the rest of the world, but the concept of the "provisional" patent application that serves as placeholder for a full, or nonprovisional, patent application is still prevalent in practice. The full (nonprovisional) application needs to be filed with 1 year of filing the provisional. If the patent is also filed in other countries before the 1-year deadline, the filings in those countries will have a priority date of the first, "provisional" filing in the United States. This gives U.S.-based inventors 1 year during which their invention is protected, and they can safely publicize their invention while deciding whether to file a nonprovisional application in the United States and the rest of the world.

However, any subject matter or claims that are not in the original (provisional) filing will be viewed as new subject matter, and in some countries, the provisional patent content may become prior art for this new subject matter. Therefore, many firms recommend to file complete, nonprovisional applications from the beginning.

One year from this submission, if you decide not to file a full (nonprovisional) international (Patent Cooperation Treaty, PCT) application, then the (provisional) patent in the home country lapses without review. This process is similar to most other countries. (See Figure 4.3 and Section 4.3.10.) Many people use the 1-year period to better evaluate commercialization prospects of the invention and justify the costs of filing a global patent application. A foreign inventor also has 1 year after filing of the patent in their country to file an application in the United States and globally (priority date will be the date of their original country filing).

4.3.9 Priority date, grace period, and public disclosure

The priority date is usually the date of filing of the first patent application. That means that any public disclosure before the date of filing is viewed as prior art in many countries and can invalidate the patent application as not being original, even if the public disclosure was by the inventor. Most countries are members of the PCT, so that a filing in one country has effect in all member countries (see Section 4.2.8 for details on the PCT).

Some countries do allow for a grace period where the applicant can disclose their invention publicly before filing a patent and not have that disclosure count as prior art.

However, not all countries allow a grace period and disclosures made publicly before the patent filing date can invalidate the patent filing. See Figure 4.2. Hence, filing a patent application before any public disclosure is the best practice to protect patent rights globally without restriction (see Box 4.4 for details on grace periods).

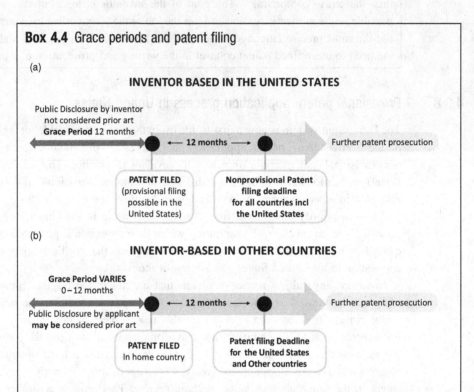

Box 4.4 Grace periods and patent filing

(a)

INVENTOR BASED IN THE UNITED STATES

Public Disclosure by inventor not considered prior art
Grace Period 12 months

⟵ 12 months ⟶   Further patent prosecution

PATENT FILED (provisional filing possible in the United States)

Nonprovisional Patent filing deadline for all countries incl the United States

(b)

INVENTOR-BASED IN OTHER COUNTRIES

Grace Period VARIES 0–12 months
Public Disclosure by applicant **may be** considered prior art

⟵ 12 months ⟶   Further patent prosecution

PATENT FILED In home country

Patent filing Deadline for the United States and Other countries

Figure 4.2 Grace periods and patent filing timelines example for an inventor (a) in the United States and (b) outside the United States

Figure 4.2 shows that the pre-application conditions for patents vary across countries. In the interest of preserving the maximum jurisdiction of rights globally, a patent should be filed *before* making any public disclosure of the invention. As explained in this chapter, a fully written, nonprovisional patent is usually recommended to be prepared and submitted at the first instance of patent filing.

Some examples of countries that allow for a 12-month grace period:

Argentina (AR)   Australia (AU)   Brazil (BR)   Canada (CA)
Columbia (CO)   Mexico (MX)   Sri Lanka (LK)   Turkey (TR)
Ukraine (UA)   United States (US)

Some countries that allow a 6-month grace period:

Russian Federation (RU)   Chile (CL)   Japan (JP)   South Korea (KR)

Some countries that do not have any grace period (or allow on a case-by-case basis a grace period where prior disclosure was made under specific, limited circumstances):

> **Box 4.4** *(cont.)*
>
> United Kingdom (UK)   China   Most Western European countries in the EU
>
> **Note:** The above lists are current as of May 2020 and are subject to change based on regulations in each country.

Why is recording the process of invention important?

In the past it was very important to keep laboratory records (in the United States) solely to establish the date of invention. However, since the United States is no longer a first-to-invent country, it is less critical to keep dated notebooks, but there is still a good case to be made for maintaining a thorough laboratory or notebook record, as it could be useful in some disputes.

Recording the process of inventing can be very important in confirming inventorship in group inventions, affirming the right to file the patent as inventor even if a project member publishes the work in the grace period allowable in some countries, and is generally useful to recognize and protect the inventor in various other circumstances. Record contributions made in group discussions to help identify claims of inventorship.

Take the example where an inventor communicates the invention to a colleague as part of a collaboration, and that colleague publishes their results before the inventor has filed a patent application. Since the collaborator made the first public disclosure, the inventor will have to establish inventorship and proof of communication of the invention to the collaborator in order to take advantage of the 1-year grace period (in some countries see Box 4.4) and obtain a patent.

Record your ideas, experiments, thoughts, and any evidence such as photos in a bound laboratory notebook with numbered pages. Invention is a conception or an idea, not just data from experiments, so ideas must be entered in the laboratory notebook. Cross out errors instead of erasing. Date each page in the book and have a witness (not a co-inventor) initial/sign each page. Typically, invention records should be corroborated by at least one person who is not an inventor, and is under an obligation to keep the invention secret before the patent application is filed. No pages should be removed from the book, and blank pages should be crossed out and dated. As electronic record-keeping increases, methods have been developed to record, date, and authenticate computer files, e.g. electronic lab notebooks. A knowledgeable attorney should be consulted regarding electronic record and document retention procedures.

4.3.10 International patent filings and the Patent Cooperation Treaty (PCT) process

As per information published on the World Intellectual Property Organization (WIPO) website (www.wipo.org), the PCT is an international treaty, administered by the WIPO between more than 125 Paris Convention countries. The PCT makes it possible to seek patent protection for an invention simultaneously in each of a large number of countries by filing a single "international" patent application instead of filing several separate national or regional patent applications. The granting of patents remains under the control of the national or regional patent Offices in what is called the "national phase."

Briefly, an outline of the PCT procedure includes the following steps:

Filing: You file an international application, complying with the PCT formality requirements, in one language, and you pay one set of fees.

International search: An "International Searching Authority (ISA)" (housed by one of the world's major patent offices) identifies the published documents that may have an influence on whether your invention is patentable and establishes an opinion on your invention's potential patentability.

International publication: As soon as possible after the expiration of 18 months from the earliest filing date, the content of your international application is disclosed to the world.

International Preliminary Examination: An "International Preliminary Examining Authority (IPEA)" (one of the world's major patent Offices), at your request, carries out an additional patentability analysis, usually on an amended version of your application.

National phase: After the end of the PCT procedure, you start to pursue the grant of your patents directly before the national (or regional) patent offices of the countries in which you want to obtain them.

The next section puts this international process in perspective with respect to the timelines and steps in the patent prosecution process in the United States.

4.3.11   Patent prosecution process

Typical patent processing timeline for U.S.-based inventor/applicant (see Figure 4.3):

Step 1: Document your invention – the invention date is important and is critical in the United States, as the person first to invent (date recorded in lab book for example) has priority to patent rights. In the rest of the world, first to file has claim to rights.

Step 2: File provisional patent with USPTO. The clock starts ticking on any patent that might issue from this "priority date." For international patents, you have 1 year from this date to submit a full application in other countries through the PCT process.

Step 3: Within one year after the provisional filing, submit a full, nonprovisional patent application to the USPTO. Simultaneously submit a PCT application to a receiving office (in this case, USPTO). PCT applications by default include all countries party to the treaty. Selection of specific countries is done later. The application is now in "Chapter I" of the PCT process. The PCT application can claim the U.S.-based priority date in most countries.

Step 4: Applications get published and are available for public review typically 18 months after application date. For PCT filing, an International Searching Authority, typically an examiner in the receiving country patent office, conducts a

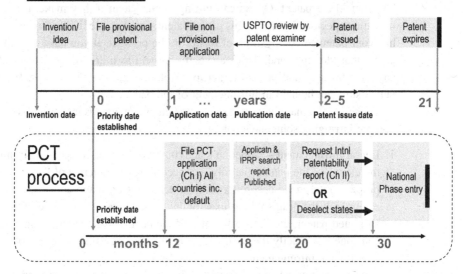

Figure 4.3 Steps for filing and prosecuting patents in the United States and through the PCT process

search and issues an International Preliminary Report on Patentability (IPRP). The search results are also published. Chapter I lasts for 30 months with no further action after unless Chapter II is selected by filing a demand. Meanwhile, at the USPTO, patents are under review and communications from the patent examiner have to be addressed and claims adjusted.

Step 5: At 8 months after PCT submission (20 months from priority date), the applicant has to select the final countries desired for final filing (national phase in each country). If demand for Chapter II is filed, a substantive examination report will be prepared, taking applicants comments (if any) into consideration. This step might be helpful in speeding up national prosecution in the receiving or home country as the examiner is usually the same person in PCT and in the receiving office.

Step 6: At 30 months from submission into the PCT process (or from the claimed priority date), the patent enters national phase filing in the countries that were finally selected. Fees for individual countries (translations, filing fees, lawyers' fees) will come due when national phase applications are filed.

Step 7: U.S. patents will issue anywhere from 2 to 5 years from initial application depending on the backlog at the Patent Office and the specific issues in the application. In recent years, the average time for first patentability review by an examiner has been about 20 months from filing, and average time for issuance of a patent has been 30 months.

4.3.12   Patent Infringement

Patents give you the right to block anyone else from selling your invention in the market. Your rights in this regard are limited to the legal boundaries of the country that has granted the patent. Others can infringe your patent if they make, use, or sell the invention specified by the claims of the patent, for example, if they sell a product that has your patented composition of matter in the product. If you have a method-of-use patent that identifies and claims a specific protein as a drug target to treat a disease, anyone who uses that protein target to develop a drug for that disease without your permission can be held in infringement of the patent.

There are two main ways to examine whether someone is infringing your patent or you are infringing someone's patent:

1. Literal infringement – someone's activities can be seen to specifically infringe on the claims of the patent, literally, i.e. according to the terms of the claims (as properly interpreted according to the patent laws for claim construction). For example, if the drug sold has exactly the formula described in one of the claims in an issued patent, then it is a literal infringement. If the protein being sold by someone has exactly the same amino acid sequence as listed in one of the claims, then it is an infringement.
2. Doctrine of equivalents – a patent claim that is not literally infringed may be infringed if the accused product or method is legally "equivalent" to what is literally claimed. This is in the gray area, where you believe someone is infringing on the spirit of the patent. If the difference between what is being claimed and what is accused of infringement is not substantial (i.e. it uses substantially the same way to achieve substantially the same result), then there may still be infringement (provided the slightly different methods or materials have not otherwise been disclaimed). In the above example, if the infringer is selling a drug with a subtle twist on the composition to skirt the formula laid out in the claim and still attains the same results, you could argue in the courts that it is equivalent to your composition and the person is infringing. This is a complex argument that relies on prior cases and should be reviewed by an experienced patent infringement lawyer. In general, the literal infringement case is much stronger, and the outcome of arguing equivalents in infringement is largely dependent on the context of the case and the perspective of the judge.

Probably the most contentious area in patent infringement lawsuits is the ability to establish intent to infringe on your patent rights, or "willful infringement," which can result in enhanced damages, up to treble the amount of the actual damage from the infringement. (Intent is not required to prove infringement; only to recover enhanced damages). Infringement lawsuits are also very expensive to fight due to the legal and technical expertise involved. However, the awards in infringement lawsuits can be very high with one of the highest being the suit between Polaroid and Eastman Kodak in which Polaroid secured a $925 million judgment against Eastman Kodak in 1990. The drug and device industries value patents and have had their share of patent disputes in the past and continuing to date.

Owing to the high cost, complexity, and level of technical expertise (both in technology and legal process) involved, taking on a patent infringement issue must be done with careful planning. It is a good idea to check the validity of the patents (held by the plaintiff) before embarking further into the process. An issued patent can still be ruled invalid and the patent which is being infringed will be tested in court by the defendant, so it is a good idea to run a validity check. If the particular action or product is in direct literal infringement, the case can be put together with more confidence. The main issues remaining then relate to the extent of damages claimed by the plaintiff and the possibility of establishing willful infringement

Conversely, if the infringement is not literal (making, using, or selling a product that is not exactly the same as that claimed or described in the patent), the next step is to put together a complete enforcement plan. In most cases (even in the case of literal infringement), this is probably a good step to take in order to clarify and get agreement on the approaches toward other potential infringers. A good legal team from one or more firms that have no conflicts with the potential infringing parties must be put together so that there is a consistency in the approach through the process.

4.3.13   "Freedom to practice" or "freedom to operate"

A patent allows the holder of the patent to stop others from commercializing products based on the patented invention. However, a single patent may not be sufficient to allow one to develop and sell a product based on the single patent (see Box 4.5). For example, a patent holder may have a claim to a drug compound with a known mechanism of action in the body to treat immunology diseases. If someone else discovers a new mechanism by which that drug can treat cardiovascular diseases, they can patent that invention as a novel finding. The primary patent holder now does not have freedom to operate in the cardiovascular area, but can continue to sell the drug for immunology diseases. The holder of the patent for cardiovascular application of the drug also does not have freedom to operate as they cannot sell the drug until they obtain a license for the drug compound from the primary patent holder.

This means that the holder of a key patent has to make sure they have the freedom to practice the patented technology in all commercial applications claimed. In order to perform an FTO search, typically you would look at issued globally issued patents in the field of application or technology area and also look at published applications that have yet to issue as patents. In order to make sure that the path to market is clear and protected for the final product, the manufacturer may need to license or own additional patents, in addition to the original invention patent on which basis the product was developed. This review of FTO is recommended to be done before the drug product progresses into commercial manufacturing, as discussed further with examples in Chapter 7.

Typically, when licensing in or buying or selling patents, a "freedom to operate" statement should be sought from a lawyer, establishing the ability to commercialize and capture the value of the patent. Note that this is not risk free, because no freedom to operate search can be guaranteed to uncover every unexpired patent which may be

---

Box 4.5 Freedom to operate

An example in drug development: If a drug targets the active site of protein X and thus treats a disease or its symptoms, the inventor of the drug can file a patent for composition of matter of the drug, protecting its chemical structure and claiming it as a likely treatment for the disease. That patent may issue on the merit that the novel composition of matter showed the inventive step and the data in the patent showed the useful step of disease treatment.

However, if someone has found earlier that specifically targeting a ligand to the active site of protein X can successfully treat the disease or its symptoms, they too can potentially be issued a patent claiming that *any* specific ligand for protein X can treat that disease, with some example drugs used to validate and reinforce the claim (not including the novel drug mentioned above). Therefore, the developer of the newly patented drug compound cannot sell their drug in the market to treat this disease unless they have licensed or otherwise obtained rights to the method of use patent described above. Another option would be for either party (typically the holder of the composition of matter patent) to challenge the validity of the other patent in court.

Thus, someone buying the company would be well advised to review existing patents or patent applications and check whether the value of the drug compound patent may be compromised by any other patents.

---

relevant. As a matter of prudence, a pragmatic and reasonable search and analysis should be done, in order to support a business case for going forward.

Drug development: A patentability search for novelty against prior art is usually done before patenting and then an FTO search is usually done once the drug is further advanced toward product with clear use cases.

Devices: A patentability search for novelty against prior art is recommended but is usually skipped or done on a cursory basis before filing a patent. FTO search is usually done when the device gets further toward a clear commercial product application.

Diagnostics: An FTO search on broad patents around the invention of new diagnostic biomarker is preferable to have done earlier along with the patentability search.

If you know about a patent that may "block" your commercialization plans, U.S. law requires a diligent inquiry to investigate whether that patent is infringed and/or is invalid. Typically, this involves obtaining the opinion of an attorney that the patent is not infringed or that it is an invalid patent, before going forward. Although an attorney opinion is not strictly required, it is recommended as a traditional way to address potentially complex and difficult patent issues for your business. Going forward without a good faith inquiry, or ignoring the results, could cause serious problems of liability and exposure for patent infringement later on.

4.3.14   Types of applications

- *Provisional application:* This is available in the United States as a potentially low-cost placeholder that is not examined and is usually used to set the priority date. However, the provisional must include a detailed description of the invention, or else the subsequent regular application submitted in a year loses the priority date for any material outside the provisional. In best practices today, the provisional application is usually the final fully drafted (same as nonprovisional) application with claims and figures, etc.
- *Nonprovisional (regular) application:* A full and final patent application with all required components (diagrams, claims, descriptions) filled in to best of inventor's knowledge
- *Continuation application:* Identical copy of original application refiled to continue prosecution on subject matter already disclosed – usually a continuation application is filed to argue to granting of denied or modified claims.
- *Divisional application:* A divisional application is the identical application refiled with a subset of the claims directed to a "different" invention in response to the "restriction requirement" by the examiner, who determined that the patent contains more than one invention. One patent = one invention.
- *Continuation in part (CIP) application:* This a new application that adds subject matter to the original application. However, the claims based on the new subject matter get a priority date as of the CIP filing and the claims covering the original subject matter get the original priority date so in effect you could be reducing the term of the new subject matter in certain claims.

4.4   Trademarks

Trademarks are usually some combination of words, phrases, symbols or designs that identifies and distinguishes the source of the goods of one party from those of others. For example, the colors, font, placement, and other features of the Becton, Dickinson (BD) logo ✹ BD (https://1000logos.net/lipitor-logo/) are owned by the Becton, Dickinson and Company. Similarly, the combination of words that make the product name BD Fluorosensor^{TM} is a trademark owned by Becton, Dickinson. Another example is Lipitor$^®$, a drug name that is trademarked by Pfizer for cholesterol lowering pharmaceutical preparations. Any representation of this specific product must use the trademarked name precisely as specified by the owner (usually in correct colors, combinations and order of letters, words or images).

A service mark is a similar combination of words, phrases, symbols, or designs that identifies and distinguishes the services of one party from those of others.

To claim rights in a mark, use the "TM" (trademark) or "SM" (service mark) designation to alert the public to the claim, regardless of whether an application has been filed with the USPTO. You do not have to register a trademark to use one or have legal claims. Essentially, the first person to either use a mark in commerce or file an intent-to-use application with the USPTO has the ultimate right to the use and registration of a mark. While not required, a trademark registration provides important legal benefits. The federal registration symbol "®" can be used only after the USPTO actually registers a mark, and not while an application is pending. Also, the registration symbol with the mark can be used only on or in connection with the goods and/or services listed in the federal trademark registration. A trademark is renewed every 10 years indefinitely, as long as it is still in use. An intent-to-use application can also be filed to safeguard the trademark before putting it in use.

4.4.1   Why register your trademark?

- Establish ownership, usage, and date of original use
- Enhanced national exclusivity for use in registered categories – easier to expand business
- Better legal protection for infringement actions
- Ability to license trademark

4.4.2   Filing a trademark with the USPTO

It will cost about $275 (for electronic filing) to file an application for use of the trademark in one class of goods or services and additional fees for use in each additional class. If a lawyer is used to file an application, fees can range from $300 to $1,500 for initial application and more for legal fees if the initial application is challenged by the USPTO.

There is good reason to look for legal assistance on registering trademarks and service marks, because the prosecution process can be quite complex. A trademark is always given for a specific category of goods or services. A trademark holder can use the trademark for multiple categories but each category of goods would typically require a separate trademark filing. Box 4.6 contains a famous example of a trademark dispute.

4.4.3   International filing of trademarks

The Madrid Agreement and Protocol (effective in the United States from 2003 onward) allows the U.S. trademark registration to serve as the basis for international coverage in countries party to the Madrid Agreement. A pan-European registration with the European Union is also advisable, as the EU did not sign the Madrid Agreement (although individual countries in the EU did).

4.5   Copyrights

From the USPTO website (https://www.uspto.gov/ip-policy/copyright-policy/copy right-basics): "A copyright is a form of protection provided to the authors of 'original

---

> **Box 4.6** Trademark dispute example
>
> A famous example of litigation on trademark is described briefly here. Apple Corps, the record company set up by the Beatles in 1968, took Apple Computer (established 1977) to court over the trademark violation of the use of the name "Apple." The two settled in 1981, with the Apple Computer company agreeing not to sell music and settling the dispute with Apple Corps for $25 million. The agreement gave the Apple Computer company rights to the Apple trademark in relation to electronic goods, computers, telecommunications equipment, data processing equipment. It also allowed for trademark use related to "data transmission services" and "broadcasting services," as well as related promotional merchandising. An additional dispute in 1991 when Apple started making music recordings possible on its computers ended with another settlement payment to Apple Corps. In 2006, since the Apple Computer company started offering music for downloads on its iTunes website, the Apple Corps took the computer company to court insisting that they were now using the Apple logo and trademark for music, in violation of the 1991 agreement. This case was finally decided in favor of Apple Computer in May 2006.

works of authorship' including literary, dramatic, musical, artistic, and certain other intellectual works, both published and unpublished. The 1976 Copyright Act generally gives the owner of copyright the exclusive right to reproduce the copyrighted work, to prepare derivative works, to distribute copies or phonorecords of the copyrighted work, to perform the copyrighted work publicly, or to display the copyrighted work publicly. The copyright protects the form of expression rather than the subject matter of the writing. For example, a description of a machine could be copyrighted, but this would only prevent others from copying the description; it would not prevent others from writing a description of their own or from making and using the machine." Copyrights are registered by the Library of Congress' Copyright Office.

Copyright protection exists from the time the work is created in fixed form. The copyright in the work of authorship immediately becomes the property of the author who created the work and lasts for 70 years after the death of the author.

This topic will not be discussed further here, as copyright protection has limited application in biotechnology. For further details on copyrights and registering copyrights, go to www.copyright.gov/.

4.6 Trade secrets

Trade secrets are also commonly termed as know-how. Even in a patent or scientific paper where details of a process have to be laid out for anyone to repeat, experienced scientists know that there is an art, a know-how that is not described in the details. This know-how, if it can be kept a secret can then build value for a company as its

> **Box 4.7** Trade secrets in drug development
>
> As an example, it is well known that when companies develop crystal structures of drug-targeted proteins, they may publish or patent the crystal structure, but the multiple structures with different proprietary compounds that help them understand how to structure a more potent compound are never published. This is the know-how or trade secret held by the scientists or inventors in the company. If the project is to be licensed to a partner for commercialization, this data and tacit knowledge remains a trade secret within the company and has great value as a form of intellectual property – know how or trade secret (data that can help quickly design a potent compound that binds the target protein correctly, eliminating months to years of experimentation).

trade secret. However, there are few commercial products where you can sustain a competitive advantage for long by keeping the process or composition of the product a trade secret. The most famous example of a trade secret is the Coca-Cola formula, which has never been patented and workers at the production factories are bound by elaborate agreements and arrangements to keep them from disclosing the single part of the process they know about.

However, in rapidly moving technology areas like biotechnology or devices, it is rare to be able to sustain an advantage of greater than a year or so by keeping a trade secret (see more discussion in Box 4.7). While trade secrets are not commonly used as the sole means to protect commercialization rights for discoveries or inventions, their use as know-how is increasingly useful in the detailed processes of biopharmaceutical manufacture. Some biopharma companies develop processes that are not disclosed except in FDA filings which are kept confidential for a number of years. Note, however, that the "best mode" of practicing an invention must be included in any patent filed for that invention; it cannot be held back as a trade secret.

4.7   IP commercialization and technology transfer

4.7.1   Commercial use of intellectual property

Predominantly in the biotechnology industry, and increasingly so in the information industry, business models have revolved around the management of intellectual property (IP) rights and the appropriation of maximum value from acquired or developed intellectual property. The limited monopoly granted by a patent gives some degree of assurance of possible revenue streams for a new product, especially products that cost a lot of money and time to bring to market. This reduction of market risk gives rise to a value for the IP itself, a value that is difficult to calculate accurately, but is nonetheless assigned by outside market forces.

A patent is used to generate revenues (and profits) by building products or blocking others from selling products based on that invention, or in the shorter term to raise financing, by:

1. Creating a limited monopoly for the patented product and thus set a favorable pricing for the duration of the patent term.
2. Licensing the patent to others who wish to sell the patented product and collecting royalties and payments from them, retaining ownership of the patent
3. Selling the patent to someone else (assigning or giving up ownership and all future rights to the patent) in a one-time transaction
4. Using the patent position to drive strategic partnerships
5. Attracting investment by reassuring investors of future revenues with a strategic patent position
6. Getting debt financing on the strength of the value or future revenues to be gained from the patent portfolio

4.7.2 Technology transfer in academic research institutions

While the role and goals of non-profit institutions in commercializing biomedical patents may be questioned, a significant number of breakthrough inventions that have changed and will change markets and technology trajectories continue to be developed at universities or research institutes. Industry has to work with these institutes to gain access to this know-how and turn it into revenue-generating products. Embodying this knowledge (tacit or explicit) in a patent makes one aspect of technology transfer dominate the process – the licensing of patents. However, although easy to overlook, it is important to understand that in many cases, there exists a know-how that goes beyond the content of the patent. Particularly in those cases, it becomes clear that technology transfer is a "contact sport," wherein scientists from the industrial licensee must interact closely with the inventor-scientist in non-profit organizations. This is the total aspect of technology transfer, not only the transfer of rights to the patented invention but also transferring the more tacit understanding and implementation of all aspects of the invention/technology.

Non-profit academic institutions and for-profit licensees can have inherent misunderstandings as each has a very different context that must be kept in mind as they work together to build value and commercialize inventions (Figure 4.4). It is thus critical for both parties to be accommodating, understanding and above all focused on the common goal to successfully work together. Figure 4.4 highlights the different perspectives between the two organizations that come together – one (non-profits) with public good as the driving motive and the other (for-profit small or large corporations) with shareholder profit as the driving motive.

4.7.3 The Bayh–Dole Act

In the United State, the 1950s–1970s saw an expansion of government R&D and a growth in R&D procurement from universities with increased competition for government research grants. By the late 1970s, there was a recognition of unrealized potential from government inventions – of 28,000 government patents, only 5% had been licensed. There was little or no uniformity in government intellectual property policies.

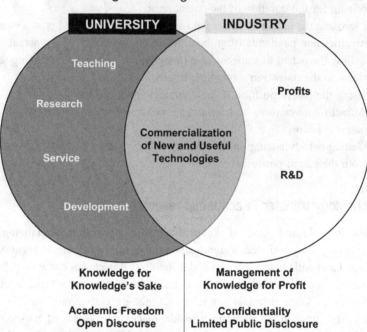

Figure 4.4 The difference in perspectives and common interests that intersect in a licensing negotiation between university and industry. Accommodation for each others' perspectives, compromise in terms of engagement, and an understanding of a common goal are required for the two parties to successfully work together. (Adapted from: Louis Breneman (2003), University-industry collaborations: partners in research promoting productivity and economic growth, Research Management Review 13(2).)

In 1980, the Bayh–Dole Act made a fundamental change in this process: Title to inventions made with government funding by small businesses, universities, and other non-profit entities would henceforth belong to those entities, and not to the government. This act also created a uniform intellectual property policy for all government agencies.

However, with these rights to the research institutes came certain duties and obligations:

- Research institutes must file patents on inventions they elect to own.
- There must be a preference for small business licensees.
- Licensed products must be manufactured in United States.
- Royalties must be
  - shared with inventor and
  - used for research and education.
- Government retains non-exclusive right to use the patented technologies for the government's own internal use. (This clause may concern some; However, the government seldom manufactures its own products, and the government can be a

large customer for the manufacturer of the patented technology as they will pay for the products to be made.)

- The university has some detailed reporting obligations (e.g. iEdison) on all intellectual property generated with government funding.

Since 1980, many universities and non-profit research institutes have set up offices of technology transfer. Inventions from these institutes have been the source of many successful products that have improved life and health for humans and also contributed to local economic development through the growth of startup ventures that were built to commercialize specific inventions. Thus, the Bayh–Dole Act has become a landmark act that many other nations are now emulating to reap the same benefits internally.

4.8 Licensing

Patents are assets. Just as land ownership can be converted to cash by leasing rights for development or habitation or mining, patent rights can be given out. A license is a commercial and legal transaction to transfer patent rights from licensor to licensee. This grant of patent rights is described by a licensing arrangement. Typical key terms in a license agreement are discussed in detail below.

A commonly asked question: Is a contract manufacturer a licensee of my patent?

Since a patent right covers the rights to block someone from making, using, or selling technology based on your patented technology, giving a contract to a manufacturer to make your patented product does imply a limited grant of a license to "make" or "use" your patented technology. However, the terms of this type of license are usually governed by a manufacturing or R&D contract. This type of contract agreement usually explicitly states that all rights to the patent are retained by the contractor.

However, if you wish to ask another party to sell your product in the marketplace and share proceeds with you, it is particularly important to have a written agreement, setting forth the rights of all parties. Such a formal legal agreement would typically grant them a license of your patent rights which are appropriate to the joint goals of the parties. For example, the license might be an exclusive grant to a distributor, who would then share in the ability to block commercialization of the patented technology by others. Otherwise, they could technically be in infringement of your patent. Thus, a license to a patent is a way to control the sale of your invention in the market.

4.8.1 Key nonfinancial terms of license agreements

The terms of license (commercialization) agreements are influenced significantly by the type of invention/technology and by the needs of the owner of the patent.

In general, a license agreement gives the licensee one or more of the following: the rights to make, to use, to have made or to sell the products based on the patented invention. In consideration of this *grant of rights*, the patent owner is paid by the licensee. The payment terms are the subject of much discussion and debate as discussed below, but the most important terms in the license agreement are the terms involved in the grant of rights. These include a carefully worded description of exactly *what rights* are being licensed (patents, trademarks, know-how, etc.), the *field of use* for which the rights are being granted, the *territory* in which the rights are being granted and whether or not the license is to be *exclusive, nonexclusive, or sole*. The sidebar contains several examples of license grant terms.

The grant of rights describes the scope and subject matter terms as shown in Table 4.3.

It can also be important for a license to specify how the patent rights that are granted can be policed and enforced, for example who can bring a lawsuit for infringement, who pays for litigation, etc.

Typically, the subject and scope of the license also includes terms that specify rights to any improvements in technology made by the licensee or licensor. -

Other terms in the body of the license agreement are important, but these terms are the main defining terms of the license agreement and are also strategically the most important terms for the licensor to decide on a priori. Licenses can be simple or very complex, depending on the circumstances; the examples in Box 4.8 and Box 4.9 are

Table 4.3 The grant of rights

Exclusivity of rights

Exclusive license	*Nonexclusive license*	*Sole license*
Only licensee and no one else (including licensor) can commercialize patent	Licensee can commercialize but so can any other licensee or the licensor	Licensor and licensee can both commercialize but licensor will not grant rights to anyone else
		Partially exclusive license (Usually limited by Territory or Field of Use) Licensee may have rights to commercialize only in certain applications (e.g. for cancer and not any other disease;) or certain countries

Territory/geographic distribution of rights

All-inclusive (worldwide)	*Conditional or geographically limited*
The licensee has rights in any geographic region (country) in which the patent has issued. This is typically the case for exclusive licenses	Licensee has rights to commercialize or practice the patent in a defined, limited set of countries. Partially exclusive or non-exclusive licenses can use this as one criterion to better control and distribute their patent rights

Field of use

All applications/disease areas	*Limited applications/disease areas*
All applications included in the patent are typically covered in an exclusive, worldwide license, but the licensor can get creative where appropriate	The licensor can get creative and use segmentation of the license terms in specific therapeutic areas based on licensee expertise or other interests to gain maximum value out of the patent

Box 4.8 Example grant of rights from university to company

An exclusive license from University of Texas Southwestern Medical Center (actually Board of Regents of University of Texas, referred to as 'BOARD') to Myogen Inc ('LICENSEE') granted for all discoveries related to cardiovascular research from a particular sponsored research agreement has the following *grant of rights*:

BOARD hereby grants to LICENSEE a royalty-bearing, worldwide, exclusive license under LICENSED SUBJECT MATTER to discover, research, develop, make, have made, use offer for SALE, SELL and import LICENSED PRODUCTS and IDENTIFIED PRODUCTS for use within LICENSED FIELD. This grant is subject to the payment by LICENSEE to BOARD of all consideration as provided herein, and is further subject to rights retained by BOARD to:

a. publish the general scientific findings from research related to LICENSED SUBJECT MATTER and IDENTIFIED PRODUCTS subject to the terms of Article 13, Confidential Information, provided however, INVENTOR shall disclose pending publications to LICENSEE in accordance with Section 6.1 of the SPONSORED RESEARCH AGREEMENT; and

b. use LICENSED SUBJECT MATTER and IDENTIFIED PRODUCTS for research that has not been sponsored by a commercial entity, teaching and other educationally-related purposes, provided, however, that any such use with respect to IDENTIFIED PRODUCTS shall be limited to IDENTIFIED PRODUCTS which are either (1) owned by BOARD, or (2) provided to BOARD by an authorized party. Any transfer of material embodiments of LICENSED SUBJECT MATTER pursuant to this Section 4.1b shall be governed by a material transfer agreement substantially in the form attached hereto as Exhibit 3.

Here, the Licensed Subject Matter (covering any intellectual findings that may arise in the research area of interest) means

"inventions, discoveries, assays and processes covered by PATENT RIGHTS and/or TECHNOLOGY RIGHTS within LICENSED FIELD;"

where Licensed Field (defined to restrict the rights of the licensee) means

"(i) treatment, prevention, diagnosis and/or prognosis of cardiac hypertrophy, heart disease and heart failure; and (ii) determination of predisposition to cardiac hypertrophy, heart disease and heart failure."

and Patent Rights (specifically written to cover multiple definitions of intellectual property) means

"BOARD'S rights in information or discoveries covered by a VALID CLAIM ... in patents, and/or patent applications, whether domestic or foreign, and all divisionals, continuations, continuations-in-part, reissues, reexaminations or extensions thereof, and any letters patent that issue thereon, ..."

Box 4.8 (cont.)

Additionally, Licensee has rights to sub-license:

LICENSEE may grant sublicenses consistent with this AGREEMENT if LICENSEE is responsible for the operations of its sublicensees relevant to this AGREEMENT as if the operations were carried out by LICENSEE. . . .

Box 4.9 Example grant of rights between companies

In 1997, Angiotech Pharmaceuticals executed a co-exclusive license granting both Cook Inc and Boston Scientific rights to use the Angiotech drug, paclitaxel, on their vascular stents. It retained certain rights for itself. This agreement is also a well-known example of the first major convergence of a drug and a device into a combination product. The following grant of rights illustrates a co-exclusive license with all the three parties tied together for successful commercialization.

Grants. Subject to the terms and conditions hereof, the following licenses are granted hereby, each effective as of the date of this Agreement:

BSC Technology License. . . . Angiotech hereby grants to BSC an exclusive (subject only to the rights granted to Cook and reserved to Angiotech in paragraphs (b) and (c) below . . .) worldwide right and license to use, manufacture, have manufactured, distribute and sell, and to grant sublicenses to its Affiliates to use, manufacture, have manufactured, distribute and sell, the Angiotech Technology in the Licensed Field of Use solely for use in the Licensed Applications (the "BSC License").

Cook Technology License. . . . Angiotech hereby grants to Cook an exclusive (subject only to the rights granted to BSC and reserved to Angiotech pursuant to paragraphs (a) above and (c) below . . .) worldwide right and license to use, manufacture, have manufactured, distribute and sell, and to grant sublicenses to its Affiliates to use, manufacture, have manufactured, distribute and sell, the Angiotech Technology in the Licensed Field of Use solely for use in the Licensed Applications (the "Cook License").

Reservation of Rights. Angiotech reserves all rights to the Angiotech Technology for (i) any use or purpose outside the Licensed Field of Use and Licensed Applications and (ii) noncommercial research purposes in all fields and applications, including the Licensed Field of Use and Licensed Applications.

The following terms were defined in the contract:

"Angiotech Technology" shall mean (a) the Patent Rights, license rights and existing technology set forth on Exhibit A hereto, (b) any New Angiotech Technology which Cook or BSC, as the case may be, elects to have included in the Angiotech Technology pursuant to Section 2.3, (c) any and all

> **Box 4.9** (*cont.*)
>
>   improvements to the foregoing developed by Angiotech, or, subject to
>   limitations and restrictions on Angiotech's rights to technology licensed from
>   third parties, for Angiotech, during the term of this Agreement (including
>   those arising under the CRADA to the extent solely owned by Angiotech),
>   and (d) Technical Information that is useful or necessary to practice
>   the foregoing.
>   "Licensed Field of Use" means endoluminal vascular and GI applications
>   "Licensed Application" means the use of Angiotech Technology in the Licensed
>   Field of Use on or incorporated in Stent Products and Endoluminal Products,
>   but specifically excluding systemic treatments and pastes, microspheres,
>   films, sprays and similar formulations in circumstances where such are not
>   applied to or incorporated in either a Stent Product or an Endoluminal
>   Product, as the case may be.

just the tip of the iceberg. Box 4.8 is an example of some terms from a licensing
agreement between a university and a company and Box 4.9 contains an example of
some terms of a licensing agreement between two commercial organizations.

4.8.2   Financial terms in a license

The various payments by licensee(s) need to be worked through and coming to
agreement on these figures can be a rather difficult negotiation in the licensing
process. Underlying these discussions of financial terms is the perception of value
of the technology/patent. However, this has to be tempered with current market trends
for the technology or application.

What is the value of the patent(s)?

The value of the patent and invention is a perception, and the subjective view of the
beholder will change this perceived value. In the end, both parties have to come to a
common subjective context within which they have agreement on the value.
A calculation on value varies by the defined scope of the application area, resources,
and time required to develop the invention, current needs of each party, and other such
subjective and objective factors, thus making it hard to accurately fix a value outside
of a context. Even quantitative tools such as discounted cash flow and option pricing
models are subjective in the assumptions that have to be entered to run these models.
Finally, a close look at comparables in the licensing marketplace is used to give a
lower and upper range for the financial terms and is useful in setting expectations on
both sides. Thus, valuation of intangible assets such as intellectual property is a
context-dependent exercise, where output from any analytical or empirical model
must always be tempered by the market forces relevant to the business context
for valuation.

A general *list of financial terms* typically included in a license agreement includes the following:

- Technology transfer or access fees or up-front payments
  - This lump sum payment, due at signing, recognizes the investment made to date by the licensor both in developing know how and the technology itself, and also includes consideration that some licensor preparation and effort may be necessary to allow access to the technology.
- Patent prosecution and maintenance fees
  - The licensee can be asked to pay legal and USPTO fees for maintaining the patent.
- Milestone payments
  - If the technology/invention succeeds in further stages of development, the licensee is asked to acknowledge the increased value (lowered risk) of the patented technology by making payments to the licensor.
- Royalties
  - Royalty rates, paid on commercial sales of the products, will vary by the state of the patented technology at execution of the license agreement and will typically be based on industry or market rates prevailing at that time. A rule of thumb sometimes used by many licensors is the 25% rule – it is often accepted that a royalty (a percentage of the sales revenues) that is equal to 25% of the expected pre-tax net profits is a fair rate. The royalty rate in the license agreement will then depend on the market for forces of each particular product. For example, if the licensee has profit margins of 60%, the royalty paid to the licensor should be (1/4 of 60% =) 15% of sales revenues. If profits of 4% are expected, the royalty rate should be around 1% of net revenues. Adaptive royalty rate – many licensing agreements include royalty rates that change as the annual revenues reach certain thresholds. Also see Box 4.10 for a legal viewpoint on what might constitute "reasonable royalty rates" and the "Georgia-Pacific factors."
- Sublicensee and sublicense fees
  - If the patent is sublicensed to another party by the licensee (if they have been given that right in the license agreement), then a portion of the payments to the licensee are typically passed through to the licensor – this arrangement can be a flat fee or also a staged royalty as appropriate.
- Minimum annual royalties
  - A minimum annual royalty payment starting with first commercial sales can be used to ensure that the licensor exercises a good effort in generating maximal revenues for the product.
- Equity considerations (also warrants and options)
  - A small licensor may find it beneficial to have a well-established licensee purchase equity as part of the payments, for validation of the company as a whole and to increase long-term investors. On the other hand, a university as licensor may accept equity in lieu of cash payments from a startup licensee that has limited resources to move the patented technology forward. Warrants and options are other methods of acquiring or selling equity in a license agreement.

> **Box 4.10** What is a reasonable royalty rate? A legal perspective on claiming infringement damages and valuation for licensing
> *Contributed by Robert Schaffer, Attorney at Darby and Darby PC, New York*
>
> If your patent is successfully enforced against an infringer, the infringer may be enjoined from further infringement, e.g. he will have to withdraw the infringing product from the market. An injunction is not automatic, and recent cases have made it harder for the patent owner to get an injunction. The other remedy for infringement is that the patent owner can recover money damages. This can be all of the profits reasonably lost by the patentee to the infringer, if the patentee has a directly competing product. By statute, the damages cannot be less than a "reasonable royalty" for the invention. In making a "reasonable royalty" determination, courts typically look to a hypothetical negotiation between a willing buyer and seller of the invention rights, and they consider the factors that typically go into such a negotiation.
>
> These factors are laid out in a landmark court case called Georgia-Pacific, and the royalty factors are called the Georgia-Pacific factors. These same factors can be used as a guide during actual licensing negotiations. The calculation of damages in a patent infringement case can be complex and highly contentious. As may be expected, the patentee tends to claim as high a lost profit or royalty as the facts arguably support, while the infringer argues for the lowest possible amount. Both parties are in the hands of the court, and there is often no reason for either party to think it will get a better deal from the judge than they could have gotten from each other in a good-faith business negotiation. Valuing a patented invention, including the reasonable royalty it can command, is something of an art form, and for a court case typically includes help from business and economics experts. Because of the high stakes involved, it may be a good idea to undertake a "reasonable royalty" evaluation before seeking to license an invention, and before bringing a lawsuit. This can particularly help to provide a ballpark "low ball" or "worst case" value for the invention, when planning to commercialize an invention or enforce a patent.

- Royalty anti-stacking provisions
  - If a licensee has to also license in other patents in order to get the product to market, then royalties payable to all licensed-in patents may stack up to make it economically unfeasible for the licensee to retain any profit from the sale of the product. An example would be a licensed-in drug compound which also requires the licensing-in of a delivery system technology in order to enable the product to be made and sold. In this case, the licensee will want to reduce the burden of the stacked royalties by requesting a subsequent reduction in each royalty while the licensors will want to minimize this reduction. There are several mechanisms and formulae that have been used to calculate the amount of royalty reduction. These terms would be placed under an anti-stacking royalty reduction provision.

4.8.3   "Boilerplate" clauses in the license agreement

There are several standard (boilerplate) parts of almost any business/legal contract that are also part of the licensing agreement contract and these are mentioned here to give an overview of the entirety of the licensing agreement. However, these 'standard' terms should also be carefully perused and negotiated if any of them have impact on the specific context of the business. These clauses will not be discussed further here, as there are many available resources that describe these terms in detail.

- Termination provisions
- Best efforts
- Warranty and indemnification
- Arbitration and applicable law
- General provisions
- Assignment
- Severability
- Entire agreement (- contents represent the agreement in entirety)
- Force majeure, contingencies
- Notices

4.8.4   Pros and cons of various types of licenses

Table 4.4 summarizes types of licenses and their pluses and minuses.

Table 4.4 Licenses, pros, and cons

Type of license	Pros	Cons
Exclusive license	Greater commitment from licensee Higher fees Higher royalties Closer monitoring of development	Higher risk of failure of product commercialization May be wrong partner/strategy Licensee will typically have single focused market approach thus not reaching max potential of technology before patent runs out Licensee can use exclusivity to block other potentially "good for humanity" type development work which in many cases could be the goal of the inventor
Nonexclusive or partially exclusive licenses	Multiple paths to market Multiple markets addressed simultaneously Increased chances of final commercialization Partial exclusivity is usually geographic	Multiple market carve-outs may raise problems down the road – a poorly executed clinical study can affect every licensee. Manufacturing rights and controls need to be spelled out. Managing licensees and multiple partnerships may be a challenge Usage is difficult if encumbered by other licensees' efforts/data/liabilities Lower fees and royalties

4.9   Biotech business models and IP licensing strategies

Once a patent is licensed, depending on the agreement, the activities of the licensor can range from "collect payments and royalty checks" to active participation in the development. Typical biotech-pharma licensing arrangements (especially at a stage before Phase III trials) tend toward the latter, with the licensor (biotech company) contributing resources and funding in a co-development agreement, and the experience-base of the pharma company people and processes helping to build expertise in the biotech company. Especially with later-stage licensing deals, the biotech company may want to have a co-marketing and sales roles so they can work alongside the large pharma experienced people and grow a commercial team to support their internal pipeline.

Established, market-leading medical device companies will usually buy the startup company or product outright rather than license it, whereas a few innovator device companies may form marketing partnerships for better market access. New diagnostic companies usually license out their technology to a commercial partner in order to gain access to the established technology platform and markets. Thus, several considerations enter into the structure and terms of the licensing agreement depending on the strategic needs of the licensor.

In general, the licensor can wait for royalties, or co-develop technology in a partnership all the way to market, or choose to co-develop only in strategic functional areas. The licensees also have similar choices, where they can develop the IP themselves to the next step, and then sublicense to a third party to complete commercialization and similarly wait for royalties, or co-develop or develop the patents themselves to market. The licensee can also bundle various in-licensed patents together to make a more attractive and comprehensive portfolio package for the next sublicensee. There are many companies that operate on a business model that works this way: license patents from one party; take the patent/technology to the next value creation step; and then license it out to someone else, collecting the incremental value created as its revenue/profits. Two examples of very successful U.S. companies from the early 1980s that actively licensed in patents from universities and licensed out to larger corporations are RCT (Research Center Technologies) and BTG (British Technology Group). There are other licensors who follow a business model that focuses on controlling and enforcing rights to patents without actually carrying out physical value-adding development beyond the invention and patent prosecution step.

Thus, intellectual property management strategies (and subsequent licensing agreements) are closely linked with the specific business model chosen by the company. Figure 4.5 is a schematic of the general business model choices that a biomedical product development company can take with its licensing and subsequent business strategy. Business models and related company strategies are also discussed in detail in Chapter 2. If a company has built itself to focus on initial R&D as per the functional area blocks in Figure 4.5, then it may choose to outsource the remaining downstream functions such as clinical trials, manufacturing, sales, etc. by either paying contractors

	R&D	FDA Process & Clinical Trials	Manufacturing	Marketing & Sales	
				US	International
BUILD Grow Internally					
BUY Strategic Partners/ Contractors					

Figure 4.5 Functional look at a business model. This schematic can be filled in to give a good idea of where the company is currently and where it should invest functionally over time.

or expert consultants from these functional areas, or may seek out a licensee who has already established themselves in those functional areas.

However, within each segment of the functional elements above lie deeper questions (are we developing services or products? are we making components or do we want to make systems? should we build sales and marketing?) that further define the business and operations. Above all, the dynamic nature of the context of business model must be recognized. Continuous changes in the markets, technology, and state of maturity of the sector require adaptability in the business model to sustain success.

4.10   Summary

In this chapter, intellectual property concepts have been introduced with a clear understanding of their value in capturing value while commercializing biomedical technologies. Product development is influenced greatly by these strategies and by the nature of intellectual property protection chosen to gain exclusivity in the market. Choices of projects and investment decisions such as investing in innovation on a new technology platform are governed by the ability to capture value in the commercialization process through the subsequently created intellectual property. This chapter has covered the main tool (licensing) and business models used to capture and use intellectual property to further product development.

Exercises

4.1. Describe types of intellectual property needed for commercialization of this product for the specific indication chosen.

4.2. Describe how specific patent claims in filed patent applications will provide market protection for the products.

4.3. If you have an initial idea, search the U.S. or EPO patent databases to determine whether a patent application or issued patent covers your idea.

(cont.)

Search scientific and general publications for keywords representing your invention/idea to determine the patentability of the invention.

4.4. Identify future IP that will/might be needed in commercialization path to access chosen market. Describe strategy for IP acquisition (assumptions are fine, just state them clearly up front). State key licensing terms for in-licensing agreement.

4.5. What might be a business strategy/model based around managing your starting IP assets?

4.6. State key licensing strategies in the context of value chain positioning with a brief rationale. Identify key assumptions.

4.7. Based on your IP strategy, plan the costs and timeline of patent prosecution, licensing or acquisition.

References and additional readings

Goldscheider, R. (2003). *Licensing Law Handbook: The New Companion to Licensing Negotiations*. St. Paul, MN: West Group Publisher.

Lerner, PL, and Poltorak, AI. (2011). *Essentials of Intellectual Property*. New York: John Wiley & Sons. ISBN: 978-0470888506

Mowery, DC, Nelson, RR, Sampat, BN, and Ziedonis, AA. (2004) *Ivory Tower and Industrial Innovation: University-Industry Technology Transfer Before and After the Bayh-Dole Act in the United States*. Stanford, CA: Stanford Business Books. ISBN: 9780804795296

Smith GV, and Parr, RL. (2005). *Intellectual Property: Valuation, Exploitation, and Infringement Damages*. New York: John Wiley & Sons. ISBN: 047168323X

U.S. Department of Commerce. (2008). *Patents and How to Get One: A Practical Handbook*. BN Publishing. ISBN: 0486411443

Useful website links

www.uspto.gov/patents-getting-started/patent-basics/types-patent-applications/general-information-about-35-usc-161 – contains general information about plant patents at the USPTO

www.uspto.gov – U.S. patent and trademark office

www.epo.org – European Patent Office

www.ladas.com/Trademarks/tmprot.html – Trademark-related blog link

www.uspto.gov/about-us/performance-and-planning/uspto-annual-reports – USPTO patent processing times

www.iphandbook.org/ – An excellent resource for reports, market trends, sample agreements and other reports on licensing and intellectual property management.

They also have a series of informative videos on various topics – www.iphandbook.org/handbook/globallearning/videos/

www.iphandbook.org/handbook/ch14/ – Further reading on FTO (Freedom to operate)

https://autm.net/ – For U.S. university technology transfer and general licensing resources

www.wipo.int/wipo_magazine/en/ – The World Intellectual Property Organization (WIPO) has an online magazine and other resources

Visit https://shreefalmehta.com/csbtbook for additional enriching readings around the topics covered in the book, topical updates on the content and for industry viewpoints and news.

5 New product development (NPD)

Plan	Position	Pitch	Patent	Product	Pass	Production	Profits
Industry context	Market research	Start a business venture	Intellectual property rights	New product development (NPD)	Regulatory plan	Manufacture	Reimbursement

Learning points

- Define target product profile (TPP) characteristics at the beginning of a biomedical product development plan using clinical study endpoints and indication use cases.
- How do drug, device, and diagnostic development processes differ?
- Why do many drugs, devices and diagnostics fail in development stages?
- How do you build a product development plan for drugs, diagnostics, and devices?
- When should you kill a project?
- How to prepare for clinical trials – what are specific issues for diagnostics and devices?
- Successfully make a pitch for a project to senior management for funding.
- What ethical issues must be recognized during product development activities?
- When should you outsource?
- How do you comply with specific certifications and laboratory regulations when setting up a new laboratory?

Read this chapter going through the book sequentially, and then revisit the exercises in this chapter after reading Chapters 6–8. The approach to new product development has to be an integrated, multidisciplinary approach, taking input from all other chapters into account (see the roadmap in the Preface).

This chapter focuses on the development of novel products with some "new-to-the-world" features, requiring a full review by the FDA before reaching market. In particular, drug or diagnostic products that do go through abbreviated product development paths (generic or repurposed drugs, and diagnostics or devices that are largely equivalent to existing approved and marketed products) are addressed in greater detail in Chapter 6. The principles presented in this chapter apply generally to new biomedical product development.

5.1   Why have a new product development (NPD) process – just get it done!

Developing new products requires input from, and interaction with, almost all functions in a company (Figure 5.1), as new products build revenues on which the entire company depends for growth and sustenance. Most large companies will have an integrated multidisciplinary product development team or a review board that has people from multiple functional areas in the company. Each functional area in the company has its useful and significant role to play in getting new products to market.

Multidisciplinary reviews are as essential as multidisciplinary teams. Product development must constantly and iteratively take input from multiple departments or corporate functions during the planning and development stages, as described in Table 5.1. If you are in a small company and don't have all these functions, you still need to consider the product from the viewpoint of all these disciplines.

Therefore, a structured process of development is needed to help organize multiple agencies and functions within an organization. The goals of investing in creating and maintaining a process for new product development (NPD) are:

- At some stage, a defined process is required and reviewed by the FDA in order to approve the product (see Chapter 6).
- A standardized design and manufacturing process is also critical to get quality certifications like ISO (International Organization for Standardization, Geneva, www.iso.ch).
- Defining the product characteristics by writing out a target product profile (TPP) document brings focus and alignment and makes trade-off discussions more objective among all stakeholders.
- To bring products from concept to market in an organized, efficient manner.
- Status of stage of development can be communicated easily.
- To improve management of the process and coordination with other resources.
- To minimize time, effort, cost.

Figure 5.1 Multiple interfaces of the new product development (NPD) process

Table 5.1 Input and interactions with various departments into the NPD process

Functional input	Purpose
Basic R&D	To transition the project
Reimbursement	To provide input into the economics of the disease state treated (and any associated procedure); to provide input into clinical trial design so that insurers will be willing to pay for the product
Marketing and market researchers	To develop product specs and possibly test early concepts/prototypes; to work with product cycle planning and to differentiate product with respect to competition
Finance division	To get agreement on budgets and adjustments
Sales division	To hear customer needs
Senior management	To review at level of portfolio and set strategy and funding,
Manufacturing	To make sure development is scalable and designed with manufacturing in mind
Regulatory affairs division	To guide the results through the regulatory gate keeper and give critical feedback on product development planning.

- To ensure quality and safety in the process and final product.
- An NPD process itself is a valuable asset of a company where learning can be captured and efficiencies transferred among all new product development projects.

5.2 Planning and preparing an NPD process for biomedical technologies (drugs, devices, and diagnostics)

The development of new biomedical products typically involves three functionally different stages, as shown in Figure 5.2. These functional stages vary in execution and time span between drugs, devices, and diagnostics as examined further below.

The first few steps in developing a new biomedical product are schematically shown in Figure 5.3. The preliminary product development plan is typically drafted by engineers, scientists, and project managers for presentation to higher level managers or committees in large companies. Part of step 1 in Figure 5.3 is the creation of a target product profile (TPP), which collects in one document the market needs, competitive advantages, and specific problem being addressed by the product so that the product characteristics and context for development are clearly defined. More on this target product profile in Section 5.7.

Another level of strategic planning and thinking about new product development takes place at the corporate level, where data on the technology roadmap of the industry, longer-term product portfolio planning, finance, marketing, etc. are all considered in the context of an industry and organizational development plan. A development scientist or engineer can better design a successful product if they have an understanding of the target market and its needs (see Chapter 2), use or mode of delivery of the product, the needs of the company, the company strategy, and an overview of the process of final delivery into the market (Chapters 6–8). As described above, keeping these points in mind during the design and development process also helps in planning the product life cycle (patent protection) and subsequent versions of the product. These perspectives will also build a greater appreciation of the significance of the role of a product development engineer/scientist in the entire NPD process.

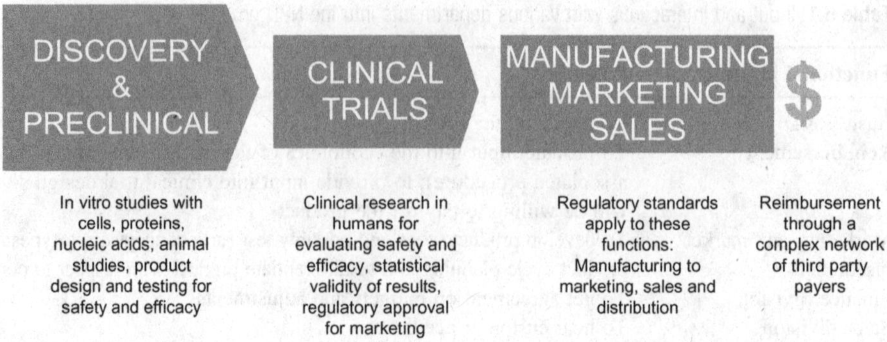

Figure 5.2 A generalized functional and value chain in biomedical product development showing three main stages of product development

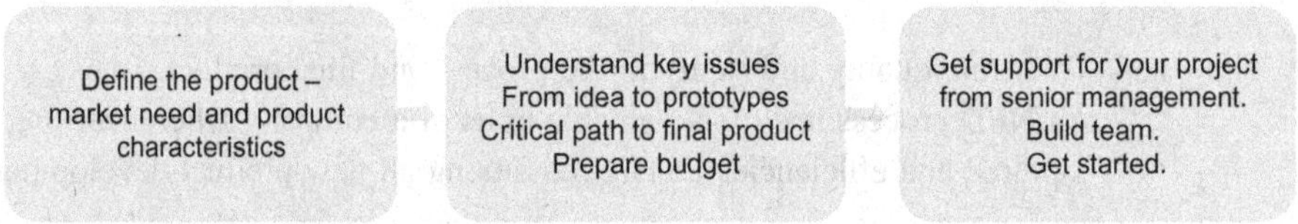

Figure 5.3 Preparatory steps by project manager and team

5.2.1   The project proposal document

The project proposal contains at least the items listed in Figure 5.4 and makes a convincing case for the project, specifically showing the risks involved, and describes how early testing will reduce the risks before the product enters clinical testing on human subjects. Some aspects of the planning process and the inputs are described in the schematic in Figure 5.4.

5.2.2   Strategy and competency of the company and goal of the project

Most companies have both a short-term and a long-term business plan on how to grow or sustain their businesses. While planning a new product, the context of the business' strategic plan and the available internal competency in technology and functional areas must be kept in mind. It also helps to understand the company's business model and current position in the value chain of the specific industry (refer to Chapter 1, Section 1.1 for more details).

5.2.3   Product life cycle planning

Each product has a reasonably well-defined life cycle from concept to development, to market, to peak sales, and finally to declining sales and obsolescence. The product life cycle from market introduction to peak annual sales for specific industries is well

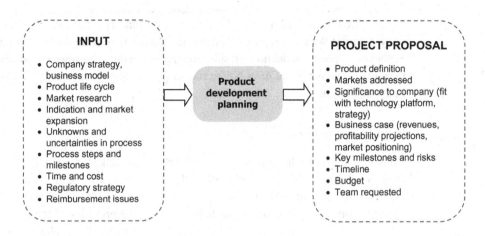

Figure 5.4 Planning for product development and proposal inputs

known as a general rule (innovative devices – ~2–3 years; drugs – ~3–5 years assuming patent status is valid; and diagnostics – ~5 years; note that these lengths will vary for specific products). The new product must fit into the current product line and market-based development cycles to help address internal and external issues and interests, particularly with respect to the long product development cycles for most biomedical products. For example, if a drug is going to go off patent after 4 years of market launch, then a parallel track or post-approval (Phase IV trial) development of slow-release or inhaled formulation could be part of an NPD plan, so that the new (patented) form of the drug product is ready for release at or before the time of entry of generic competition into the market. A life cycle planning strategy for a device might be to develop a broader, technically complex, fully configured platform at the first run, and then introduce various parts/features of the device onto the market in a sequential measured fashion, preempting competitors who will not be able to match the pace as redesigning their products will take too long.

5.2.4 Market research inputs

Components of market research are described in Chapter 2. The outcome of the market research should identify the product characteristics and define the product and target market in greater detail. One other specific outcome of market research is the identification of a primary indication and indication expansion strategies. The inputs from market research (voice of customer or customer needs [VOC] is one such input) are consciously and quantitatively synthesized into product characteristics by using various design tools such as a "house of quality" (where design and functional inputs from external and internal sources are weighed against each other by assigning numerical significance, arriving at a relative importance of each identified product characteristic) or some other design matrix that brings a display of market research inputs to design/development teams. This approach is discussed in Section 5.10.1. The product development and design process for drugs (small molecules and biologicals) typically involves taking a higher-level view of the market as NPD design decisions in early drug development typically revolve around mode of administration (oral, intravenous) or specific reduction of side effects by dosing adjustments. However, the business decisions in the drug NPD process involve significant market research and in-depth analysis.

5.2.5   Identify key unknowns and risks

Knowing where the hurdles lie in the NPD process makes it more likely that the choices to be made along the NPD path will enable a successful product launch. Major risks such as toxic side effects must be addressed as early as feasible. Sometimes, the identification of key unknowns can help to clarify the priorities of the development testing and studies that need to be done. For example, if the pharmacodynamic behavior of the drug compound is identified and listed as an important parameter that dictates many other product development steps, then specific tests to evaluate and optimize those pharmacodynamic parameters should be prioritized in the product development plan. Or if the long-term toxic effect of a material is identified as a key risk for a device, then parallel to other development studies, a long-term study with the material in vivo (may not be the final version of the material, but close enough that the result would be indicative) could be initiated. Having a list of these specifically identified risks or unknowns can be helpful in building an NPD process with greater chances of success. More to the point, *the product development plan should show how the key risk factors will be tested early in the project.*

5.2.6   Build a milestone-based plan for product development

Milestones are results of tests that show key reductions of risk in NPD. For example, a milestone for a device could be the successful mechanical testing of the device functions in repeated cycles to failure, showing expected or better-than-expected device usage lifetime or an animal model study that shows the expected benefits of using the device against the current alternative gold standard. A milestone for a drug could be the demonstration of efficacy in an animal model showing improvement or disease modification against the current standard of care. A milestone for diagnostic development could be the satisfactory repetition of a sensitive level of measurement, establishing a level of reliability for the diagnostic test. The more specific the milestones, the easier it will be to build a convincing NPD plan. It is important to plan to do the knock-out or go/no-go tests early. However, *planning tests to weed out early failure points can be a conceptual challenge for scientists or engineers in product development, who are focused more on understanding the scope of the technology or on getting the technology to work rather than on figuring out risk factors and potential failure points.*

Typical example milestones that reduce risk in product development are presented in Table 5.2 for each subsector of biomedical product. An attempt has been made to segregate milestones to correspond with typically discussed product development stages.

Table 5.2 Example milestones in product development for various sectors

Stage	Pharmaceutical	Medical device	Diagnostics	Software/ healthcare IT
Stage 0: Concept	Healthcare need / problem clearly identified (indication defined) Solution to address need and business case for investment clearly presented			
Stage 1: Proof of concept or feasibility demonstration	For new therapeutic, lead compound identified and shown to work against target in vitro tests In vivo efficacy	First functional device prototype shown to work as expected (in vitro, in vivo)	Primary functional verification of assay in laboratory	Algorithm concept and demo proof with partial algorithm or small set analysis (for AI or software-based product/service) showing basic functional benefit over state of the art
Stage 2: Alpha prototype	Formulated product manufactured under Good Laboratory Practices (GLP) shown to be safe Animal efficacy, early Absorption, Distribution, Metabolism, Excretion and PharmacoKinetics (ADME/PK) characteristics demonstrated	Advanced prototypes built using pilot manufacturing processes Device effective in in vivo tests and basic safety issues identified or addressed	Diagnostic platform firmed up Efficacy and benefit over state of the art shown in small patient groups	First free or paid customers with Minimum Viable Product (MVP) to learn and reiterate.
Stage 3: Beta prototype	Formulated product manufactured under good manufacturing practices (GMP) used for completing toxicity testing FDA meeting with agreement on indication Investigational New Drug (IND) application filed	Device manufactured with design and quality controls Safety tests done and Investigational Device Exemption (IDE) submitted to FDA	All system components (e.g. H/W, S/W, reagents, consumables) are finalized and transferred to manufacturing Product optimized with clinical feedback	Fully functional software version with user interface ready
Stage 4: Clinical testing and regulatory approval	Phase I: safety in normal humans shown FDA meeting for trial protocol review and approval of endpoints Phase II: efficacy in target patient	Pilot studies in patients shows efficacy 510(k) approved OR Pivotal study in patient population shows efficacy and safety	Clinical studies complete with acceptable sensitivity and specificity of test showing clinical benefit.	Clinical studies showing positive comparison with current alternative practices or state of the art practices

Table 5.2 (*cont.*)

Stage	Pharmaceutical	Medical device	Diagnostics	Software/ healthcare IT
	population shown Phase III: meets agreed on endpoints with statistical significance Regulatory approval	PreMarket Authorization application (PMA) approved (MDR – Medical Device Review) review completed in EU)		
Stage 5: Commercial launch	Manufacturing scale-up plan in place Market adoption growth in market share Post-marketing studies for expansion of indications			

Note: For specific considerations for diagnostics development for low- and middle-income countries, see Mugambi, ML, et al. (2018). How to implement new diagnostic products in low-resource settings: an end-to-end framework. *BMJ Global Health* **3**, e000914.

For specific considerations for medical device development for low- and middle-income countries, see Vasan, A, Friend, J. (2020). Medical devices for low- and middle-income countries: a review and directions for development. *Journal of Medical Devices* **14**(1),010803.

5.2.7   Specific risks known to occur frequently during the development of biomedical products

Biopharmaceutical/drug products that fail in development typically fail because of one or more of the following reasons (data from R. Lipper [1999]).

Drug candidate molecules in development fail to get to market because of the following reasons (percent compounds failed):

- (41%) Poor ADME characteristics (absorption, distribution, metabolism, or excretion in the human body; desired product characteristics could not be met)
- (31%) Lack of efficacy – compounds did not show the benefit expected
- (22%) Toxicity
- (6%) Market or business reasons

Diagnostic products that fail in development typically fail because of one or more of the following (see Box 5.1 for examples):

- Lack of clinical utility = no correlation with clinical outcome.
- Needed sensitivity/specificity of the assay not verified in subsequent clinical studies.
- Wrong test principle chosen (genetic, expression, protein, metabolite).
- Wrong test format chosen – (centralized vs. point of care). Perhaps in the in vitro diagnostics (IVD) industry more than the others, there is major issue with setting market standards and testing platforms. The example in Box 5.1 illustrates this point.

Box 5.1 Examples of failures or problems in diagnostics development
Sensitivity-related failure
An IVD test being developed by a large diagnostics company for sepsis was aimed at locating a surrogate marker so that an intervention decision could be made in a timely fashion. Approximately 100,000 in vitro samples were analyzed and studied with extensive molecular mechanisms of sepsis and cell culture studies. When the designed assay was tested in a clinical setting, the assay was just not sensitive enough on real clinical samples and could not deliver results in the time frame that a clinical decision was required to be made. The exquisitely designed diagnostic test thus failed to show clinical utility due to lack of clinical sensitivity. According to Dr. Christoph Hergersberg (then Global Biosciences Leader at GE Central R&D, quoted from a personal interview) – "This is the 'so-what?' factor. You have a great diagnostic test – so what?? It has to have clinical utility and be tied to clinical outcomes to be accepted and successful."

Failure due to incorrect choice of test principle
HER2/neu is an epidermal growth factor tyrosine kinase that is known to be over-expressed in breast cancer. This overexpression can be due to the amplification of the gene as measured by fluorescence in situ hybridization (FISH) or in the circulation of breast cancer patients by immunohistochemistry (IHC). Both tests have been developed commercially but which one is more predictive? Even if you come up with a better principle, is the distinguishing part of the test (which may be scientifically a more rational test principle – e.g. chromosomal-level detection of rearrangement) more clinically relevant and will it be competitive in cost and utility to the other? This is again, the "so-what?" question in diagnostic development . . .

Market standard related test format incorrectly chosen
If you are developing a test for use in critical care patients, who are in the intensive care unit, point of care (POC) testing is de rigueur. Therefore, one would start developing and creating a POC test. If, however, a centralized lab could beef up the workflow to run this test with a faster turnaround time, the effort and time to develop a new platform might not be worth it. The competition would be driven by economic value and the time spent to develop a new POC test might be better served in developing a diagnostic test that could be licensed to a large diagnostic that already has a POC testing system in place. For example, Luminex is a market leader in multiplexing, where several assays are run in one test, making it much more cost-effective.

Complicated and sophisticated test that may be more accurate but does not add much more value to final clinical outcome
A cascade of activated proteins related to a systemic pathology is discovered and a FISH (fluorescent in situ hybridization) assay on the entire cellular signaling path is developed to get a multiparameter diagnostic test. This new multiparameter test is potentially more sensitive or more specific. If the existing clinically used test is a

---

Box 5.1 (*cont.*)

simple blood test to get the same diagnosis, then the more complex (and expensive) test will have to show not only better sensitivity and specificity but also a *much* better predictive value to successfully replace the current clinical blood test.

Lack of patent protection

Lack of adequate patent protection can result in economic failure for a manufacturer, even though their test works. A strong intellectual property position is sometimes difficult to achieve and protect in the diagnostics industry.

The example of a patent dispute between Metabolite and LabCorp highlights a unique patent protection issue for the diagnosis industry: University scientists discovered and patented a way of diagnosing vitamin deficiency by measuring levels of an amino acid called homocysteine in the blood. Metabolite (small startup company) licensed the patent from the university and then sublicensed it to LabCorp (a large company). LabCorp paid royalties to Metabolite for a few years while it sold the tests and conducted assays according to the methods described in the patents. When a better assay process was invented by Abbott, LabCorp switched to that and stopped paying royalties to Metabolite. Metabolite took LabCorp to court; LabCorp pleaded that the relationship between homocysteine levels and vitamins was a law of nature and could not be patented, thus the Metabolite patent was invalid. Metabolite won an award against LabCorp (LabCorp failed in appeal to the U.S. Supreme Court on a technical matter), but this case raises some important issues – the diagnosis industry is made up of these kinds of correlations that occur in normal and healthy processes. Will these numerous patents also be challenged? It will become very difficult to develop new diagnostic product and take them to market without some certainty and protection from the patent laws.

- Test is too complicated versus existing format. This is a value proposition issue where in the new test in development may be more precise and sensitive than needed.
- Reliability – repeatability and precision not achieved.
- Nonlinear response of assay in clinical use.
- Patents are not comprehensive or valid.

Medical devices that fail in development typically fail because of one or more of the following:

- Failed to meet efficacy
- Safety, toxicity or instability in device behavior/mechanics
- Biocompatibility
- Business or market reasons

See Box 5.2 for examples of medical device failure in development.

Box 5.2 Examples of failures or problems in device development

This is an example of a device that failed to meet its claimed efficacy after market clearance. Curon Medical sold a device called Stretta^{TM} for treatment of gastro-esophageal reflux disease (GERD). Cleared under a 510(k), it had problems with establishing long-term efficacy. The device was rejected by the technology assessment boards of various payers and the company filed for bankruptcy in late 2006 and closed its doors, taking the product off the market.

Another example of failure of a marketed product due to inadequate studies in product development is Boston Scientific's Enteryx product for GERD, which had problems with mechanics of delivery, causing complications and deaths. Injections of the polymer (used to thicken the esophageal wall) were frequently wrongly delivered by physicians and caused serious complications, even death in one case of injection into the aorta. This product was taken off the market in 2005.

5.3   Killing the project early or try some more?

It is important to understand that, despite good technical progress and data, projects get killed in development. Usually, multiple considerations enter into that decision, including market and reimbursement issues, competitive pressures, newly discovered patent issues, lack of patentability or competing, regulatory changes, financial constraints, or simply a change of strategy in company due to a management change or merger or acquisition of the company. However, a counterintuitive point frequently brought up in biomedical product development is discussed here: *Why is early failure a (relatively) good thing in biomedical product development?*

5.3.1   Early failure is better than late failure in biomedical product development

A high cost and rapidly rising rate of expenditure in the later stages of medical product development makes it attractive to recognize potential product failure points at the earliest possible stage of development. There are always new product ideas that could be tackled by the resources freed up by failed projects in a large company (although that is not always the case for startups). Also, exhaustive and early tests for potential side effects or toxic effects might help prevent deaths or morbidities for patients, and for the sponsor, reduce the liabilities of legal lawsuits (see Chapter 8, Section 8.12) in the future when the product is marketed to a broader population. Examples are given in Box 5.3.

Product development in the clinical testing phase can cost tens of millions of dollars. Therefore, there is a significant interest in testing for known safety concerns or for specific problematic aspects of delivery or dosage as early as possible. The goal is to identify the risk correctly as early as possible and then either kill the project or fix the issue and thus reduce the chances of failure later on in the project.

Box 5.3 Examples of desirable early failure milestones
Device example
A device was known to have a problem with chronic fibrosis forming around it, leading to premature failure of the device functions. In developing a next generation device, an in vivo test was carried out very early in the material selection and design process to determine the amount of likely fibrotic reaction. The goal was to help learn about the process of fibrosis. One such material, when tested in this way, had to have its edges reshaped to make them more rounded, resulting in much lower buildup of the cells involved in the fibrous deposits and thus showed a way to make the final device less prone to failure due to fibrosis in the body.

Drug example
Chronic drugs may likely fail in late development or regulatory review if they have significant side effects on cardiovascular function. A new class of compounds being developed for a chronic disease should incorporate in the NPD plan, acute (short-term) ion-channel binding and other cardiovascular studies on molecules in the early design process (lead compounds) so that further investment in certain classes/types of compounds is halted in the NPD process (early failure). The typical process would be to test drug candidate compounds (at a much later stage of development) in expensive long-term cardiac toxicity studies in large animals. Several such efforts, to look for in vitro markers that might help to predict specific drug toxicities, are in progress in academia and in industry.

5.3.2   When to kill a project?

The decision to stop investing in a project is one of the most challenging issues in biomedical product development for two reasons: (1) people are passionate about their projects, as the projects have a higher goal toward helping better human health; and (2) because biology is so complex, it is far easier to kill a project at the first sign of toxic side effects or equivocal data than to allow it to continue to try and reach efficacy. Conversely, there are many stories of projects that were killed and discarded by one company but were subsequently developed into very successful products by other companies (see Box 5.4 for case example).

Management is concerned with final outcomes, such as product differentiation (with respect to competitors or existing alternatives), pricing, reimbursement, market acceptance, user convenience, and compliance, and has to show eventually (to the FDA and market payers) that the balance between safety versus benefits tips toward benefits to the patient and that the product has real-world clinical utility. In fact, the clinical trial endpoints that are acceptable for regulatory approval are often not relevant to the insurers or payers, especially in the case of medical devices or diagnostics. The payers are looking for a real-world clinical outcome benefit and hence it may be prudent for the company to gather an extended set of data in the

---

Box 5.4 Old drugs never die – they get revived . . .

A story in the *Wall Street Journal*, dated August 11, 2003, described how a small company, Genesoft, successfully picked up and turned around a drug (brand name "Factive") development project that big pharma (Glaxo SmithKline [GSK]) had killed. SmithKlineBeecham (SKB) had licensed in Factive from a company (LG Group) in Korea. The drug, an antibiotic to treat respiratory infections, failed FDA review as the FDA was concerned about a rash that developed on patients after its use. But the product development team had found the rash to be intermittent and carried out more studies to demonstrate this. However, at the time SKB had merged with GSK. GSK dropped the project, never submitting the last findings to the FDA (after having spent 5 years and $200 million on its development), as they were increasingly focused on billion-dollar drugs and did not feel the last safety hurdle would be overcome. The lead scientist from the Factive project left GSK as a result of the merger and was working for GeneSoft. When he heard over a newswire about the opportunity to license the drug out again, he convinced his CEO to pick it up. They struggled but were successful in raising the funding and obtaining the license from the Korean company. They continued development and in 1 year filed for FDA approval and successfully put the drug on the market, transforming Genesoft's future and improving patient health.

---

pivotal approval trial or additional clinical trials so that a real-world clinical outcome benefit can be quantified for the commercial success of the product.

Sometimes a project will get killed by management because of changes in regulatory or market environment, frustrating the scientists and engineers, who have achieved the technical milestones. In addition, there are budget allocations and resource constraints that usually factor heavily into a decision to kill a project. For example, if a project has been funded for a number of years and the team still has major "red flag" issues to overcome, it is more likely that the project will be killed during the next review. The product development team, focused on showing efficacy, should attempt to gain better appreciation of these external and internal business issues and monitor them as best as possible.

While it may be easier to evaluate and quantify external factors (markets, regulatory, economic factors, competition) as criteria to drop projects, the decision to drop a project for technical reasons alone is the hardest. There are numerous examples in many industries (typically in larger companies) where a project was declared officially over due to a scientific or technical hurdle but a passionate scientist or engineer with insight kept tinkering with the concept to overcome the hurdle. They succeeded in overcoming the technical problem after numerous months/years and that resulted in renewed funding, development, and launch of the new product. Examples of such products exist in various industries and include GE's digital X-ray, DuPont's BioMax biodegradable polymer, IBM's silicon geranium devices, Analog Device's accelerometer, Corning's Pyrex glass at the turn of the century, and many others. There are also many well-known drugs that have been redirected or reformulated from "failed" status and have become successful and useful drugs. For example, Viagra, now prescribed for erectile dysfunction, was a failed cardiovascular drug; AZT, a successful treatment for AIDS, was a failed cancer drug; thalidomide, a successful multiple myeloma treatment, was previously a sedative and leprosy treatment; and many successful rare disease treatments have been developed out of drugs originally developed or marketed for other indications.

Breakthrough ideas will continue to emerge as one-off cases but can also result in significant angst among management, as the product might have reached market faster if the company had continued its investment into the project. There is no simple answer to this quandary. However, several approaches and concepts from management theory, which relate to better management of product development in high uncertainty and emerging technology domains, are discussed in the next section.

5.4   Uncertainty-based view of product development processes

There are four main uncertainties in all product development – two that management can control are *resource uncertainty* and *organizational uncertainty*, and two that management has to adapt and adjust to are *technical uncertainty* and *market uncertainty*.

Technical uncertainties are addressed by engineers and scientists in product development, whereas market uncertainty – in terms of competition and market segmentation, pricing, etc. – can be factored into the product development processes and product definition/characteristics. Resource and organizational uncertainty may occur because the project may not have the right amount of funding assigned to it at the right times, the team in charge of the project may disband, a senior manager supporting the risky project may leave, expertise in that technical area may not exist in the organization, or market channels may be new or unknown to the organization. However, these types of high internal uncertainty projects are also means for the company to renew itself from within as these breakthrough launch the company growth trajectory into a steeper curve. Examples include the Avastin drug product at Genentech, which made it a world-leading oncology company in a few short years; the digital X-ray project at GE; the first drug-coated stent at Boston Scientific; and Viagra at Pfizer.

As the uncertainties in product development rise, so also does the possibility to give the company tremendous advantage and market share or revenues by radically changing the markets or the technology base for competition. These radical innovations are either supported through alternate channels of product development within a company that has well-established incremental innovation systems, or sometimes these projects need support from management with long-term views not driven by quarterly performance. If these risky projects do not find funding, they may go "underground," calling on resources obtained through informal networks and volunteer efforts to continue to make progress. At some point the project emerges from the dark, undergoes examination, and competes for resources. This pattern has been seen repeatedly in several new products as described in the books listed in additional reading section at the end of this chapter. Why do these potentially breakthrough projects not receive management attention sooner? Because funding and organizational commitment requires a business plan or business case that reflects a level of certainty that simply does not exist throughout much of the life cycle of a breakthrough innovation project. Some management books that discuss this particular problem of radical or breakthrough innovation management (see References and Additional Readings at end of this chapter).

In all the three approaches that deal with high uncertainty projects – the stage-gate, discovery-driven, and learning-plan approaches – the most important assumptions are clearly defined, and tests are developed to prove or disprove the assumptions. However, the purpose of these different approaches is typically as described here: milestone-based planning is usually better suited for an incremental product development; discovery-driven planning is used to develop a new business concept and related product(s); and innovative solutions to address an identified problem or market need are explored and better defined in the learning plan approach, as shown in Figure 5.5.

The sequence of milestones is clearly formulated in the stage-gate and discovery-driven processes as the end goal is quite clear in these approaches, as opposed to the learning plan. The clarity of objectives in discovery-driven planning allows the team to work backward from the goal, recognizing assumptions and evaluating their validity. Further details on these various approaches can be studied in the management papers and books listed in the References and Additional Readings for this chapter.

This book focuses on the stage-gate process, as this process is particularly well-suited for biomedical product development. The regulatory gatekeeper (FDA) forces specificity and focus of application (the indication) very early on in the development

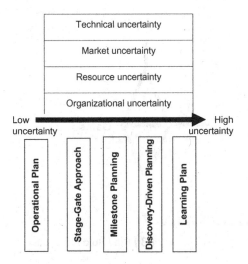

Figure 5.5 Uncertainty-based view of product development planning

process making it imperative to develop a defined process. Over the last decade, most companies (in diverse industries) have adopted some version of the stage-gate process first introduced by Robert Cooper in 1988. The stage-gate process consists of activity stages, between each of which is a decision point or a gate for a go/no-go decision. The content and format of the "gates" are usually based on perceptions of risk at that stage of the project. The actual decision-making (go, no-go, or repeat stage again) process depends on both the context of the project and the culture and processes of the company but should usually include the scientific team leader (or members).

5.5　Stage-gate approach

5.5.1　Stages and gates

Stages are key areas of activity that define a functional area or focus. Moving from one stage to the next also generally correlates to major steps along the value chain (see Chapter 1). Gates are the decision-making evaluation points for the results of that stage. Each stage could have multiple evaluation criteria within a gate. If the outcome from that stage does not meet the predefined parameters, then the project does not move on to the next stage of activity. For example, if a diagnostic test cannot demonstrate reproducibility to a certain preset level, then it cannot move on to clinical testing. An example gate for a drug would be the demonstration of efficacy in an animal model of the disease or the demonstration of adequate stability in formulation; without showing which the drug will not be allowed to move onto formal preclinical toxicity testing studies.

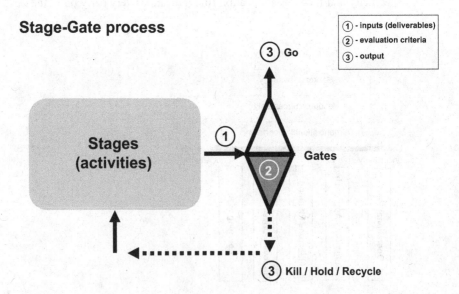

Figure 5.6 Stage and gate detail model

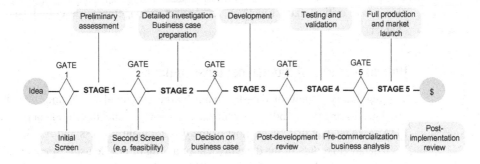

Figure 5.7 Stage-gate process as described by Robert Cooper. (Adapted from Cooper, 1988.)

Cooper, in his original paper (1988), defined six stages of activity in a product-development process and it is conceptually useful to see those original six stages and screening gates as shown in Figure 5.7.

5.5.2 How to configure a stage-gate process plan for my biomedical product

Each stage should be focused on one key functional area with the gates defined by key risk-reducing results of that stage. Within each stage, you can have multiple substages and gates, representing key areas of evaluation of the project. For example, in a combination product that involves live cells implanted on a bioresorbable material coated with a drug, the first stage of preclinical testing could have substages that evaluate each of the individual components with regard to cross-reactivity (e.g. will the implanted cells behave in the right manner when placed on the drug-coated resorbable device?).

A brief discussion of the various stages of development for drugs, diagnostics, devices, and combination products is given below, with a conceptual description of the types of evaluation gates. Each company will have different processes and variations of the stage-gate process in place and the reader must apply the general principles described here in the specific relevant context.

While constructing a stage-gate plan, it is important to be as quantitative as possible in defining both the specific parameters in a gate and the interplay among various parameters. With regard to the interplay between parameters, for example, drug solubility may have a set threshold, but a lower value could be acceptable if other pharmacokinetic parameters like half-life are better than expected. Or a device life span may be reduced if the weight threshold is reached. These can be mapped in a matrix with some weights and multipliers on each parameter interaction box. For example, in diagnostic development, the interaction box between specificity and sensitivity gates might have a multiplier of 0.8, so that if the specificity is higher than the expected level, the sensitivity could be lower than current threshold and still yield an acceptable product.

The stage-gate plan for product development, identifies activities and key milestones, and is useful for generating milestone-based budgets, resource-allocation plans, and critical path timelines for project management (described later in the chapter).

5.5.3   Unique features of biomedical development

The regulatory review of data to obtain market approval is a unique feature of all regulated biomedical products and impacts the new product development process. The FDA reviews data that spans most of the development process and it is important to keep detailed records of the development of a drug or device using specific guidelines as suggested by the FDA (Chapter 6). This requirement to store early product design and development data for reporting to the government is unique to this industry. The various multiple iterations in product design and development also need to be well documented and presented with explanations to the FDA. The processes of product testing and design are regulated by the Quality System required to be followed by global regulatory bodies including the FDA (QSR regulations; see Chapter 6). In particular, once the product is in final preparations to enter human clinical testing, the design is in effect "locked in." Any change in key product characteristics during that phase of testing requires added notifications to be submitted to the FDA and in some cases, further progress of the development will depend on FDA approval. This "design lock" relatively early in the product development process is another unique feature of biomedical product development.

There are benefits to the public and eventually to the manufacturer, from this onerous regulatory process, as explained in Chapter 6.

5.6   Ethical requirements in biomedical product development

If product development proceeds without regard to general ethical principles and the specific checks and balances described here (also Box 5.5) and in Section 5.16 (certifications, licenses, and regulations needed by biological research facilities), the company can face negative consequences including a complete halt of all product development activities ordered by the regulatory body.

Ethics considerations formally enter the NPD process at the time of *preclinical testing* of the drug or device and continue through the clinical development and post-approval stages of development. Specific embodiments of these considerations are described in the following sections.

5.6.1   IACUC – Institutional Animal Care and Use Committee

In the United States, any study being done on vertebrate animals, or on tissues or cells taken directly from an animal, requires the establishment of an animal care and use program and policy at the company/institute. This Institutional Animal Care and Use Committee (IACUC; vernacular pronunciation "I-a-cook") will review the specific study protocol. An animal care program must be managed in accordance with

---

Box 5.5 Some notes on regulations

IACUC resources are available online at: http://grants.nih.gov/grants/olaw/refer ences/outline.htm.

The *Animal Welfare Act* has provisions to regulate and ensure that animals used in research, for exhibition, or as pets receive humane care and treatment. Regulatory authority is vested in the secretary of the U.S. Department of Agriculture (USDA) and implemented by USDA's Animal and Plant Health Inspection Service (APHIS). Rules and regulations pertaining to implementation are published in the Code of Federal Regulations, Title 9 (Animals and Animal Products), Chapter 1, Subchapter A (Animal Welfare). Available from: Regulatory Enforcement and Animal Care, APHIS, USDA, Unit 85, 4700 River Road, Riverdale, MD 20737-1234. File Name 9CFR93.

The Public Health Service (PHS) Policy on Humane Care and Use of Laboratory Animals was updated in 1996 and contains guidelines on what constitutes humane and ethical treatment and care of animals used in research. Information concerning the policy can be obtained from the Office for Protection from Research Risks, National Institutes of Health, 6100 Executive Boulevard, MSC 7507, Rockville, MD 20892-7507

IRB related: FDA regulations that apply to clinical investigations and govern the development of drugs, biologics, and devices are contained in Title 21 of the Code of Federal Regulations (CFR), which can be purchased from the Superintendent of Documents, Attn: New Orders, P.O. Box 371954, Pittsburgh, PA 15250-7954; (202-512-1800, fax: 202-512-2233)

---

applicable federal, state, and local laws and regulations, such as the federal Animal Welfare Regulations and Public Health Service (PHS) Policy on Humane Care and Use of Laboratory Animals (PHS 1996). The IACUC reviews the animal care program and approves any experimental procedure that involves vertebrate animals. The IACUC committee must include a veterinarian. The IACUC reviews facilities and study protocols for the following:

- Minimum standards of care and treatment.
- Research facilities meet required standards of veterinary care and animal husbandry, dogs' exercise, primate psychological well-being, etc.
- Minimize the pain or distress caused by research as best as the experiment allows (anesthesia or pain-relieving medication). Note: The Animal Welfare Act also forbids the unnecessary duplication of a specific experiment using regulated animals.

5.6.2   IRB – institutional review board

An institutional review board (IRB) review and approval of any study protocol that involves interactions with human subjects is required by the FDA. IRBs that approve studies of FDA-regulated products must be established and operated in compliance with 21 CFR (Code of Federal Regulations) part 56. An IRB review is mandatory before starting human clinical trials with any biomedical product, even a diagnostic or software program that may not involve direct interaction with the patient. The main purpose of an IRB is to ensure that the rights and welfare (safety) of the subjects participating in a clinical trial are protected. Thus, the IRB rigorously reviews (in addition to the submitted study protocol) the "Informed Consent Form" that a subject signs to confirm the subject understands the procedures and risks involved in the clinical trial (see Box 5.6). The IRB also verifies that the sponsor (product developer) has obtained all necessary permissions from the FDA before beginning the trial.

Box 5.6 Informed consent

Informed Consent Forms – minimal requirements for protection of clinical trial subjects as per the Code for Federal Regulations (21CFR50)

Title 21 – FOOD AND DRUGS

CHAPTER I – FOOD AND DRUG ADMINISTRATION, DEPARTMENT OF HEALTH AND HUMAN SERVICES

PART 50 – PROTECTION OF HUMAN SUBJECTS – Table of Contents

Subpart B – Informed Consent of Human Subjects

Sec. 50.25 Elements of informed consent

(a) Basic elements of informed consent. In seeking informed consent, the following information shall be provided to each subject:

(1) A statement that the study involves research, an explanation of the purposes of the research and the expected duration of the subject's participation, a description of the procedures to be followed, and identification of any procedures which are experimental.

(2) A description of any reasonably foreseeable risks or discomforts to the subject.

(3) A description of any benefits to the subject or to others which may reasonably be expected from the research.

(4) A disclosure of appropriate alternative procedures or courses of treatment, if any, that might be advantageous to the subject.

(5) A statement describing the extent, if any, to which confidentiality of records identifying the subject will be maintained and that notes the possibility that the Food and Drug Administration may inspect the records.

(6) For research involving more than minimal risk, an explanation as to whether any compensation and an explanation as to whether any medical treatments are available if injury occurs and, if so, what they consist of, or where further information may be obtained.

(7) An explanation of whom to contact for answers to pertinent questions about the research and research subjects' rights, and whom to contact in the event of a research-related injury to the subject.

Box 5.6 (*cont.*)

(8) A statement that participation is voluntary, that refusal to participate will involve no penalty or loss of benefits to which the subject is otherwise entitled, and that the subject may discontinue participation at any time without penalty or loss of benefits to which the subject is otherwise entitled.

(b) Additional elements of informed consent. When appropriate, one or more of the following elements of information shall also be provided to each subject:

(1) A statement that the particular treatment or procedure may involve risks to the subject (or to the embryo or fetus, if the subject is or may become pregnant) which are currently unforeseeable.

(2) Anticipated circumstances under which the subject's participation may be terminated by the investigator without regard to the subject's consent.

(3) Any additional costs to the subject that may result from participation in the research.

(4) The consequences of a subject's decision to withdraw from the research and procedures for orderly termination of participation by the subject.

(5) A statement that significant new findings developed during the course of the research which may relate to the subject's willingness to continue participation will be provided to the subject.

(6) The approximate number of subjects involved in the study.

(c) The informed consent requirements in these regulations are not intended to preempt any applicable Federal, State, or local laws which require additional information to be disclosed for informed consent to be legally effective.

(d) Nothing in these regulations is intended to limit the authority of a physician to provide emergency medical care to the extent the physician is permitted to do so under applicable Federal, State, or local law.

5.7 Define the product and process = indications, endpoints, target product profile, and minimum viable product

The goal of all development work in any company is to get a product to market as quickly and as efficiently as possible. For biomedical products to get to market, there is an additional consideration, as the market is defined through very specific language in the approval from the FDA. The disease for which the product is approved to be marketed, is called the *indication* (discussed in greater detail in Section 2.5.5). All data collected in carefully designed experiments (clinical studies with human subjects) and submitted to the FDA have to satisfy rigorous evaluation by the FDA and must show that the product will perform as stated *for the specific disease condition* that it is being indicated for. Each new indication for the same product needs a separate application to the FDA by the product developer (follow-ons to the primary approval are shorter). On approval, the FDA issues a "label" letter that very specifically defines the indication for the product; and the manufacturer cannot market or make any claims outside of treating or diagnosing the disease condition/ indication on the label. Therefore, the product development process must have the selection of the "indication," or the specific state of the disease for which the product is going to be developed as an early milestone, perhaps even at concept stage.

Product development studies have to be focused and designed carefully to finally convince the FDA, with statistical analysis, that the product works safely to treat or diagnose that specific indication or disease compared to either placebo or current comparable products. The final statistical analysis of the study outcome, on which the approval is based, is called the *endpoint(s)* of the study. Clinical endpoints are distinct measurements or analyses of disease characteristics reflecting the effect of a therapeutic intervention in a clinical trial or study. The product development plan and in particular the clinical studies are designed and structured to try and reach a statistically significant endpoint. Endpoints can be grouped as single, multiple, or composite and the final claim (marketing, efficacy, etc.) made on the device or drug will depend completely on the evaluation of this set of endpoints. Hence study design must be carried out keeping the endpoints in mind. Examples of simple and composite clinical trial endpoints for drug products are given in Box 5.7, and Box 5.8 gives a device example on how to choose a clinical endpoint. Box 5.9 provides an outline of a typical target product profile. Market research input is frequently used to help identify the endpoints and specific indication (see Chapter 2) especially in the context of competitive products. Surrogate endpoints, typically used in drug development, are discussed in Section 5.8.3.

While statistically significant endpoints are useful for gaining market approval from the FDA, an equally relevant data point is the clinical outcomes analysis. Clinical outcomes measures and real-world data-driven clinical trials are used for economic analysis and decisions by insurance companies and payers on reimbursement the new product. The clinical outcomes measures are usually additional to the endpoints collected for FDA approval. These outcomes provide measures of actual patient-life benefits, such as number of surgical follow-up procedures, systemic cost reductions for hospitals, or reduced rehabilitation visits. Companies are now increasingly including such additional data collection points in their clinical trials in order to show value and get favorable commercial considerations for their new products. The company may carry out additional or extended clinical trials to track such clinical outcomes after their main pivotal trial for approval. The planning for these clinical trials based on identified patients' and payers' needs is usually indicated in the target product profile developed at the start of a project.

What is a target product profile (TPP) document and what is it useful for? The TPP is a detailed analysis of a potential new product in comparison to competitors and the existing standard of care. This document is typically owned by the project leader or product manager but has inputs from multiple stakeholders such as regulatory group, marketing, engineering/pharmacology, clinical, etc. Most important part of the TPP is the identification of the primary indication planned to addressed by the product. The TPP also defines key characteristics of the product and its development context – such as specific desired dosing characteristics (once a day vs thrice, oral vs injected, etc.),

---

Box 5.7 Primary and composite endpoints for drug studies
Material adapted from an FDA presentation, 2004

Primary endpoints comprise a set of clinical endpoints based on which clinical benefits are assessed. Clinical studies with the product have to show a statistically significant effect (compared to a control group) to satisfactorily conclude that the primary endpoint has been met. Primary endpoints usually provide characterization of various aspects of a disease and are used to describe clinical benefits.

Examples of primary endpoints in clinical trials:

Anti-hypertension drug trial primary endpoints:
　　Supine diastolic blood pressure
Congestive heart failure drug trial primary endpoints:
　　Reduction in incidence of all-cause mortality
　　Reduction in incidence of stroke
　　Reduction in incidence of myocardial infarction
Alzheimer's drug trial primary endpoints:
　　Alzheimer disease assessment scale – cognitive subscale
　　Clinician's interview-based impression of change
Epilepsy device trial primary endpoints:
　　Percent reduction in seizure rate
　　Percent reduction in drop attack rate
　　Parental global evaluation of seizure severity

Secondary endpoints
In a clinical trial, secondary endpoints form a set of clinical endpoints that are intended for possible inclusion in the label, after efficacy has been demonstrated by the primary endpoints.

Composite endpoints
Composite endpoints are a combination of several primary endpoints, used when the disease manifestation is complex.

Example
A study of Losartan (COZAAR) vs Atenolol in 9193 hypertensive patients had three primary endpoints – reduction in incidence of cardiovascular (CV) death, stroke, or myocardial infarction (MI) – which were summarized in a composite endpoint as below:

Composite endpoint (CV death, stroke, MI) = The time to the first occurrence of either CV death, stroke, or MI.

Box 5.8 Challenges in designing clinical endpoints (device example)

This example is an extract from a FDA Guidance document for percutaneous tissue ablation in atrial fibrillation on clinical trial design and endpoints (www.fda.gov/cdrh/ode/guidance/1229.html). This minimally invasive procedure involves ablating/selective removal of a segment of the tissue that gives the heart muscle its synchronizing pulsation signals.

Clinical study designs for percutaneous catheter ablation for treatment of atrial fibrillation

Introduction and scope

Atrial fibrillation (AF) is a complex arrhythmia; its precise mechanisms remain unclear, and the clinical presentation, arrhythmia characteristics, and underlying pathophysiology are variable. This guidance document addresses study design issues associated with catheter ablation devices intended for treatment of atrial fibrillation. These devices (product code: LPB; Electrode, Percutaneous, Conduction Tissue Ablation) are class III, requiring premarket approval applications before marketing (section 513(a) of the Federal Food, Drug, and Cosmetic Act (21 U.S.C. 360c(a))).

Study endpoints: primary effectiveness endpoint

In the future, it may be feasible to demonstrate that ablation therapies for AF positively affect disease outcomes. At the current time, it is probably most appropriate to evaluate ablation therapy for AF as a palliative therapy and to select endpoints that have the potential to clearly demonstrate a reduction in symptoms caused by AF. FDA believes that evaluation of reduction of AF burden (or reduction in the incidence of AF) is problematic as the primary endpoint for a study designed to evaluate therapy for paroxysmal AF. Measurement of this endpoint post-ablation could be strongly influenced by various, non-therapy-related factors. . . . *For a primary effectiveness endpoint, FDA recommends the relatively unambiguous endpoint of freedom from symptomatic atrial fibrillation at one year* [emphasis added]. This outcome should be in the absence of antiarrhythmic drug therapy ... A one-year follow-up period both minimizes the confounding effects of the clustered, non-random AF recurrence pattern that was previously discussed and provides sufficient time to evaluate adverse events, e.g., pulmonary vein stenosis, that may be manifest or progressive only at late time points in some patients.

Primary safety endpoints

In considering primary safety endpoints, FDA acknowledges that an ablation intervention arm and a drug intervention arm may have different safety criteria.

Box 5.8 (*cont.*)

Ablation procedure safety endpoint

For an ablation procedure safety endpoint, FDA recommends for the devices addressed in this guidance document, a composite serious adverse event endpoint that includes, but need not be limited to, the following:

- Transient ischemic attack
- Cerebrovascular accident
- Major bleeding
- Cardiac tamponade
- Pulmonary vein stenosis
- Pericarditis
- Myocardial infarction
- Diaphragmatic paralysis
- Death.

Composite serious adverse event endpoint

For a drug intervention arm, FDA recommends a composite serious adverse event endpoint, which includes, but need not be limited to, the following:

- Life-threatening arrhythmia
- Transient ischemic attack
- Cerebrovascular accident
- Anaphylactic reaction
- Pulmonary hypertension (if amiodarone therapy)
- Death

Box 5.9 Outline of a typical target product profile

Typical biopharmaceutical TPP includes the following sections:

Molecule (NBE/NCE) and General Product Information: Brief description of the molecule, product name, general information

Mechanism of Action: The mechanism by which the product produces an effect on a living organism

Clinical Pharmacology: Pharmacokinetic information, distribution, and pathways for transformation

Indication for Use: Targeted disease and population

Target Manufacturing Profile: Formulation, shelf life, storage conditions, delivery system

Primary Efficacy Endpoints: The primary clinical outcome measures. *Note: endpoints are usually proposed as 3 different scenarios: minimal, base, optimal*

Box 5.9 (*cont.*)

Secondary Efficacy Endpoints: Additional endpoints that are not required to be met in a clinical trial.

Expected Safety Outcomes: The primary safety outcome measures

Contradictions: Known or expected contraindications

Commercial Landscape: Description of the competitive landscape at launch time, prospects, positioning, competition, and business case for investment in the product.

Regulatory: Expected BLA/NDA approval date

R&D Go/No-Go: Criteria for upcoming decision points or milestones

From NIH website: www.ninds.nih.gov/Funding/Apply-Funding/Application-Support-Library/CREATE-Bio-Example-Target-Product-Profile-TPP.

Example TPP for a therapy for pain associated with diabetic neuropathy

Product properties	Minimum acceptable result	Ideal results
Primary product indication	Relief of pain symptoms in diabetic neuropathy	Relief of symptoms in neuropathic pain syndromes
Patient population	Adults with diabetes who experience moderate to severe pain	Adults with diabetes who experience moderate to severe pain
Treatment duration	Chronic	Chronic
Delivery mode	Subcutaneous injections	Subcutaneous injections
Dosage form Regimen	Prefilled vials with liquid Once every month	Prefilled vials with liquid Once every 2 months
Efficacy	A 40% decrease in pain score in 30% of patients	A 70% decrease in pain score in 50% of patients
Risk/side effect	Devoid of local injection effect and clinically significant CNS side effect	Devoid of local injection effect and any CNS side effect
Therapeutic modality	Antibody	

An example of a detailed TPP for improved diagnostic products is found at www.finddx.org/wp-content/uploads/2016/01/HCV-TPP-Report_17July2015_final.pdf (last accessed Nov 2020).

permissible level of side effects. The TPP also specifies key claims or benefits (safety/ efficacy) over competitors or the existing standard of care (key differentiations), details mechanism of action, and finally lists the desired clinical outcomes or end-points for human trials for registration. Economic value of indication and claims are also usually included in the TPP, which can be thought of as a business plan for the product but with more technical details than might be typical in business plans. The TPP document is a snapshot in time, and updates are expected during the development pathway. Many drug and device development companies, in particular, submit a TPP to the FDA (optional, not required) to support and enhance their dialogue with the FDA.

The TPP is thus a key strategic document, capturing the value proposition of the proposed product and directing the development of a regulatory strategy. The TPP provides a framework that ensures that a company's product development program is efficient, by defining all relevant medical, technical, and scientific information required to reach the desired commercial outcome.

As part of a product development plan in the TPP, certain products may also include the first launch of a "minimum viable product" (MVP) (e.g. a first prototype of a decision support software solution as a trial, a first prototype working medical device provided under humanitarian exemption, or a lab-developed diagnostic test launched before FDA approval). MVP is a concept from the information technology industry where an early version of an actual fully functioning product is released (not partial prototypes) in order to learn more from customers about their usage, wants and needs. Eric Ries, in his book *The Lean Startup* (2011), on page 96, defines the MVP as that "version of a new product which allows a team to collect the maximum amount of validated learning about customers with the least effort." Releasing early product versions for final users on an experimental basis to learn and iterate to a better product is not really possible to do on a short turn-around time or low cost in the biomedical industry. An MVP concept could put patient health at risk. Regulatory approval is needed just to have a test run with subjects. Product versioning in drugs and devices is a common strategy, but in years or decades, not in the "days/weeks" time frame espoused for iteration of MVP product versions for software.

However, with the recent emergence of software products as medical devices (SaMD), there is greater import to this kind of product development method in the biomedical industry. The FDA recently (2019) launched a Digital Health Software Precertification (Pre-Cert) Program. The Pre-Cert Program for SaMD products allows companies with "a robust culture of quality and organizational excellence" to apply for company-wide precertification. Precertified companies are free to innovate more rapidly – without applying for 510K clearance with every new product release.

Also, in the United States, in vitro diagnostics products are frequently launched into commercial use before FDA approval by a certified laboratory as a laboratory-developed test (LDT). Thus the LDTs are like an MVP launch of the in vitro diagnostic test under development, allowing early interaction with the market (users and payers) to learn more about the desired final product performance.

5.8   Typical drug development process

Drug development takes over 10 years, from the definition of the problem and identification of drug target to market approval from the regulatory body. The process typically costs over $800 million (includes the cost of failed prototypes); while varying estimates exist, it is clear that the majority of the costs arise from human clinical testing of the drug (Figure 5.8a). Figure 5.8a shows the approximate time for each stage of development along with the approximate cost for each stage of drug development not including the failures and infrastructure fixed costs for each stage. These cost estimates are based partly on the author's experience developing an inhaled orphan disease drug with a novel small molecular entity. The economic analyses on drug development costs must, of course, include the costs of failures along the way as it is only a foolishly optimistic entrepreneur who would think that their single molecule effort will be successful all the way to market (the statistics in drug development show that less than 10% of molecules that start development get approved and get to market). Figure 5.8b shows, in slightly greater detail, the relative ratio in distribution of costs of drug R&D across various functional stages of development.

The functional segments of the value chain for drug development are schematically outlined in Figure 5.9 (also discussed in Chapter 1) and the stages for each functional segment of the value chain are discussed in greater detail in the following sections.

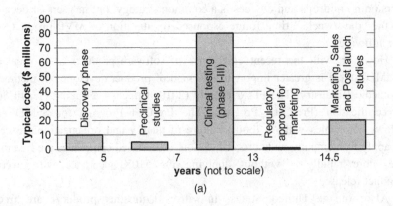

(a)

Figure 5.8a Typical time and costs for development of a new drug

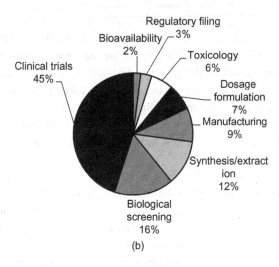

(b)

Figure 5.8b Pie chart illustrates, in percentages, the allocation of development costs. (Data from Mathieu, *Parexel's Bio/Pharmaceutical R&D Statistical Sourcebook* (2005)).

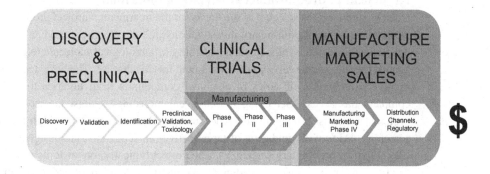

Figure 5.9 Activities in drug development

5.8.1 Discovery and preclinical testing

Target discovery and validation

A drug development project often starts with the identification of a disease problem that is lacking adequate treatment (a "medical need") and the discovery of a pathological mechanism that appears to influence the disease, its symptoms, or its progression. An example would be the initial idea to develop a treatment for Alzheimer's disease, as the drugs currently available are not adequate. A little research on etiology and pathology of the disease will reveal that there are three possible ways to prevent disease progression or possibly reverse the disease – block glutamate neurotoxicity, prevent amyloid plaques, or attenuate neuro-inflammation. Tissue cellular processes that are altered from normal health behavior have some changes in the quantity (decreased or increased levels being made), the form (mutation), or function (activity) of the proteins that carry out these processes. Either one of the above approaches to developing a new treatment for Alzheimer's would point to a specific enzyme or cell receptor whose activity would have to be blocked, enhanced, or modulated (typically the former is the easier to do with a drug). These enzyme or receptor proteins then become known as *drug targets*. Target discovery usually takes place in academic settings, where such basic and fundamental phenomena are studied, but with the advent of industrial-scale genomics and proteomics and now, artificial intelligence–based software approaches, discovery of new targets is likely to happen in corporate research settings as well.

Target validation: Once discovered, the target has to be validated – experiments must be carried out to show that changing the activity of the target protein will affect the disease outcome positively. In this preclinical stage, studies are done using knockout models (where expression of the target protein is completely knocked out by genetic manipulation at the formative stage) or other disease models and interventions. These experiments involve fundamental biological principles and practices and are expensive and time consuming, and can last a number of years.

Hit to lead to drug product candidate optimization

Once a valid target is selected as the thesis for the treatment, a project proposal can be written up to launch preliminary investigations into developing a product that would develop either a new synthetic chemical entity or a biological molecule (e.g. antibody against the selected target) as a drug treatment for this disease. The next steps typically involve examination of series of chemical compounds from libraries to which the company has access. Screening the chemical molecules against the isolated target protein, or against a relevant cell type, (using techniques known as high throughput screening, HTS) will yield a set of target-active molecules that are designated as *hits* (the output of preliminary screening). The project typically moves to its next level of funding and resourcing, as medicinal chemistry is now involved in refining and optimizing the hit compounds into *lead compounds*. This *hit to lead* optimization and refinement process can take from months to years to yield an acceptable *drug product candidate*.

There are many tools available today to help speed this process up, including computational screens and high throughput crystallography, which can lead to very detailed understanding of the optimal molecular structure. More discovery tools such as artificial intelligence/machine learning will continue to be developed and applied to this area of development, as there is need to make this part of the process more efficient and productive. As seen in Figure 5.10, the lead optimization process from hit to drug candidate involves, in addition to medicinal chemistry, a variety of biological tests (efficacy screening in cells or whole animal models, pharmacokinetics, metabolism, early toxicity) to evaluate the properties and biological activity of the compounds being synthesized. The process is inherently iterative (Figure 5.10). Box 5.10 indicates some preferred product characteristics for an orally ingested chemical drug product. Box 5.11 shows how to transform a compound into a drug.

Box 5.10 Sample gate parameters for lead compounds

Lipinski's "rule of five" is typically one set of parameters used as cut-off characteristics to reject or select compounds when good intestinal absorption is important (most oral drugs).

The Lipinski's rule of five states that poor absorption or permeation is more likely when:

- Molecular weight (MW) is over 500.
- LogP is over 5 (or MLogP is over 4.15).
- More than 5 H-bond donors (expressed as the sum of OHs and NHs).
- More than 10 H-bond acceptors (expressed as the sum of Ns and Os).

The rule of five is so called because the cutoff for each of four parameters is either five or a multiple of five(Lipinski et al. [1997]).

Other parameters used could include characteristics such as solubility (above a certain level); potency against the specific drug target, measured in IC_{50} (the concentration at which the compound shows 50% of maximum theoretical activity). For example, a cancer compound may be expected to kill all cancer cells in a culture; if it kills 50% of the cells at a 10 micromolar concentration, the IC_{50} of that compound for that cell line is 10 micromolar.

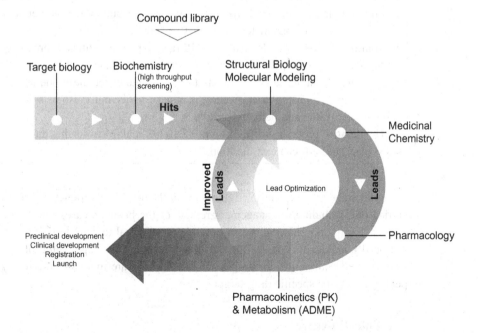

Figure 5.10 Preclinical stages of developing a drug product

Box 5.11 Transforming a lead compound into a drug

There are several issues in product development that young companies, and in particular inexperienced staff, sometimes fail to consider, during preclinical testing.

In general, the small molecule or biologic must be druggable. Since issues in developing biologics into products are discussed in Section 5.8.2, this sidebar note will focus on small molecules. "Druggable" is a term which means "must be capable of becoming a drug product," but this term really covers many characteristics of the small molecule such as the Lipinski rules mentioned above, manufacturability (if it is a 50-step complex synthesis with low yields, or involves toxic reagents, it might not be possible to scale up; or if polymorph crystal forms emerge in the final scaled-up process with different properties than the original compound), stability, protein binding, solubility, and other such parameters.

Two examples of missed opportunities due to poor preclinical work were quoted in an interview by the head of a global preclinical contract research organization. The first example was that of a small molecule that was being tested in experimental animals – the compound was actually degrading into a new compound while stored overnight and the scientists were making their (patent and IND) claims on the wrong molecule. The scientists in the startup company had not known to do a stability check on the small molecule in its final formulation. Another example was that of a group that had developed an anti-cancer agent that was quite effective but had to be stored and transported at –70°C to remain stable. This group spent a few million dollars on developing the drug compound and on proof of concept clinical trials, but the drug did not get taken up commercially, as not everyone could store the drug in –70°C freezers, limiting distribution and market adoption.

There are various parameters that must be satisfied before a drug candidate can be confirmed. The technology of the API (active pharmaceutical ingredient) must be transformed into a product by considering all of the above and also confirming appropriate packaging and excipients that are suitable for the mode of administration – e.g. for inhaled dry powder, the inhaler device must also be selected to fit the size of capsule and delivery profile.

The iterative process shown in Figure 5.10 has multiple tests that the compound must pass before it can exit the loop.

At this stage gate, typically, a full review of the optimized product characteristics is carried out, including a management review of the business case, and the design is "frozen," with only minor changes possible after formal toxicity testing (next stage). The lead compound(s) passes through many tests in this stage, some of which are standard and common to all drugs, while other tests are more specific to the therapeutic area or the specific drug target.

Preclinical toxicity

If the drug candidate has passed the screening gates on specific characteristics such as good pharmacokinetic parameters, then it enters a formal toxicity testing program and scale of production under regulatory guidelines. From this stage onward, *the drug molecule (active pharmaceutical ingredient or API) is usually tested as a product formulation (solid, liquid, suspension) with excipients (other non-active components that do not change biological activity of the API and are usually selected from those already approved for human use).* Toxicity testing usually requires studies on at least two animal species (typically rodent and canine) to identify the doses at which toxic effects are seen. Additionally, specific effects on various tissues or organs may be indicated based on the mechanism and characteristics of the drug (e.g. neurotoxicity screening if the compound crosses the blood-brain barrier or lung toxicity if the compound is inhaled). The manufacturing methods and related quality-control parameters such as lot-to-lot reproducibility, product stability, trace element analysis per lot, yield consistency, etc., are also established at this point, as the API has to be scaled up. These quality-controlled manufacturing processes are termed GMP (good manufacturing practices; see Chapter 7 for details) and all product for human use must be made using these regulated methods and documented for submission to the FDA, usually with three production lots being evaluated for the submission. Only FDA approved and inspected facilities can manufacture the drug compound for human testing. The compound formula is typically locked in prior to these preclinical toxicity studies and any change, however minor, must be communicated to the FDA and approved before proceeding further. Laboratory experiments must be carried out using processes that follow the strict FDA guidelines known as the current good laboratory practices (cGLP). The GLP guidelines outline the processes and recording requirements to be followed in data generation, handling, and analysis. This term and related regulations for promoting the generation and exchange of high-quality and valid data are used in most other countries around the world. These data are submitted to the FDA as part of a petition called an IND (investigational new drug application), requesting initiation of testing in humans. These interactions with the FDA, including the contents of an IND application, are described in more detail in Chapter 6.

5.8.2   Distinctions in preclinical development of biotechnology drugs (large molecule biologicals)

The biological drug is usually a protein or a glycoprotein, made by biological processes in a bioreactor full of living cells (the current preferred production cell is a Chinese Hamster Ovary or CHO cell line) or in other whole organisms (secreted in milk of transgenic goats or in tuber of transgenic plants for example), rather than a chemical molecule made by synthetic chemical processes. Subtle changes in the composition of the biotherapeutic molecule can cause significant changes in biological activity, making it important to work with as close to a final molecule as possible, even in the early stages of drug development. The cell line, production and processing conditions, and the product itself must be well characterized early on in the process. Investing in scalable, well-characterized processes early on in the development process needs to be balanced with having a stable, reproducible source of the compound, and while smaller batches or production are possible even with synthetic drugs, (when it is not known if the project will progress to clinical studies), biological drugs will usually need more planning and investment at this stage.

A risk factor that is specific to the use of biological drugs is the development of antibodies in humans against the synthesized protein drug, leading to drop in efficacy or severe side reactions. Strategies to test and reduce the potential for antibody formation in humans must be evaluated and applied early in the development process.

5.8.3   Drug candidate clinical testing to market approval

The FDA suggests that the company arrange its first formal interaction visit before starting human clinical testing. This is not mandatory but highly recommended as further described in Chapter 6. All clinical studies must be carried out using current good clinical practice (cGCP) guidelines. Planning for clinical studies usually must incorporate the following considerations (also see Box 5.14):

- Clinical endpoints must be carefully selected (see Chapter 6, Section 6.1 for details). Input from marketing departments/research (Chapter 2) and reimbursement specialists (see Chapter 8 for details) must be incorporated into the study design so that the product is developed with the appropriate data and characteristics
- Patient inclusion and exclusion criteria must be defined – examples of such criteria are found in Box 5.12.
- Clinical study design and protocol. The measurements made during the study are defined in this document, including details on the dosing schedule. The number of patients to be recruited, group assignment, protocol for drug dosing/delivery, data collection and archival, reporting of adverse events, and many other criteria are defined clearly before the study begins. See Box 5.13 for examples and details.
- Size and length of the clinical study must be determined by a robust statistical analysis, as the final data must show a statistically significant effect of the drug in patients over placebo to be approved by the FDA and must show a better effect than the existing standard of care. This analysis must be done as early as possible in the planning process, as this will give an estimate of the time and cost involved. For example, a 500-patient study that measures the effect of a drug therapy on overall survival of the patient groups will take 1.5 years to recruit patients, 2–3 years for the drug dosing period and patient data recording, and half a year for analysis of the collected data, at a medical cost of approximately \$12,000 per patient included in the study. This hypothetical study will cost \$180 million (medical costs per patient is about 30% of trial costs as a general rule of thumb) and take 5 years to complete.
- Data analysis techniques have to be laid out a priori, as the FDA will want to review the prospective trial and the specific hypothesis the clinical studies will be testing.

Human clinical testing for drugs typically proceeds in four phases:

Phase I clinical studies

The drug (active pharmaceutical ingredient (API) and its formulated carrier compounds) is first tested in a set of normal healthy patients to determine the toxic

Box 5.12 Patient selection criteria for clinical trials of a new asthma drug

Inclusion criteria

1. Males and females between the ages of 18 and 50.
2. FEV1 of 80% or greater than the predicted value. [Note: FEV1 is the Final Expiration Volume-1 and a measure of pulmonary function that classifies the severity of asthma.]
3. PC20 FEV1 < 8 mg/mL on methacholine challenge test. [Note: another measure of pulmonary function that classifies the severity of asthma.]
4. Blood pressure ≥ 110/70 mm Hg.
5. Pulse rate ≥ 60 beats/min.
6. No significant health issues.
7. Non-smoker or ex-smoker < 10 pack/year.
8. Able to complete diary cards and comply with study procedures.
9. Females of childbearing age may participate only if they have a negative pregnancy test, are nonlactating, and agree to practice an adequate birth control method (abstinence, combination barrier and spermicide, or hormonal) for the duration of the study.

Exclusion criteria

1. History of upper/lower respiratory tract infection or asthma exacerbation within 6 weeks of first baseline visit.
2. Currently diagnosed with chronic obstructive pulmonary disease (COPD).
3. Used any oral or inhaled corticosteroids within 4 weeks of the first baseline visit.
4. Used short-acting antihistamines within three days of the first baseline visit.
5. Used chromones (i.e. Intal, Tilade) or long-acting antihistamines within 1 week of first baseline visit.
6. Used leukotriene modifiers within 14 days of first baseline visit.
7. Used methylxanthines (i.e., theophylline) within 72 hours of first baseline visit.
9. Used ipratropium bromide (Atrovent) or other anticholinergics within 8 hours of first baseline visit.
10. Used short-acting beta2-agonists within 8 hours of baseline visit.
11. Taking any beta-blocker medications.
12. History of any cardiovascular, neurological, hepatic, renal, or other medical conditions that may interfere with the interpretation of data or the patient's participation in the study in the investigator's opinion.
13. Smoked within 1 year prior to first baseline visit.
14. Any clinically significant deviation from normal in either the general physical examination or laboratory parameters as evaluated by the Investigator at the screening visit.
15. History of alcohol or drug abuse.
16. Participation in another research trial within 30 days of starting this trial.

Box 5.12 *(cont.)*

17. Known allergy or sensitivity to the study drug.
18. History of adverse reaction to beta-blocker medication.
19. Inability to give consent and/or unwillingness or inability to comply with study procedures.
20. Inability to swallow the study medication tablet.
21. History of life-threatening asthma.

Box 5.13 Reconfiguring drug indications

A clinical study protocol is usually written by multidisciplinary team that will typically include a clinically trained physician, a biostatistician, a regulatory affairs person, marketing person, and researchers from the team that discovered the drug. Increased data collection will lead to increased costs, but taking measurements outside of the expected effects has resurrected many drugs that were failing to achieve efficacy in their main development indication.

A classic example is that of Viagra, which was developed for angina but in Phase I studies the clinician noticed that majority of the drug-treated volunteers had an unusual side effect (penile erection), while the majority of the placebo subjects did not have any such side effects. Since this was a drug with vasodilatory effects, it made them wonder whether the vasodilation was predominantly in the penile area. The rest is history. A new blockbuster class of drugs for sexual dysfunction emerged from this finding.

Other studies that have specifically tried to include additional endpoints to see whether rescue indications emerged (in case the drug failed the primary endpoint) have not always been successful. Examples include the Phase III anti-obesity studies conducted by Eli Lilly, Pfizer, and Regeneron for an antidepressive drug (Regeneron's drug was actually developed for Lou Gehrig's disease) that showed weight loss effects in Phase II/III studies. All of these three drugs with reconfigured indications failed to show sustained and significant effect in obesity.

---

thresholds and thus define the therapeutic window for dosage. The most difficult product characteristic to determine (due limited sets of tests possible and high cost of testing in humans) but also the most important, is the dosage or therapeutic window definition, that is, how much drug to give or how frequently. Usually, lower bounds of the therapeutic window are set by the early cellular or in vivo animal experiments and upper bounds are set by these Phase I studies.

Each participant is given a single dose of the drug and is closely monitored for adverse drug reactions. If none occurs, the dose of the drug is progressively increased until a predetermined dose or serum level is achieved or until some event marking toxicity occurs.

Note: Phase I studies in cancer therapeutics are sometimes carried out in terminally ill cancer patients who have failed other therapies.

Phase II clinical studies

The purpose of the Phase II studies is to determine the correct dose–response range (optimizing dosage characteristics) for the new drug and to verify its efficacy for the intended disorder. These studies are carried out in 20 to 100 patients randomly placed in placebo- or drug-receiving groups – called a randomized study – with specific *endpoints* defined in the clinical trial protocol for proving that the drug is effective. These endpoints are statistically determined differences from the placebo group in the parameters measured. For example, one endpoint of an anti-hypertensive medication can be that the blood pressure in the drug group must be 20% lower than the pressure in the placebo (or standard of care) group with strong statistical power to the result. For scientifically more rigorous experimental design, neither the patient, the physician, nor the drug company should know which randomized group is receiving placebo or drug until the completion of the study or upon emergence of a (predetermined) significant difference between groups. This type of experiment is called a *randomized, double-blinded study*, where there is little chance of even unconscious bias or manipulation of data or patients. Box 5.14 contains a brief discussion and suggestions from the FDA on clinical study design. The data from this phase are crucial in determining whether to proceed with more extensive studies in large populations (Phase III).

Note: Sometimes Phase I/II trials may be combined into one set of studies, or each Phase I and Phase II trial can have subcomponent studies titled Phase Ia, Ib IIa, IIb etc. It is advisable to have the FDA review the data at the end of Phase II to get their agreement in writing on the design (size and approvable endpoints) of the Phase III clinical trial. Often, the direct clinical outcome desired by the patient is not measurable in the clinical trial or may take too long to determine (e.g. improved survival endpoint) and a set of "surrogate endpoints" are used to seek regulatory approval. These surrogate endpoints might also be biomarkers, which are defined characteristics (e.g., molecular, histologic, radiographic, or physiologic

---

Box 5.14 Drug clinical study designs
Extract from FDA guidance

Before a new drug or biologic can be marketed, its sponsor must show, through adequate and well-controlled clinical studies, that it is effective. A well-controlled study permits a comparison of subjects treated with the new agent with a suitable control population, so that the effect of the new agent can be determined and distinguished from other influences, such as spontaneous change, "placebo" effects, concomitant therapy, or observer expectations. FDA regulations [21 CFR

Box 5.14 (*cont.*)

314.126] cite five different kinds of controls that can be useful in particular circumstances:

(1) placebo concurrent control
(2) dose-comparison concurrent control
(3) no-treatment concurrent control
(4) active-treatment concurrent control, and
(5) historical control

No general preference is expressed for any one type, but the study design chosen must be adequate to the task. It is relatively difficult to be sure that historical control groups are comparable to the treated subjects with respect to variables that could effect outcome, and therefore use of historical control studies has been reserved for special circumstances, notably cases where the disease treated has high and predictable mortality (a large difference from this usual course would be easy to detect) and those in which the effect is self-evident (e.g., a general anesthetic).

Placebo control, no-treatment control (suitable where objective measurements are felt to make blinding unnecessary), and dose-comparison control studies are all study designs in which a difference is intended to be shown between the test article and some control. The alternative study design generally proposed to these kinds of studies is an active-treatment concurrent control in which a finding of no difference between the test article and the recognized effective agent (active control) would be considered evidence of effectiveness of the new agent. There are circumstances in which this is a fully valid design. Active controls are usually used in antibiotic trials, for example, because it is easy to tell the difference between antibiotics that have the expected effect on specific infections and those that do not. In many cases, however, the active-control design may be simply incapable of allowing any conclusion as to whether or not the test article is having an effect. ...

For certain drug classes, such as analgesics, antidepressants or antianxiety drugs, failure to show superiority to placebo in a given study is common. This is also often seen with antihypertensives, anti-angina drugs, anti-heart failure treatments, antihistamines, and drugs for asthma prophylaxis. In these situations, active-control trials showing no difference between the new drug and control are of little value as primary evidence of effectiveness and the active-control design (the study design most often proposed as an alternative to use of a placebo) is not credible.

It is often possible to design a successful placebo-controlled trial that does not cause investigator discomfort nor raise ethical issues. Treatment periods can be kept short; early "escape" mechanisms can be built into the study so that subjects will not undergo prolonged placebo-treatment if they are not doing well. In some cases, randomized placebo-controlled therapy withdrawal studies have been used to minimize exposure to placebo or unsuccessful therapy; in such studies, apparent responders to a treatment in an open study are randomly assigned to continued treatment or to placebo. Subjects who fail (e.g., blood pressure rises, angina

---

Box 5.14 (*cont.*)

worsens) can be removed promptly, with such failure representing a study endpoint.

Institutional review boards (IRBs) may face difficult issues in deciding on the acceptability of placebo-controlled and active-control trials. Placebo-controlled trials, regardless of any advantages in interpretation of results, are obviously not ethically acceptable where existing treatment is life-prolonging. A placebo-controlled study that exposes subjects to a documented serious risk is not acceptable, but it is critical to review the evidence that harm would result from denial of active treatment, because alternative study designs, especially active-control studies, may not be informative, exposing subjects to risk but without being able to collect useful information.

For further reference, also visit the guidance document on conducting complex and innovative clinical trial designs at www.fda.gov/regulatory-information/ search-fda-guidance-documents/interacting-fda-complex-innovative-trial-designs-drugs-and-biological-products *[last accessed Dec 2020]*

---

characteristics) that are objectively measured as indicators of normal or pathologic processes, or responses to an exposure or intervention.

Phase III clinical studies

The purpose of the Phase III studies is to verify the efficacy of the drug in much larger populations that better reflect the general population of patients/customers, and to identify any side effects that may not have occurred during Phases I and II, so that the sponsor and FDA can determine that the drug is safe and effective for its intended use. When sufficient data are collected, a rigorous statistical analysis is carried out to prove the efficacy and safety improvements over existing treatments, and that data and completed analysis are submitted to the FDA with a request for approval to market the drug (NDA: new drug approval). Contents of the NDA are discussed in Chapter 6.

Phase IV (post-market) clinical studies

This title is given to clinical studies that are conducted after the drug is marketed, typically, clinical studies in large or select populations. Often, special subpopulations, such as pregnant women, children, or the elderly, are included, with the company's goal being to expand the indication or prove efficacy in new indications. The FDA's concern in sometimes requiring Phase IV studies is to monitor specific safety concerns that arose earlier. Reports from ongoing Phase IV studies must be sent to the FDA every 3 months during the first year, every 6 months during the second year, and annually thereafter. The sponsor must notify the FDA of any unexpected adverse effects, injury, and toxic or allergic reactions.

5.8.4   Manufacturing, marketing, sales, and reimbursement

The manufacture of the drug (major considerations and processes discussed in Chapter 7) is carried out under strict quality-control guidelines and regulations in approved facilities that are regularly inspected by the FDA. The API may be made at one site and combined with the excipients at another location, where the formulation of the final form of the drug – in pill, liquid, or powder from – is completed and packaged for distribution. The drug company has to send in labels and product brochures to the FDA for review. Reimbursement, sales, and distribution are discussed in greater detail in Chapter 8.

5.8.5   Keeping a record for the FDA

The FDA requires the submission of a *drug master file* (see Chapter 6), which contains all significant preclinical and clinical product development data and records, in order to consider the drug product for market approval. The details of this record, its content and use are discussed in Chapter 6.

5.8.6   General stage-gate process for new drug development

The stage-gate criteria shown in Table 5.3 for drug development are generalized and will likely differ in specifics either based on the particular nature of the drug being developed or based on the company's internal processes (see Figure 5.11).

5.9   Typical diagnostics development process

Diagnostics are mostly regulated as devices, but since the clinical usage aspects of diagnostics are more closely related to drug therapies, their development is discussed first in this section and device development (which covers several points relevant to diagnostic development) is discussed in Section 5.10.

In vitro diagnostics are developed in two stages –front-end investigative or exploratory research leads to the discovery of a protein up-regulated or down-regulated in a disease as a potential diagnostic marker, followed by extensive validation studies to confirm the diagnostic correlation. The discovery is validated in in vitro studies that may include analysis of various stratification markers from tissue or serum samples to examine and optimize the accuracy, specificity, sensitivity, and reproducibility of the assay.

Further assay development and optimization will typically continue on a chosen clinical testing technology platform. At or before this assay product optimization stage, the product characteristics and TPP parameters are defined. Before beginning clinical trials, a diagnostic assay and kit is developed that is user friendly and includes clinical sample handling processes validated in a clinical environment. The development of a commercial test method (with any needed change in the established commercial technology platform) is completed before clinical trials are

Table 5.3 Stage-gate criteria for drug development

Stage 1: Target discovery and validation
Biological studies to identify target and indication of interest. If carried out in a company, this is typically part of a larger discovery effort, without a specific project necessarily assigned at this stage.

Gate 1:
Primarily a business case review. Potency of target and druggability (accessibility to drug compound) of target are other criteria. Validation data for target must have broad scientific acceptance. Can the target be put into the company's processes? For example, if the company's technical foundation for drug design is based only on crystal structures of the target and crystal structure of the target is not available, that target might not pass this gate.

Stage 2: Hit to lead
Lead identification and optimization, includes biology activities. Formal project management in place.

Gate 2:
Specific product physico-chemical characteristics, like Lipinski's Rule of Five, lack of toxicity in early tests, efficacy in animal models of the disease, other project- or disease-specific parameters (e.g. blood–brain barrier penetration; oral bioavailability). For example, Gate 1c: The new drug compound must show potency of at least 50% inhibition of cell proliferation at 10 micromolar concentration against a chosen cell line.

Stage 3: Lead to drug candidate
Formal cGLP studies, toxicity studies, advanced efficacy studies in animal models if needed; also includes formulation and further medicinal chemistry optimization of drug compound; or if biological drug, could include fine-tuning of separation procedures, full characterization of post-translational modifications that are acceptable, etc.

Gate 3:
Drug candidate meets or passes preferred pharmacokinetic, physico-chemical, pharmacologic, and toxicology parameters. No objection by FDA to IND package approved by FDA to start clinical studies.

Stage 4: Phase I clinical studies
Determine toxic reactions and toxic dose in small group of healthy normal humans.

Gate 4:
Lack of significant adverse side effects; identification of therapeutic window that is suitable; adequate pharmacokinetic parameters in human subjects.

Stage 5: Phase II clinical studies
Dose-finding studies, verification of efficacy in 80–100 (or more) patients with disorder.

Gate 5:
Efficacy and lack of toxicity; dose-response determined for Phase III approval study.

Stage 6: Phase III clinical studies
Include large-scale studies for efficacy and safety with a final report submitted for market approval to the FDA.

Gate 6:
Statistical analysis of results from studies supports the claims of efficacy and safety. FDA approval of NDA after review.

Stage 7: Phase IV – post-market approval clinical studies
Carried out in larger or selected population groups, the FDA reviews data from this ongoing trial periodically.

Gate 7:
No significant adverse event and no added toxicity finding. New indication added with selected patient group.

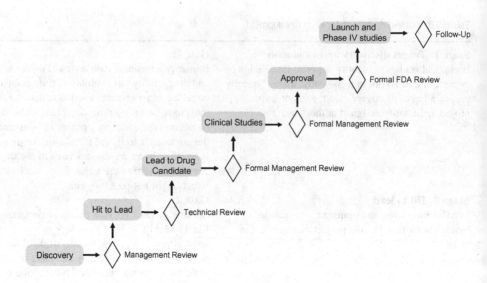

Figure 5.11 Conceptual stage-gate process for developing a drug

launched. A strong correlation with clinical outcomes is validated using the configured clinical assay. The testing platform can be changed at this point, as long as the new technology platform can be shown to produce equivalent results. Larger companies may choose to first configure the assay on their already commercialized test instrument platform.

The typical development stages and gates for development of in vitro diagnostics are highlighted in Table 5.4.

Stage 6 may not be necessary if the test is cleared under a 510(k) (see Chapter 6 for details on 510(k) process), but a prospective clinical trial may be needed by the payers (see Chapter 8).

Some manufacturers sell analyte-specific reagents (ASRs) to established certified clinical laboratories (especially startups, as the ASRs are the early proprietary portion of their in vitro diagnostic product). These laboratories can offer the diagnostic test commercially as their own laboratory developed test before FDA approval. ASRs are defined by the FDA as "antibodies, both polyclonal and monoclonal, specific receptor proteins, ligands, nucleic acid sequences, and similar reagents which, through specific binding or chemical reactions with substances in a specimen, are intended for use in a diagnostic application for identification and quantification of an individual chemical substance or ligand in biological specimens." ASRs are typically sold as specialized reagents to other IVD manufacturers, CLIA (Clinical Laboratory Improvement Amendments, used by CMS to regulate all laboratory testing in the United States) labs, and non-clinical labs as "research reagents" without any claims on clinical performance. However, due to low regulatory hurdles to get to market (as a laboratory developed test), ASRs can be a good early source of revenues for some startups.

Table 5.4 Stages and gates for diagnostic development

Stage 1: Biomarker discovery and validation Biological studies to identify a marker linked to a disease. This discovery research can be carried out inside a company as part of overall basic research activities but is usually discovered in academic medical research centers, which have access to large pools of annotated clinical samples and basic science researchers.	**Gate 1:** A business case review and a feasibility or proof of concept review. Validation data for biomarker must have strong scientific evidence that links marker to disease. The biomarker can be a protein, gene sequence, or specific pattern of genetic changes.
Stage 2: Clinical test development Development of the assay procedures and the technical method.	**Gate 2:** A reproducible assay that can be performed on clinical samples. Test should require minimal sample preparation, should be portable across technology platforms with simple operation, show operator independent reproducibility, etc.
Stage 3 Analytical assessment optimization Assessment and optimization of the analytical methods.	**Gate 3:** Gates can include good reproducibility, high sensitivity, high specificity, predictability, and a linear response over the relevant range of measures.
Stage 4: Scale-up, manufacturing, and early sales as nonregulated test	**Gate 4:** Manufacturing cost per test must be under identified threshold. Reagent sales can proceed for "research use only" without going through FDA review.
Stage 5: Retrospective clinical trials, registration, and commercialization Partnerships with access to clinical samples must be set up if needed, registration path defined and data collected from analysis of clinical samples from previously completed study (typically for a therapeutic).	**Gate 5:** Clinical relevance must be clearly shown by good sensitivity and specificity, with diagnostic correlated to clinical outcome. Sales of diagnostic test can proceed in Europe and the United States, respectively, after CE mark and 510(k) FDA clearance (if applicable; see Chapter 6).
Stage 6: Prospective clinical trials, registration, and commercialization Diagnostics with significant risk (as per FDA regulations) must submit an IDE to start trials. Prospectively designed studies for efficacy with a final report submitted for pre-market approval (PMA) to the FDA. Companion diagnostic approaches must partner with a therapeutic in development to leverage clinical trials and coordinate time to market.	**Gate 6:** (See Chapter 6 for regulatory process details.) The IDE (investigational device exemption) submitted to the FDA to get permission for clinical trials must have the following: (i) IVD cutoff value(s); (ii) preanalytical and analytical studies designed to demonstrate the reliability of the assay, particularly around the cutoff value(s) (iii); other analytical studies that support the conclusion that use of the IVD does not expose subjects to unreasonable risk of harm; and (iv) clinical trial protocol. Statistical analysis of results from studies supports the claims of efficacy and correlation to clinical outcome. FDA approval of PMA for marketing.

If an instrumentation product is being developed as a technology platform for diagnostic assays and not a specific diagnostic test, the product development stages are quite different and can be summarized as below:

Stage 1: Definition of product specification with input from market needs and product standards.

Stage 2: Manufacturing cost analysis and business case are generated, which become the basis of a go/no-go decision. Cross-functional, multidisciplinary team is formed.

Stage 3: System and subsystem design. Create a working automated instrumentation process flow for the chemistry and analysis. The generalized assay configuration will flow into the product specifications. Formalized software and hardware specifications are developed.

Stage 4: Validation. Make sure the test system reports the correct results and conforms to customer needs. Protocol testing and beta testing results are used to fine-tune or change elements of the instrumentation system.

Typical development of a new type of clinical diagnostic test can take 3–8 years and costs about $15 million to $75 million, depending mostly on the clinical trial and subsequent investment in new instrument development, commercial production activities and market infrastructure. Figure 5.12 points out specific stages in the development of a diagnostic test, with a recognition that the length of the development period depends largely on the type of product and the type of regulatory gateway the product will enter (see Chapter 6 for more details). In essence, some diagnostic products can be sold on the market with minimal clinical development (e.g. ASRs), while others will need data from a full-fledged prospective clinical trial in order to be approved for sales and acceptance in the market.

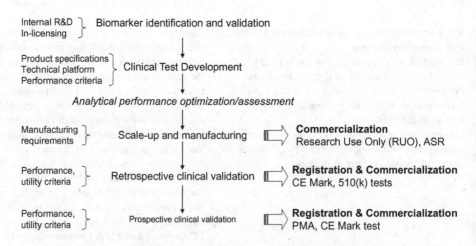

Figure 5.12 Development steps for a new diagnostic. This diagram recognizes that not all diagnostic products will have to go through a prospective trial and that the development pathway is shorter for some types of products that require minimal regulation.

These figures will change significantly as co-development of drugs and companion diagnostics become more common (see Section 6.7 for more details), as the cost of the clinical trials for the diagnostic may be reduced by leveraging the diagnostic clinical testing into the pharmaceutical clinical trial. One potential use for a companion diagnostic would be to identify predicted-responsive patients for selection into the drug trial (patient enrichment strategies), increasing the chance that the drug will show a high benefit-to-risk ratio with a smaller clinical trial. Figure 5.13 lays out the key diagnostic development stages in a timeline comparison to the drug development stages. Biomarker validation is completed in a series of preclinical studies that usually includes human samples. The multiple validation studies for the diagnostic test are typically carried out during preclinical phases of drug development (Figure 5.13). A technically validated test with a valid biomarker will then be evaluated for clinical utility, performance, and significance during the clinical testing phases for the drug.

The process of co-developing drugs and companion diagnostics is also postulated to bring more effective and safer drugs to market in a shorter time frame. Pharmacogenomics (or personalized medicine) is the term used to describe the identification of such personalized drug therapies, and related business and regulatory issues are discussed in more detail in Chapter 6, Section 6.7.

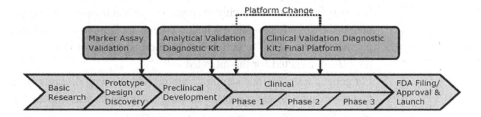

Figure 5.13 This schematic describes the timing of various studies and functions involved in diagnostic development, compared to drug co-development steps. Figure taken from the FDA website content on co-development of drugs and diagnostics (see Chapter 6 for more details).

Box 5.15 General considerations for planning and evaluating clinical studies for in vitro diagnostic (IVD) devices

Extract from FDA guidance: www.fda.gov/cdrh/oivd/guidance/1549.html

The following are some general recommendations that may be used when planning and evaluating clinical studies. An additional resource to consider when seeking guidance on reporting clinical and/or method comparison studies is the STARD (Standards for Reporting of Diagnostic Accuracy) statement [Bossuyt et al., 2003], which is a roadmap for improving the quality of reporting of studies of diagnostic accuracy.

1. Plan studies to support the intended use claim for the device with data that are representative of the population for whom the device is intended. Include a

Box 5.15 (*cont.*)

diversity of ethnic groups if the marker/mutation varies according to ethnicity. Use investigational sites appropriate to the intended use and claims being sought. Clearly outline efforts to define population sampling bias when this issue may impact performance.

2. Describe all protocols for internal and external evaluation studies. Clearly define the study population and inclusion and exclusion criteria and the chosen clinical endpoint. If literature is to be used to support your intended use, you should clearly explain the study population, inclusion/exclusion criteria, and endpoints in the publication and reflect how the device will be used in practice. Establish uniform protocols for all external evaluation sites prior to study and follow them consistently throughout the course of data collection.

3. Determine sample size prior to beginning the clinical study. The sample size should have sufficient statistical power to detect differences of clinical importance for each marker, mutation, or pattern. Consider other approaches in cases with a small available sample size, for example, a disease allele having a low prevalence in the intended use population.

4. Describe the sampling method used in the selection and exclusion of patients. If it is necessary to use archived specimens or a retrospective design, provide pre-specified inclusion and exclusion criteria for samples, and adequate justification for why the sampled population is relevant to the patient population targeted for the intended use.

5. For genetic tests, include samples from individuals with diseases or conditions that may cause false positive or false negative results with the device (i.e., within the differential diagnosis), if appropriate.

6. Analyze data for each individual test site and pooled over sites, if statistically and clinically justified. Justification of data pooling over sites should address variation between sites in prevalence, age, gender, and race/ethnicity.

7. Describe how the cut-off point (often the distinction between positive and negative, or the medical decision limit) will initially be set, and how it will be verified, if appropriate. If a cut-off is specified for each of multiple alleles, genotypes or mutations, describe the performance characteristics of each cutoff as it relates to its respective allele, genotype or mutation. The description of how each cut-off is determined should include the statistical method used [e.g., receiver operating characteristic (ROC) curve].

8. Diagnostic devices that assay the presence of a particular pattern (e.g., single nucleotide polymorphism (SNP) set, haplotype pattern), should ideally be validated in a prospective clinical trial. An example of such a device would be a test using a defined SNP set to discriminate between patients who may or may not experience an adverse event associated with a particular drug. Since it is statistically problematic to validate discrimination patterns in the same study in which they were defined, the simplest way to address this is to validate the pattern with an independent data set. Determination of the statistical

> **Box 5.15** *(cont.)*
>
> significance of a retrospectively determined feature pattern may not be possible or minimally would call for careful use of complex statistical procedures, such as bootstrapping, or an explicit cross-validation scheme. Given that it can be easy to obtain a low misclassification rate for a retrospectively determined feature pattern even on random data, a valid procedure for obtaining the statistical significance of such a pattern should be provided. The simplest approach statistically is to evaluate the pattern on an independent data set from a prospective clinical trial, if that is feasible.
>
> 9. Account for all individuals and samples. Perform appropriate data audits and verification before submitting to FDA. Give specific reasons for excluding any patient or test result after enrollment.
> 10. Perform studies using appropriate methods for quality control. Describe the materials and methods used to assess quality control.

5.10   Typical device development process

Due to the diverse nature of medical devices, development cycles are highly variable across the industry. One of the first steps in your non-drug or combination product is to assess whether the product is going to be recognized by the regulatory body as a device, drug, or diagnostic. This determination of classification will drive the regulatory pathway and could significantly impact the development costs for approval.

Devices are defined in the Medical Device Amendments of 1976 to the Food Drug and Cosmetic Act [21 U.S.C. 321(h)] as:

An instrument, apparatus, implement, machine, contrivance, implant, in vitro reagent, or other similar or related article, including any component, part or accessory, which is

(a)   recognized in the official National Formulary, or the United States Pharmacopoeia, or any supplement to them,
(b)   intended for use in the diagnosis of disease or other conditions, or in the cure, mitigation, treatment, or prevention of disease, in man or other animals, or
(b)   intended to affect the structure of any function of the body and which does not achieve its primary intended purpose through chemical action and which is not depended upon being metabolized for the achievement of its primary intended purposes

Devices can be looked upon as being durable, implantable, or disposable, and perform either therapeutic, diagnostic, or monitoring functions. Examples of each type include: durable devices – lithotripsy machine used to break kidney stones in vivo; implantable devices – pacemaker; disposable device – adhesive bandage. Most IVD products are also looked upon as devices, and product development considerations have been discussed in Section 5.9.

The product development process for a medical device also depends largely on the FDA *classification of risk* (there are three categories of risk), which will dictate the

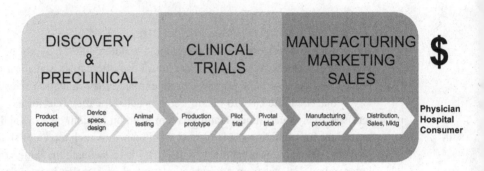

Figure 5.14 Schematic of device development value chain / process

extent of safety and clinical testing required. Therefore, it is important to understand the classification schema for devices discussed in Chapter 6. Additionally, devices that carry out equivalent functions and are similar to other approved devices will have a shorter development path, also discussed later in Chapter 6. This section specifically covers new medical devices that have new characteristics or applications and require a full development path with rigorous clinical studies.

Depending on the technology, novelty, application, and other factors, product development may take from a few months to a few years and may cost from a few hundred thousand to millions of dollars. However, most devices are developed through the key schematic steps in Figure 5.14.

5.10.1   Discovery, feasibility, and optimization – design and preclinical testing

Product concept

The idea for a product usually stems from an invention or discovery, or from examining the needs of caregivers in specific diseases, disorders, or treatment of trauma. Problems in the current methods of care give rise to product opportunities, but the problems must be carefully analyzed to obtain the right product characteristics. Market research is a key activity in this phase (Chapter 2).

The *indication* for the product and the *specific mechanism of action* are two key factors that will dictate the FDA classification and also the focus of the development efforts, affecting the cost and timeline for development of the device. It is thus important to identify clearly, from market research and technology evaluation, the exact indication and primary mode of action of the device technology.

Usually, a business review is sufficient for a gate decision at this point.

Device design and specifications

Medical device design and product development is controlled and regulated by the FDA as described in detail in Chapter 6. Most importantly, the project team must take input and guidance from their regulatory affairs officer in the company (or from an outside consultant) and have a quality system in place for collecting appropriate documentation that can be compliant with the FDA Quality System Regulation (QSR, 21CFR 820) or similar regulation requirement in other countries. At this point, it is imperative to read the requirements for the design process and related documentation in the Quality Systems Regulation Manual (website reference in References and Additional Readings at end of the chapter) and, in particular, the Design Controls section of the Quality System Regulation (21 CFR Part 820, Subpart C, Sec.820.30), which outlines the requirements that each manufacturer of any Class II or Class III device, and certain Class I devices, must meet when designing such products or related processes, and when changing existing designs and processes.

The design process for devices also requires input from marketing, sales, manufacturing, and other corporate functions. In particular, the design process must incorporate the customer requirements, gained through market research (interviews, surveys, customer input through the sales and marketing organization, third-party market research), and design and evolve the product characteristics to satisfy those customer needs. Box 5.16 details some aspects of design inputs. In general, design inputs must be formally recorded for regulatory compliance.

Typical steps in device design are:

- Identify the need
- Define the problem
- Set design objectives/constraints
- Search for necessary background info
- Devise alternative solutions (synthesis)
- Analyze solutions
- Evaluate solutions, make decisions, communicate

The *product design specifications (PDS) document* is generated after careful evaluation and after striking a balancing act among various criteria and characteristics to arrive at design specifications (specifications). At the end of the process, as various choices are made, evaluated, and discarded iteratively in the product development stages, it is important to go back to the starting needs and ensure the adherence of the final output to the initial criteria that were identified as important to the customer (Figure 5.15). This focus on meeting customer needs seems like a simple step but is often forgotten in the process between early development and final stage production. Aggressively pursuing customer inputs for identifying true needs or collecting feedback to new features is one of the most rewarding tasks in product design but one that is usually hardest to do for analytical engineers. Taking initiative in this task is a good way for young engineers to stand out in a team.

A popular analytical approach that transforms customer needs (the voice of the customer [VOC]) into engineering characteristics of a product or service is called *quality function deployment* (QFD) or "house of quality." The QFD was originally developed in Japan and then imported to the United States by car-manufacturing firms and is now used in virtually all industries. The QFD process prioritizes each product/ service characteristic based on customer needs, while simultaneously setting development targets for product or service development. A thorough review of customer

Box 5.16 Design Inputs

Design inputs (device requirements) for the selected design project must employ procedures and documents must demonstrate that the design inputs established for the device considered factors such as (not a comprehensive list):

- The intended uses of the device

    The needs of the user and the patient
    Compatibility with environment of use and with accessories/auxiliary devices
    Safety and performance characteristics
    Limits and tolerances
    Risk analysis
    Toxicity and biocompatibility
    Environmental issues
    Electromagnetic compatibility
    Human factors (see FDA guidance documents on human factors on FDA/CDRH
        website)
    Labeling/packaging
    Reliability/stability
    Voluntary standards
    Manufacturing processes
    Sterility

In addition, the design input sources must be well-documented. These design input sources could typically include:

    Customer input through meetings, focus groups, surveys, trade shows etc.
    Comparison testing of competitor products or benchmarking surveys/activities
    Internal manufacturing and service department inputs
    Risk analysis (including hazards analysis and, as appropriate, Design Failure
        Mode and Effects Analysis)
    Review of literature, FDA reports, histories of similar products
    Input from regulatory, quality assurance, R&D, marketing departments

Finally, these design inputs must be reviewed, approved and subsequently reviewed again as they evolve. All these processes and steps must be well documented in the design history file (DHF).

Side note on managing design projects: The most common parameters to be desired by product managers include (i) fast speed to market, (ii) high quality of design of product, and (iii) low cost of the project and/or product. It is a well-known rubric in product design in general (not just medical products) that these represent three essential corners of a product design process (triangle) and product managers (or customers) can optimize any two of the three factors but must then compromise on the third.

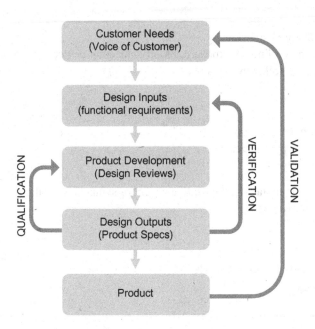

Figure 5.15 Incorporating customer or market-driven design specifications in the design process. (Figure adapted from talk by Mr. Laurence Roth, Percardia.)

needs, through observation, interview, and focus groups, is incorporated into various quantitative measures of product characteristics as deemed important or demanded by the customer. If a certain characteristic is rated very high by customers, the product characteristics related to that function or characteristic of the product get higher weights in the decision matrix. Additional weights may also be added if no competitor product has that characteristic (competitive advantage for the product under development). Various other combinations of considerations can be brought into the matrix to result in a prioritization of design and development activities and product characteristics with the end user in mind. Identifying the right customer and then carrying out primary market research (interactions at trade shows, in-person site interviews, email surveys, phone calls, focus groups) with that group of customers is the best way to bring a successful product to market.

Recognizing that variability in product quality characteristics is a major cause of customer dissatisfaction and development inefficiencies, a set of statistical, engineering and design tools called "design for six sigma" have increasingly been adopted by many industries. The successful application of "six sigma" processes in bringing quality into all manufacturing and product development processes in various industries has created an interest in applying these practices and tools to early stages of product development in the biomedical industry. Further details of the six sigma processes can be accessed on many websites through the internet and through the additional readings at the end of this chapter.

Box 5.17 Toxicity testing

A catheter, which is a limited-exposure, blood-contacting device, has to be tested for toxicity with the following parameters:

- Cytotoxicity
- Sensitization
- Irritation/reactivity
- Systemic toxicity/pyrogenicity
- Hemocompatibility

A permanent, blood contacting device like a stent or synthetic vascular graft has to go through the above tests, plus:

- Subchronic toxicity
- Genotoxicity
- Implantation study (special controls)
- Chronic toxicity
- Carcinogenicity

Hardware and software design modules must also be specified with similar inputs and formal specifications documents. In addition, validation and verification of software modules are an important area for device manufacturers, as seen by the increasing use of computing parts, software, and microprocessors in many devices. Other components in the biomedical device design process include *human factors engineering*, which looks at user interactions with the device in order to design-in ways to prevent human/user error.

Specifying and designing the characteristics of the device, The design phase begins with a risk analysis, safety analysis, assessment of human factors requirements, and formulation of project planning documents (see Box 5.17) which lead to specifying and designing the characteristics of the device.

Animal and toxicity testing

The device prototype is evaluated in animal models before the final production run. These studies are used to optimize product characteristics. Specific biocompatibility and toxicology or safety studies will depend on the type of device (listed under special regulatory controls for that device type), the materials used, and the application. For example, a neurological implantable device that contacts cerebrospinal fluid and brain parenchyma may need an animal implant test to determine local effects on brain tissues and fluids, the susceptibility of seizures, and other physiological functions. In general, biocompatibility/toxicology tests include the following:

- Acute, subchronic, and chronic toxicity
- Irritation to eyes, skin, and mucosal surfaces
- Sensitization
- Hemocompatibility

- Genotoxicity
- Carcinogenicity
- Effects on reproduction including developmental effects
- Specific organ toxicity/effects as needed

Risk and reliability testing

Medical devices that are not disposable should also undergo a series of reliability tests, typically run in laboratory settings. Reliability is a measure of the potential for failure of a device – mechanical, electrical, material failures could be involved in the reliability testing depending on the particular device. Potential hazards or risks of failure mechanisms are critical to review during design and development of all medical device products.

Production prototype

The results of these tests are all factored into the final, optimized design of the device. The development of a production prototype is carried out after senior management meets for another formal design review. At this stage a design "freeze" is implemented. Some aspects of the device design might already be fixed and defined as unchangeable by the biocompatibility tests. The device design must also include consideration for ease of manufacturing scale up – called "design for manufacturing." The business case is also analyzed and reviewed again, before the commitment to go to production is made. Quality control and engineering divisions should be involved in making production prototypes so that the scale up can be carried forward seamlessly into manufacturing stage. The design lock in has to occur before the production prototypes are taken into clinical trials.

5.10.2 Special considerations for device clinical trial design

Clinical trials

The preferred study design in regulated product trials is a double-blind randomized trial, in which the patients are randomly assigned to a treatment or control group with neither the care provider nor the patient knowing which group is which. However, in device trials it is usually not possible or ethical to mask the device, especially in an implanted or surgical device. Therefore, device clinical trials are typically run as randomized trials with parallel groups in which patients are randomly assigned to only one device or treatment regimen. The control against which the device efficacy or safety is compared is typically an active control, where another currently accepted device or treatment regimen is applied. Since a placebo group cannot usually be included, some trials have no choice but to compare results against historical data. Another design in clinical trials is a crossover study in which a patient sequentially receives more than one device or treatment regimen in the clinical trial and effectively serves as their own control. A further discussion on trial design is found in the guidance document referenced in Box 5.18.

Box 5.18 Guidance for clinical trial design for medical devices

The following are extracts from FDA guidance document Design Considerations for Pivotal Clinical Investigations; *quoted material appears in italics*

Found at https://www.fda.gov/regulatory-information/search-fda-guidance-documents/design-considerations-pivotal-clinical-investigations-medical-devices

In the foreword to the document, the FDA states the scope of the guidance document as follows: *"This guidance describes principles for the design of pre-market clinical studies that are pivotal in establishing the safety and effectiveness of a medical device. Various factors are important when designing any medical device clinical study, including general considerations of bias, variability, and validity, as well as specific considerations related to study objectives, subject selection, stratification, site selection, and comparative study designs."*

While the manufacturer may submit any evidence to convince the Agency of the safety and effectiveness of its device, the Agency may rely only on valid scientific evidence as defined in the PMA regulation section entitled, 'Determination of Safety and Effectiveness' (21 CFR 860.7). A thorough reading of that section is strongly recommended. The guidance also states: *"Isolated case reports, random experience, reports lacking sufficient details to permit scientific evaluation, and unsubstantiated opinions are not regarded as valid scientific evidence to show safety or effectiveness."*

Design of the clinical trial

The trial objective

The study objectives provide the scientific rationale for why the study is being performed. The objectives should provide support for the intended use of the device, including any desired labeling claims. Claims can be supported statistically by formal hypothesis testing or by point estimates with corresponding confidence intervals. For pivotal studies designed to test a scientific hypothesis, the study objectives should include a statement of the null and alternative hypotheses that correspond to any desired claim. This hypothesis should be formulated with extreme care and specificity to effectively evaluate a particular type of intervention. A question such as "Is my device safe and effective?" is far too general to be meaningful. What is the proper way to evaluate effectiveness in the target condition and population? What are the unique safety concerns of the device intervention? Is the device as effective or more effective than another intervention? If so, is it as safe or safer? Is the evaluation of safety and effectiveness limited to a particular subgroup of patients? What is the best clinical measure of safety and effectiveness?

The attempt to answer these and similar questions will provide an essential focus to the trial and should provide the basis for *labeling indications*.

For example, if a new device has been developed to treat a progressive, degenerative ophthalmic disorder for which there currently exists an alternative therapy

Box 5.18 (cont.)

using an approved device, how should effectiveness be determined? Does the new device slow or halt degeneration? If so, does it restore functions that had previously been lost? Does it reduce pain or discomfort? Is it to be compared with the approved device and is it thought to be as good as or better than the old device for some purpose? Does it have fewer adverse reactions?

Identification and Selection of Variables

The observations in a clinical study involve two types of variables: outcome variables and influencing variables. Outcome variables (also known as endpoints) define and answer the research question and will have direct impact on the claims for the device. These endpoints should be directly observable, objectively determined measures subject to minimal bias and error.*The endpoints, outcomes, or measurements should provide sufficient evidence to characterize the clinical effect of the device (for both safety and effectiveness) for the desired intended use. The endpoints, outcomes, or measurements should be clinically meaningful and relevant to the stated study objectives and desired intended use. The pivotal study should be designed to demonstrate clinical benefit to the specified subject population rather than to simply demonstrate how the device functions.*Influencing variables are any aspect of the study that can affect the outcome variables (increase or decrease), or can affect the relationship between treatment and outcome. Imbalances in comparison or treatment groups in influencing variables at baseline can lead to false conclusions by improperly attributing an effect observed in the outcome variable to an intervention when it was merely due to the imbalance. For example, blood pressure generally increases with age. If a group of individuals in the treatment group is significantly younger, and possess lower mean pressures than subjects in the control group, and are then compared using blood pressure as the outcome variable, the investigators may falsely conclude that an intervention was responsible for the observed "reduction" in blood pressure.*To ensure that the subjects in the clinical study reflect the desired target population, the protocol should specifically define eligibility criteria that match the key characteristics of the intended target population. In conducting the trial, the sponsor should ensure that only those individuals meeting these criteria are included. These are referred to as the inclusion/exclusion criteria for subject entry into the study.*

Control groups

Every clinical trial intended to evaluate an intervention is comparative, and a control exists either implicitly or explicitly. The safety and effectiveness of a device is evaluated through the comparison of differences in the outcomes (or diagnosis) between the treated patients (the group on whom the device was used) and the control patients (the group on whom another intervention, including no intervention, was used). A scientifically valid control population should be comparable to the study

Box 5.18 (*cont.*)

population in important patient characteristics and prognostic factors, i.e., it should be as alike as possible except for the application of the device.

There are many types of control groups. The regulation 21 CFR 860.7(f)(1)(iv) identifies four types of controls:

(i)   No treatment – *Where objective measurements of effectiveness are available and placebo effect is negligible, comparison of the objective results in comparable groups of treated and untreated patients*

(ii) Placebo control – *Where there may be a placebo effect with the use of a device, comparison of the results of use of the device with an ineffective device used under conditions designed to resemble the conditions of use under investigation as far as possible*

(iii) Active treatment control – *Where an effective regimen of therapy may be used for comparison, e.g., the condition being treated is such that the use of a placebo or the withholding of treatment would be inappropriate or contrary to the interest of the patient*

(iv) Historical control – *In certain circumstances, such as those involving diseases with high and predictable mortality or signs and symptoms of predictable duration or severity, or in the case of prophylaxis where morbidity is predictable, the results of use of the device may be compared quantitatively with prior experience historically derived from the adequately documented natural history of the disease or condition in comparable patients or populations who received no treatment or who followed an established effective regimen (therapeutic, diagnostic, prophylactic).*

(v)   Subject as own control (considered in the FDA guidance document but not in the regulation) – *This design is only possible when the experimental device and control intervention effects are local and do not overlap.* Subject as non concurrent own control is not adequate as comparison for most therapeutic studies.

The guidance document also covers various parameters and considerations specific to diagnostics device clinical performance studies.

Among other useful pointers about carrying out clinical studies, it is noted that the protocol includes a detailed statistical analysis plan that clearly describes the precise strategy to analyze the data.

Other useful pointers offered by the FDA in the guidance document:

-   *The rationale why other endpoints or alternative study designs with less potential for bias were not selected may also be helpful for review of the study.*
-   *It is also advisable that investigator input be sought during the study design phase.*
-   *Clinical data managers play a critical role in providing input into study design and case report form design based on past experiences running similar clinical studies.*

Most drugs are administered via well-established and widely accepted standard modes of administration. However, for devices, the method of use or implantation is a major factor in the success or failure of a device trial and *investigator training and protocol compliance* form a critical aspect of many device trials.

Device trials typically have a short safety or biocompatibility study (comparable to Phase I trials in drugs) and a pilot efficacy or feasibility study (Phase II in drug trials) followed by a larger pivotal clinical study (Phase III in drug trials) that compares the safety and efficacy of the new device to the current standard of care.

5.10.3   Device manufacturing

Device manufacture also has to be done in compliance with FDA guidelines and regulations for tracking each and every step of the manufacturing process, including raw material sourcing. These guidelines are known as current good manufacturing practices (cGMP) and each facility that is certified by the FDA is also regularly inspected. Further details on manufacture and regulatory compliance issues are discussed in Chapter 7.

5.10.4   Keeping records for the FDA

The design history file, all preclinical studies, and all clinical trial data need to be carefully documented for review by the FDA. These records are maintained in four different formats/records: The device master record (DMR), the design history file (DHF), device history record (DHR), and the technical documentation file (TDF) are discussed in greater detail in Chapter 6. In particular, the design reviews carried out during product development must be well documented, with all decisions and design changes explained and supported with technical analysis and sound reasoning.

5.10.5   Device development stage-gate process

Figure 5.16 shows the stage-gate process for device development. Table 5.5 explains each step in detail.

5.11   A few general notes on biomedical product development

Some key dos and don'ts with respect to biomedical product development:

DON'T – carry out extensive studies for alternate diseases/indications once a product has entered formal clinical development for a particular chosen indication. All data involving a product have to be delivered to the FDA. If a vertebrate animal in a study using the same product dies, revealing a new risk, the death has to be reported to the FDA and the main ongoing clinical study may be paused while the FDA re-evaluates the risks involved.

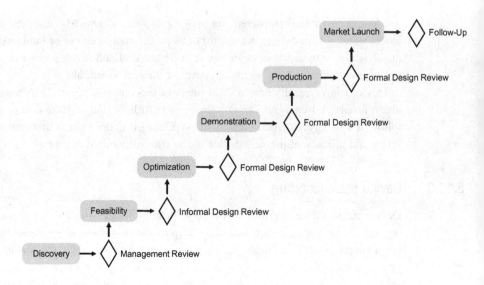

Figure 5.16 Conceptual stage-gate process for device development

DO – obtain appropriate insurance for the human clinical trials.

DON'T – forget to put the adequate ethical controls and regulatory certifications in place in the preclinical stage – animal review committees, laboratory certifications for OSHA, biosafety regulations, hazardous materials, department of health certifications, etc. Section 5.16 has a summary of these requirements.

DO – carry out a failure mode and effects analysis (FMEA, particularly relevant for device design) as described in Box 5.19.

DON'T – forget to include effective information security and privacy controls.

DO – design with empathy for the users; understand who your product's users will be and design the product with a view from their shoes

DO – consider, as a startup, putting your quality system in place as you go through the design process rather than trying to put a whole system in place all together. Only describe the processes you're confident you can execute on and select processes noting that a quality system's primary objective is ensuring that user needs are met by the product design. However, in preparation for regulatory submissions, remember that if it isn't documented, it didn't happen.

DO – write product specifications with the test in mind. For example, the Software Requirement Specification (SRS) is tremendously important because it potentially defines the largest test effort within the product. Every specification written will be painstakingly verified multiple times. A key strategy is to focus on what test the specification will motivate. Is that something we need to test? Does it tell us how to test it? Is it clear what passing is? These questions collectively embody the term "testability." Writing the specification with the test in mind encourages a deliberate focus on what's important to explicitly verify.

Table 5.5 Stage-gate process for device development

Stage 1: Discovery
Product concept and preliminary data collection
Exploration, brainstorming
Sources: R&D, marketing, internal R&D, customers,
competitors' literature, academia, conferences

Gate 1: Management Review
"Gentle" screen
Strategic alignment
Initial feasibility
Magnitude of opportunity
Output = "Development Initiation Proposal"
Initial resources allocated

Stage 2: Feasibility and concept testing
Market size, market potential, market acceptance
Technological readiness, proof of principle
IP position
Product specifications
Prototype production
Project planning
Testing plan
Regulatory strategy
Budget, timeline preparation

Gate 2: Repeat of previous screen, with more information
"Must have," "should have" features, market
opportunity, technical feasibility
Output = "development continuation
proposal"
Project contract
Spending increases, budget allocated

Stage 3: Optimization
Design controls implemented
Design refined
Preclinical (animal) testing
Design is frozen
Materials, equipment, tooling
Packaging, sterilization
Data for FDA submission gathered – preclinical toxicity

Gate 3: Formal design review
Product performance evaluation:
  Customer needs
  Cost and pricing
  Safety, efficacy, reliability
  Target market definition
Output = extension of "development
continuation proposal"

Stage 4: Demonstration (if needed)
Validation of manufacturing process
FDA submission for clinical trials
Pilot-scale production for clinical trials; may also be used
for product launch
Clinical trials – pilot and pivotal studies

Gate 4: Formal design review
Review of clinical data
Review of validation data
Output = extension of "development
continuation proposal"

Stage 5: Production
Scale-up of manufacturing process
Validation of GMP compliance
Product QA/QC validation

Gate 5: Formal design review
Final gate before product reaches public:
  Review of validation data
  Updated financial projections
  Operations and marketing plans
Output = approval to launch

Stage 6: Launch and follow-up
Marketing launch, sales team
Distribution
Monitoring of manufacturing (QA/QC)
Field support, customer training
Educational programs

Gate 6: Feedback and evaluation
Adjust programs as needed
Feedback to next cycle of development

Box 5.19 Failure mode and effect analysis

Adapted from a talk by Mr. Michael Cohen, President of the Institute for Safe Medication Practices

Step 1: Develop a process flow diagram that articulates how the product should be used.

Step 2: Use the diagram and assume the worst possible scenario at each step.

Step 3: Predict the effects of the failure.

Step 4: Rank the likelihood of the occurrence.

Step 5: Rank the estimated severity of the failure.

Step 6: Rank the likelihood of detection.

Step 7: Add steps 4, 5, and 6 and divide by 3 to develop a criticality index.

Step 8: Develop ranges which are unacceptable.

Step 9: Decide on interventions.

Step 10: Take action and assess the effects of the action.

DON'T – turn on your quality system too late. Done well, the quality system makes product execution efficient by keeping product objectives in front of you, and explicitly delivering on top-level concerns like product safety and efficacy.

NOTE: some of the above points were paraphrased from the website https://www.designnews.com/design-hardware-software/3-dos-and-donts-medical-device-software-development

5.12   Project management

5.12.1   Project management tools – Gantt charts and critical path

A Gantt chart is a bar chart that shows how project elements (stages), schedules, and other time related systems progress over time. The horizontal black bars in Figure 5.17 show the summary elements (each summary stage of development) while the shorter horizontal bars indicate substages or activities/tasks within that summary stage. The vertical bars indicate dependencies, meaning the next task cannot start until the connected previous task has completed. The Gantt chart (or any other related graphical method) is useful as it makes it easy to visually check timelines and job allocations per person so that all team members and management are unequivocal about the tasks and timeline responsibilities. Also, progress can be easily tracked against Gantt charts and the software programs available today allow for changes to cascade through the chart making revisions in timelines and responsibilities easy to communicate. A Gantt chart drawing and planning tool is available in a software package called Microsoft Project (this one was generated in MS Excel).

Another type of visual representation of the complex set of tasks in a project is the critical path method (CPM), which can also be very useful in identifying and allowing

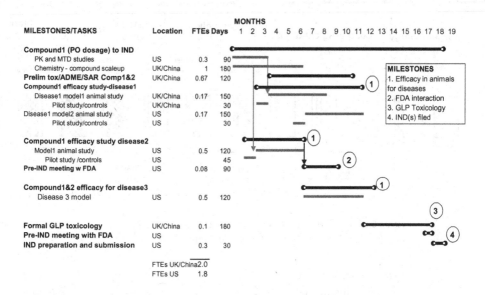

Figure 5.17 Example Gantt chart for preclinical development of a drug that has more than one indication.

specific prioritization of tasks that impact the timeline of the project. A brief description of the method is reproduced here (from http://en.wikipedia.org/wiki/Critical_path):

The essential technique for using CPM is to construct a model of the project that includes the following:

1. A list of all activities required to complete the project (also known as Work breakdown structure),
2. The time (duration) that each activity will take to completion, and
3. The dependencies between the activities.

Using these values, CPM calculates the starting and ending times for each activity, determines which activities are critical to the completion of a project (called the critical path), and reveals those activities with "float time" (are less critical). In project management, a critical path is the sequence of project network activities with the longest overall duration, determining the shortest time possible to complete the project. Any delay of an activity on the critical path directly impacts the planned project completion date (i.e. there is no float on the critical path). A project can have several, parallel critical paths. Since project schedules change on a regular basis, CPM allows continuous monitoring of the schedule, allows the project manager to track the critical activities, and ensures that non-critical activities do not interfere with the critical ones.

5.12.2   Team composition

Multidisciplinary teams must be formed for a given project, to carry out the varied tasks involved in product development, as seen in the previous sections. In almost all biomedical product development (drugs, devices, and diagnostics), there is a need to get input from sales and marketing, reimbursement, regulatory affairs, finance, and

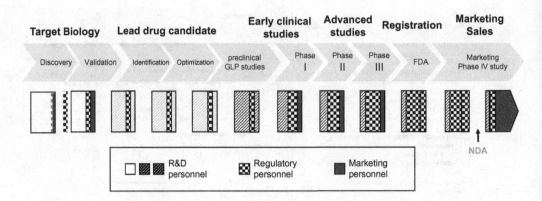

Figure 5.18 Composition of a product development team in drug development path. Device development teams will have similar configurations/composition.

general management throughout the development process (see Figure 5.18). The level and intensity of engagement with various functions varies over the course of the project. For example, the interactions with marketing, regulatory affairs, and reimbursement would increase steadily as the project went into clinical studies, with each stage seeing more involvement and feedback from marketing, regulatory affairs, and reimbursement as they work on positioning the product in a competitive marketplace. Even within a technical project team, the composition of the team is usually multidisciplinary, and the technical backgrounds of team members might change as the project moves along. A new drug project team would typically have a biologist/biochemist, medicinal chemist (protein chemist or cell biologist in the case of biological drug), and project leader in the team as the molecule went through early discovery studies. With progression into advanced preclinical stages, a regulatory affairs and manufacturing person might be added into the team to help prepare for the first interactions with the FDA. During clinical stage studies, a clinical physician, regulatory affairs, biostatisticians, and manufacturing might be part of the project team, either continuing with the same project manager or handed off to a different project manager. In a medical device development project, the engineering (commercial manufacturing) division might get involved as the prototype development stage was reached, and design review and feedback from manufacturing, marketing, and regulatory might become necessary.

5.12.3   Team management in a matrix environment

Sound personnel management practices are important given the long-term development cycles in the drug and devices industry. Matrix management structures used in many companies, still retain the hierarchical vertical management structure (seen as control of functional areas from top management down in Figure 5.19). The rigidity of the separation between line management (functional area) and project management depends on the culture and type of matrix organization existing in the company. Matrix management relies on cooperation and communication. In a project

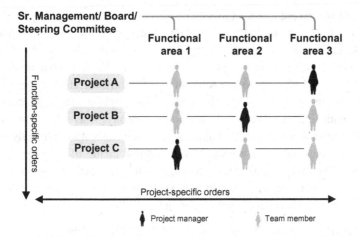

Figure 5.19 Matrix environment in a company showing the responsibilities of the team leaders and the functional group heads in a typical company

environment, the decision-making authority would usually rest with the project manager, but in a matrix environment all major decisions are reached by consensus between line managers (typically managers of functional areas) and project managers. Project staff have to be committed to both the project and their own departments (functional group).

In larger organizations, there is usually one full-time project manager who leads one project through a process. Often, projects are handed over between project managers specializing in the various preclinical, clinical, and marketing functional groupings. In smaller organizations, there may be a doubling-up of functional management and project management leadership responsibilities in the same person. There are two organizational extremes in a matrix management environment: (1) all personnel report only to line (functional area) managers and project managers will have to coordinate through the line managers; (2) all personnel in a project are put in a team and report only to project manager. In practice, a mix of the two organizational architectures is used. Generally speaking, the matrix environment puts a greater onus on managers to manage personnel, communicate expectations and effectively build consensus for their projects.

5.13   Formulating budgets

Estimate a success budget (studies planned have a successful outcome) with some extra padding for slips, error, and repeat studies. One cannot realistically budget in every possibility – for example, you could not have budgeted for the event that the expensive rodent study with a lead candidate drug had to be aborted because the drug compound would precipitate out of solution after 3 hours and could not be delivered reliably. If one were to take into account all the failures (failed hit compounds, lead compounds) in drug development, the total cost to develop a product is estimated at over $800 million.

The best way to put a budget together is to put a success budget together first and then start adding in more resources based on failures of the known highest risk events. Try to be realistic and state assumptions. In the end, if there is a technical issue that blows your budgeted amount anyway, there is not much to be done about it but to present the facts. Unless there is a solution in sight for that technical issue, there is a good chance the project will be dropped, which might be the right choice for a company with limited resources. In a startup company, where death of the project could also mean the death of the company, the management are typically willing to take more risks to continue a project, and this risk taking often gives rise to new innovative ideas and insights that could improve the final patient outcome.

A few key steps that can help in putting a quick annual or total project budget together from a product development plan are described here:

1. *Estimate the personnel resources needed for each stage/substage per year.*

This is typically done by calculating how many people are involved and estimating their time involvement using a full-time equivalent (FTE) measure. FTE is a way to measure a worker's productivity and/or involvement in a project. An FTE of 1.0 means that the person is equivalent to a full-time worker. An FTE of 0.5 would signal that the worker is only half-time on the project for 1 year, or that they are full-time on the project for 6 months. One way to calculate FTEs per stage is to list the total number of people that would be involved, multiply by number of weeks spent on the project that year, and divide that number by the total number of salary weeks in the year. Multiply this resulting number by the FTE cost, estimated per job group by your finance department. The fully loaded FTE cost rate typically includes overheads, benefits, and other costs. This number will be much higher than the actual salary of the person. For example, a medicinal chemist in the United States is typically given an FTE rate of $200,000 to $250,000, whereas their salary may be half that amount or less. But that higher number is the actual cost to the company to employ and provide an environment for that person's work.

2. *Estimate the materials and capital expenditures cost.*

Direct costs of materials and consumables (expensive antibodies, special probe tips, expensive pressure catheters) that will need to be purchased specifically for the project (even though they may be used later by other projects) must be added up for each stage. Capital expenditures ("cap ex") are usually expense items (equipment or fixed assets) that have a lifetime longer than a few years and cost more than a company-set threshold amount. Examples are gene-sequencing machines, mechanical testing equipment, automated cell sorting machines, and microscopes. Capital expenditure purchases can require submission of a formal justification of the purchase and specific authorization from some chain of command in the company (if the item is significantly expensive).

3. *Get a quote for any contracted research.*

For example, animal studies may need to be contracted to a specific animal research contractor who has developed expertise in that model. That quote, and the

Stage	PERSONNEL/FTE	Number	Time	Unit	Base Cost (over 1 year)
4a & 4b & 4c.	Project manager	1	0.8	FTE	$1,70,000
	Senior Chemist	1	0.4	FTE	$2,50,000
	Senior Biologist	2	0.75	FTE	$2,25,000
	Biology research asstnt	1	0.6	FTE	$1,15,000
	Senior Biomedical Engineer	1	0.2	FTE	$2,00,000
	CONSUMABLES/ SUPPLIES				
In vivo test in healthy animals, hypertensive animals, toxicity tests	Pressure transducers, assay kits, chemicals, materials costs				
	CAPITAL EXPENDITURES				
	Micro-centrifuge				
	CONTRACTED/ OUTSOURCED				
	Outsourced animal study				
TOTAL					

Note: Base Cost for personnel includes taxes, benefits, overhead costs, some materials

Figure 5.20 Example of a budget worksheet for a stage. Time = no. of weeks on project divided by no. salary weeks in a year.

additional costs of any further analysis of the results/tissues needs to be specifically included in the budget calculations.

Add these three pieces together for each stage of the project development plan and pretty soon you will have an estimated budget per year and a total for the project. This calculation will at least get you an order of magnitude estimate of the development cost of the project and some idea of the resources that you will need to plan to acquire. Figure 5.20 shows an example budget preparation worksheet.

5.14 How to get your project funded in a larger organization

First, do not make the mistake of thinking that just because the technology or invention is very exciting, the project will be funded. In a science-driven, pure R&D organization it may be easy to get people excited by a new idea and by the "hot" topical nature of the concept/technology, but a business is also driven by many other considerations, as described in all the chapters in this book.

Communicate, communicate, and then communicate some more. That does not mean that one should be redundant and communicate the same message over and over again! Each point of communication must adapt to the audience and to the questions and concerns heard, and the same message may need to be delivered with different perspectives each time. Box 5.20 is an example of how a message was adapted to the audience to help them understand the value and need for a particular project. Communications must cover the elements listed in Section 5.14.1.

Box 5.20 An excerpt from Conger (1998), "The Necessary Art of Persuasion"
Reproduced here by permission, all rights reserved by Harvard Business Press

Robert Marcell [was] head of Chrysler's small-car design team. In the early 1990s, Chrysler was eager to produce a new subcompact-indeed, the company had not introduced a new model of this type since 1978. But senior managers at Chrysler did not want to go it alone. They thought an alliance with a foreign manufacturer would improve the car's design and protect Chrysler's cash stores.

Marcell was convinced otherwise. He believed that the company should bring the design and production of a new subcompact in-house. He knew that persuading senior managers would be difficult, but he also had his own team to contend with. Team members had lost their confidence that they would ever again have the opportunity to create a good car. They were also angry that the United States had once again given up its position to foreign competitors when it came to small cars.

Marcell decided that his persuasion tactics had to be built around emotional themes that would touch his audience. From innumerable conversations around the company, he learned that many people felt as he did-that to surrender the sub-compact's design to a foreign manufacturer was to surrender the company's soul and, ultimately, its ability to provide jobs. In addition, he felt deeply that his organization was a talented group, hungry for a challenge and an opportunity to restore its self-esteem and pride. He would need to demonstrate his faith in the team's abilities.

Marcell prepared a 15-minute talk built around slides of his hometown, Iron River, a now defunct mining town in Upper Michigan, devastated, in large part, by foreign mining companies. On the screen flashed recent photographs he had taken of his boarded-up high school, the shuttered homes of his childhood friends, the crumbling ruins of the town's ironworks, closed churches, and an abandoned railroad yard. After a description of each of these places, he said the phrase, "We couldn't compete"-like the refrain of a hymn. Marcell's point was that the same outcome awaited Detroit if the production of small cars was not brought back to the United States. Surrender was the enemy, he said, and devastation would follow if the group did not take immediate action.

Marcell ended his slide show on a hopeful note. He spoke of his pride in his design group and then challenged the team to build a "made-in-America" sub-compact that would prove that the United States could still compete. The speech, which echoed the exact sentiments of the audience, rekindled the group's fighting spirit. Shortly after the speech, group members began drafting their ideas for a new car.

Marcell then took his slide show to the company's senior management and ultimately to Chrysler chairman Lee Iacocca. As Marcell showed his slides, he could see that Iacocca was touched. Iacocca, after all, was a fighter and a strongly patriotic man himself. In fact, Marcell's approach was not too different from Iacocca's earlier appeal to the United States Congress to save Chrysler. At the

Box 5.20 (*cont.*)

end of the show, Marcell stopped and said, "If we dare to be different, we could be the reason the U.S. auto industry survives. We could be the reason our kids and grandkids don't end up working at fast-food chains." Iacocca stayed on for two hours as Marcell explained in greater detail what his team was planning. Afterward, Iacocca changed his mind and gave Marcell's group approval to develop a car, the Neon.

With both groups, Marcell skillfully matched his emotional tenor to that of the group he was addressing. The ideas he conveyed resonated deeply with his largely Midwestern audience. And rather than leave them in a depressed state, he offered them hope, which was more persuasive than promising doom. Again, this played to the strong patriotic sentiments of his American-heartland audience.

No effort to persuade can succeed without emotion, but showing too much emotion can be as unproductive as showing too little. The important point to remember is that you must match your emotions to your audience's.

5.14.1   The art of persuasion

- Establish credibility
- Illuminate advantages
- Provide evidence
- Connect emotionally
    - Show commitment
    - Understand and adjust to audience

5.14.2   Business case

In general, certain points need to be well researched before taking the project case to senior management:

- ◆ Tie to company vision and strategy.
- ◆ Tie to current concerns of top management.
- ◆ Quantify benefits where possible.
- ◆ But do not ignore unquantifiable benefits.
- ◆ Demonstrate awareness of risks; show plans to overcome them..
- ◆ Use style and format usually preferred
- ◆ Be ready for formal presentations.
- ◆ Lobby informally beforehand.
- ◆ Options?
- ◆ Provide information and counterarguments to resisters ahead of time.
- ◆ Be prepared with other items on project proposal (see Chapter 4, Figure 4.4).

5.14.3   Valuation decision – net present value (NPV)

Management will frequently carry out an economic analysis (typically a net present value [NPV] analysis) to determine whether the project value fits current financial/ economic value thresholds. The *net present value* is calculated by first projecting all costs and revenues for this project over a certain number of years. The net cash flow for each year is then discounted to the previous year by a certain discount rate (decided by the risk perceived in the project or by a general cost of capital to the company or the expected rate of return for a project – this rate may be set by the company for all projects at that stage of development). The result is sequentially brought down to the present time. The net of the negative discounted cash flows (costs or investments by the company) and positive discounted cash flows (revenues) is typically expected to be a positive number for a project to be approved and sometimes can be compared to other projects to make portfolio decisions among projects. Therefore, if a project has revenues 10 years out from current date, then those revenues will need to be sufficiently large to balance out the closer (less heavily discounted) costs.

Other valuation methods used in the biomedical industry include risk adjusted net present value, real options, decision tree analysis, and Monte Carlo simulations (please see References and Additional Readings for more details on these techniques). In general, every economic decision is built on certain assumptions, and every project manager must pay very close attention to the basis of the assumptions. Additional considerations include health economics (discussed in Chapter 8).

5.14.4   Stakeholders

The next useful step is to carry out a formal or informal stakeholder analysis to prepare to communicate more effectively (see also Box 5.20).

List all stakeholders
- All the people who are affected by your work, who have influence or power over it, or have an interest in its successful or unsuccessful conclusion.
- These people include your family, boss, coworkers, future recruits, community, the public, your coworkers, alliance partners, trade associations, etc.

Prioritize stakeholders
- *High-power, interested people*: these are the people you must fully engage and make the greatest efforts to satisfy.
- *High-power, less-interested people*: put enough work in with these people to keep them satisfied, but not so much that they become bored with your message.
- *Low-power, interested people*: keep these people adequately informed, and talk to them to ensure that no major issues are arising. These people can often be very helpful with the detail of your project.
- *Low-power, less-interested people*: again, monitor these people, but do not bore them with excessive communication

Considerations in preparing to deal with stakeholders

- How can you help the high-power, high-interest stakeholders do *their* job better with your project?
- Use the opinions of the most powerful stakeholders to shape your projects at an early stage:
  - To support you
  - To improve the quality of your project
- Win more resources – this makes it more likely that your projects will be successful.
- Making sure they understand the benefits of your project, they are more likely to support you when needed.
- Anticipate what people's reaction to your project may be and build into your plan the actions that will win people's support.
- What financial or emotional interest do they have in the outcome of your work? Is it positive or negative?
- What motivates them most of all? What information do they want from you? How do they want to receive information from you? What is the best way of communicating your message to them?
- What is their current opinion of your work? Is it based on good information?
- Who influences their opinions generally, and who influences their opinion of you? Do some of these influencers therefore become important stakeholders in their own right?
- If they are not likely to be positive, what will win them around to support your project? If you don't think you will be able to win them around, how will you manage their opposition?

Once this stakeholder analysis is carried out, a clear picture of how communication messages need to be tailored for each particular audience will emerge. In addition, the communication effort can now be focused more effectively.

5.15   Outsourcing product development

Why outsource?

1. Does it save the company money in the short term? in the long term?
2. Does it bring the product to market faster?
3. Does the contracted party bring in expertise and tools that the company does not have the need, the time, or the money to build itself?

Outsourcing work is fraught with problems (many of which are driven by communication issues) and a project manager must not underestimate the time that will be required to successfully manage and complete an outsourced project. In general, outsourcing can add value to a company if it is managed well and integrated into the product development plans and company strategy.

	R&D	FDA Process & Clinical Trials	Manufacturing	Marketing & Sales	
				US	International
Do It Yourself					
Strategic Partners/ Licenses					

Figure 5.21 Strategic business model for planning internal and external development

Another way to look at the decision of whether to outsource or to build the capability in house is to evaluate it from a view of the business model of the company (Figure 5.21). What parts of the value chain will the firm retain vs. outsource with a long-term view? Figure 5.21 is a useful if simple tool to visualize the company's business strategy and understand internally which component of development is going to be partnered or outsourced.

If the company is never going to build the capability to do cGMP (FDA-regulated process) manufacturing, then the decision to outsource manufacturing is easy to make. If the company does not have animal facilities and only a couple of animal studies need to be done, then also, the decision to outsource these animal studies is relatively simple to consider. In these cases, outsourcing brings in expertise to the project that is not present in the company.

The debate on whether to outsource (buy) or build capabilities in house depends on many other considerations, and these considerations will vary with the context of the specific function and project under discussion. For example, if there is a need for speed, it might be faster to outsource a clinical trial to a CRO that has connections in many countries or many sites and can reduce the trial recruitment times significantly. Alternatively, when working with academic collaborators, it is likely that they are driven by different priorities and might not be able to expand resources rapidly to take on a large job. In this instance, the company might be able to hire people faster to get the job started in house (build). Other considerations include the complexity of the task and the concern or sensitivity of intellectual property rights that might be generated or that are embodied in the process. For example, if the process involves the development of new protocols in making a very high-margin, expensive product, the risk of that know-how being disseminated to other players in the industry through the contract house may be too high, as the company might lose its competitive advantage in the market. On the other hand, some jobs that are very complex – like building relevant transgenic disease models for drug testing – may best be outsourced to places where past expertise lies in building these models. If the process requires novel methodology that would require the contractor to reconfigure their infrastructure, it might be cheaper (especially if multiple products in development go through that facility) to invest in a larger space and build the capability in house. Figure 5.22

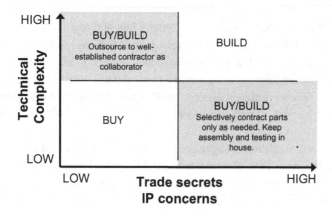

Figure 5.22 Outsourcing decision chart for projects with different levels of complexity and intellectual property concerns

demonstrates the various balances between choosing to buy (outsource) vs. build (internal investment) given certain IP and process complexity issues. When complexity is high, but IP concerns are low, outsourcing might be considered, but to a sophisticated provider with whom company personnel will tend to work more as collaborators, in order to make sure the process performs as required in a complex environment. If the IP concerns are high, but the project complexity is low, then again, an opportunity to selectively contract out only limited parts or specific functional components might be considered – for example, in outsourcing chemistry, one might want the service provider to make compound intermediates, while the final compounds are made in the company. The chart in Figure 5.22 also shows areas where it is probably prudent to build internally, in areas of high complexity and high IP concerns where expertise needs to be kept in house, or in areas where both are low and expertise can easily be built in house.

Three practical suggestions to have a successful outsourcing experience, especially when offshoring:

1. Check references and speak with customers that have contracted similar projects.
2. Conduct an internal readiness assessment to make sure existing processes, infrastructure, and assay or testing methods can support an outsourcing development initiative in terms of time commitment, responsiveness, communication tools, etc.
3. Try out new offshore development contractors with small and relatively low-risk projects first.

5.16   Summary of preclinical certifications and laboratory regulations

The following summary is not a comprehensive list, but indicates some key ethical and regulatory compliance issues that must be kept in mind while setting up and

---

Box 5.21 A note on understanding U.S. federal law and regulations

Laws are passed by Congress with relatively general language, intended to address a particular issue in a way that can be consistently applied to all instances with the issue. These laws are then passed down to a particular government agency to apply and administer. The government agency then decides how the law must be applied – resulting in regulations, which serve as the practical rules for citizens to adhere to the law.

The regulations by all federal agencies are published and updated in the Code of Federal Regulations (CFR). The reference 21CFR 280 = Volume Title 21, section 280 of the CFR. Electronic version of the CFR is at http://www.gpoaccess.gov/cfr.

---

running a laboratory. While compliance in some cases is required by federal law or agency (Employment law, Drug Enforcement Agency, etc.), in other cases the requirements and procedures vary by state (e.g. Department of Health). It is recommended to proactively peruse their areas of regulation and to determine whether company or project activities fall under those regulations (see also Box 5.21).

Pertinent regulatory bodies
- Environmental Protection Agency (EPA)
- Occupational Safety and Health Administration (OSHA), an agency of the U.S. Department of Labor
- Drug Enforcement Administration (DEA), a U.S. Department of Justice law enforcement agency that enforces the Controlled Substances Act of 1970
- National Institute of Health (NIH; if obtaining federal funds for biomedical research)
- Relevant state's Department of Health
- Local or institutional committees on environmental health and safety, animal safety, biosafety, radiation safety, and various institutional review boards (IRBs) that review protocols prior to allowing any human testing of new products

Biosafety
- The NIH requires that organizations conducting recombinant-DNA research have an institutional biosafety committee to ensure that the work meets the safety requirements set forth in the agency's guidelines.
- At many institutions, the biosafety panels, which were first established some 30 years ago, have taken on additional responsibility for overseeing research involving other potentially hazardous biological materials, such as infectious microorganisms, toxic biological substances, and biological allergens. Fear that those materials could be used by terrorists has intensified the oversight. Federal law spells out strict controls over the use of so-called select agents – that is, organisms and toxins that can pose a severe threat to human health, such as smallpox,

botulism, and anthrax. Government involvement with and oversight of local biosafety committees has increased, regarding the review of "dual-use" studies – projects with the potential for dangerous as well as beneficial applications.

- An institution must arrange for appropriate hazardous waste disposal whether of a chemical nature (solvents, toxic volatiles), biological waste (tissues, carcasses from animal studies, etc.) or contaminated materials. Certified contractors must be used. Disposal of radioactive materials must be accompanied by a sound monitoring program of employee exposure.
- Institutions must monitor the use, storage, and disposal of radioactive material. Those regulations emanate from the Nuclear Regulatory Commission. The institutional radiation-safety committee is responsible for the policies and procedures for acquiring and using such materials.

Other certifications and ethical regulatory oversight that the organization must have in place

- Certifications: In many states, if working with animals, a Certificate of Approval for Working with Live Animals is required by the laboratory. This may require an inspector's visit by the state health department.
- If using controlled substances (i.e., pentobarbital for anesthesia), a Controlled Substances license is required. Registration with the DEA is also required for use of controlled substances.
- Materials Safety Data Sheets (MSDS) must be maintained for all chemicals used in the laboratory and must be made available to all involved employees.
- To follow the NIH Guide, a Biosafety Committee, and an Institutional Animal Care and Use Committee (IACUC) must be constituted.
- Local labor laws for the state have to be followed – non-discriminatory practices, good safety procedures, appropriate training for employees, all have to be followed in order to avoid problems if an audit is ever carried out.

5.17   Summary

The new product development path listed here for drugs, devices, and diagnostics is rather generalized, as each product will have its own nuances and specific gates in the NPD process. In addition, each company will evolve its own NPD process, and the general processes discussed here describe key issues that are broadly important to consider during NPD. It is important to realize that Chapter 5 should be read in the context of all the rest of the chapters in this book. It is positioned centrally in the book because the reader is now better informed while going through the remaining chapters to develop a sound regulatory path and reimbursement strategy and then come back and revise the product development plan.

Exercises

5.1. Examine carefully the wording of the indication identified in exercises in Chapter 2. Does it match the target market population? Are the desired product characteristics incorporated in drafting the particular indication?

5.2. Identify specific endpoints for clinical trials that will meet product characteristics and address the unmet need of the target patient population.

5.3. Identify key risk-reducing milestones. Describe stages of development (use general templates in chapter) and endpoints of each stage or substage.

5.4. Generate a stage-gate development plan/chart with timelines.

5.5. Define gate assessment criteria as quantitatively as possible.

5.6. Summarize resources needed up to each milestone of development including final manufacturing – space, personnel, equipment, and consumables with high-level approximation of costs (in categories or in blocks of $50,000).

5.7. Generate a Gantt chart using milestones and stages, with all key assumptions annotated.

5.8. Generate a budget with monthly cash flow and large-item expenses outlined (make sure you justify cost assumptions for big-ticket items).

5.9. Prepare a presentation with project justification and project plan, budget, and timelines clearly described.

5.10 Carry out a risk analysis and hazard analysis considering the typical usage and selected product design specifications and then identify changes in design required to mitigate such risks or hazards and prevent failure.

References and additional readings

Block, Z, and MacMillan, IC. (1993). *Corporate Venturing: Creating New Businesses within the Firm*. Boston: Harvard Business School Press. ISBN: 0875846416

Bossuyt, PM, Reitsma, JB, Bruns, DE, et al. (2003). The STARD statement for reporting studies of diagnostic accuracy: explanation and elaboration. *Clinical Chemistry* 49(1), 7–18.

Christensen, C. (2003). *The Innovator's Dilemma: The Revolutionary Book That Will Change the Way You Do Business*. Boston: Harvard Business School Press. ISBN: 0060521996

Conger, JA. (1998). The necessary art of persuasion. *Harvard Business Review* 76(3), 84–95.

Cooper, RG. (1988). The new product process: a decision guide for managers. *Journal of Marketing Management* 3, 238–255.

Fries, R. (editor). (2001). *Handbook of Medical Device Design*. New York: Marcel Dekker Press. ISBN: 0824703995

Gad, SC. (editor). (2005). *Drug Discovery Handbook*. New York: Wiley-Interscience. ISBN: 0471213845

Galbraith, JR. (1971). Matrix organization designs – how to combine functional and project forms. *Business Horizons* 17(1), 2–40.

Justiniano, J, and Gopalaswamy, V. (2004). *Six Sigma for Medical Device Design*. Boca Raton, FL: CRC Press. ISBN: 0849321050

King, PH, Fries, RC, and Johnson, AT. (2018). *Design of Biomedical Devices and Systems*, 4th ed. Boca Raton, FL: CRC Press. ISBN: 9780429786068

Larson, E, and Gray, CF. (2020). *Project Management: The Managerial Process*, 8th ed. New York: McGraw-Hill. ISBN: 978-1260238860

Larson, EW, and Gobeli, DH. (1988). Organizing for product development projects. *Journal of Product Innovation Management* 5, 180–190.

Leifer, R, McDermott, CM, O'Connor, GC, Peters, LS, Rice, M, and Veryzer, R. (2000). *Radical Innovation: How Mature Companies Can Outsmart Upstarts*. Boston: Harvard Business School Press. ISBN: 0875849032

Lipinski, CA, Lombardo, F, Dominy, BW, and Feeney, P. (1997). Experimental and computational approaches to estimate solubility and permeability in drug discovery and development settings. *Journal of Advances in Drug Delivery Reviews* 23, 3–25.

Lipper, RA. (1999). How can we optimize selection of drug development candidates from many compounds at the discovery stage? *Modern Drug Discovery* 2, 55–60.

Lynn, GS, Morone, JG, and Paulson, AS. (1996). Marketing and discontinuous innovation: the probe and learn process. *California Management Review* 38(3).

Mathieu, M. (2005). *Parexel's Bio/Pharmaceutical R&D Statistical Sourcebook*. Parexel International Corp.

Metzler, R. (1994). *Biomedical and Clinical Instrumentation: Fast Tracking from Concept through Production in a Regulated Environment*. Interpharm Press. ISBN: 0935184503

Mugambi, ML, Peter, T, Martins, SF, and Giachetti, C. (2018). How to implement new diagnostic products in low-resource settings: an end-to-end framework. *BMJ Global Health* 3, e000914.

O'Connor, G. (2008). *Grabbing Lightning: Building a Capability for Breakthrough Innovation*. San Francisco: Josey Bass Publisher. ISBN: 978-0787996642

Ogrodnik, P. (2019). *Medical Device Design*. London: Academic Press. ISBN: 9780128149621

Pisano, GP. (1996). *The Development Factory: Unlocking the Potential of Process Innovation*. Boston: Harvard Business School Press. ISBN: 0875846505

Project Management Institute. (2000). *A Guide to the Project Management Body of Knowledge (PMBOK®)*. ISBN: 1-880410-23-0

Ries, E. (2011). *The Lean Startup: How Today's Entrepreneurs Use Continuous Innovation to Create Radically Successful Businesses*. New York: Crown Business. ISBN: 9780307887894

Sapienza, AM. (1997). *Creating Technology Strategies: How to Build Competitive Biomedical R&D*. New York: Wiley-Liss Publishing. ISBN: 0471153702

Stromgaard, K, Krogsgaard-Larsen, P, and Madsen, U. (editors). (2017). *Textbook of Drug Design and Discovery*, 5th ed. Boca Raton, FL: CRC Press. ISBN: 9781498702782

Vasan, A, and Friend, J. (2020). Medical devices for low- and middle-income countries: a review and directions for development. *Journal of Medical Devices* 14(1), 010803.

Whitmore, E. (2012). *Development of FDA-Regulated Medical Products: Second Edition, A Translational Approach*, 2nd ed. Milwaukee, WI: ASQ Quality Press. ISBN: 9780873898331

Websites of interest

For more details on surrogate biomarkers see www.fda.gov/drugs/development-resources/surrogate-endpoint-resources-drug-and-biologic-development [accessed Nov 2020].

For FDA's Quality Systems Manual: www.fda.gov/medical-devices/postmarket-requirements-devices/quality-system-qs-regulationmedical-device-good-manufacturing-practices

Tufts Center for the Study of Drug Development has various white papers and other publications which are good resources on various topics in the development of new drugs: http://csdd.tufts.edu/overview-publications

Visit https://shreefalmehta.com/csbtbook for additional enriching readings around the topics covered in the book, topical updates on the content and for industry viewpoints and news.

6 The regulated market: gateway through the FDA

Plan	Position	Pitch	Patent	Product	**Pass**	Production	Profits
Industry context	Market research	Start a business venture	Intellectual property rights	New product development (NPD)	Regulatory plan	Manufacture	Reimbursement

---

Learning points
- What are the functions of the U.S. Food and Drug Administration (FDA)?
- What is the importance of indications and endpoints in mapping a regulatory route?
- How do FDA regulations impact preclinical and clinical product development planning?
- Identify a path through the regulatory process for a drug, device, or diagnostic product idea
- What are some major differences in regulatory requirements among key countries or regions?
- What are the important contents in the submissions required by the FDA at various points?
- How does a product with a combination of diverse technologies get approved?
- How are personalized medicines developed through the regulatory process?
- How do regulatory agencies approve new technologies like AI-based diagnostic programs?

6.1 FDA role and significance for biomedical product development

6.1.1 Introduction and history

A brief history of the FDA proves useful to understanding the rationale and role of this organization in the health care system of the United States. The U.S. FDA processes are followed widely given ongoing efforts to globally harmonize healthcare product review, and hence form a good benchmark for this chapter, and in general, for regulatory bodies of various countries around the world.

Federal oversight for drugs started with the establishment of U.S. customs laboratories to administer the Import Drugs Act of 1848, as the United States had become the world's dumping ground for counterfeit drug materials. However, the founding of the modern FDA is in the 1906 Food and Drugs Act. This Act, establishing a national agency to put a stop to food adulteration and fake remedies, was passed in the context of trade concerns. Various trades and the public were concerned about the economics of varying state laws and the increasing use of new synthetic chemicals to create cheap and unsafe adulterated food (deodorized rotten eggs, revived rancid butter, substituted glucose for honey). Upton Sinclair's book *The Jungle* highlighted poor health conditions in the Chicago meat-packing plants and precipitated an ongoing discussion in Congress. President Theodore Roosevelt signed the "Food Bill" in June 1906. This Act made it illegal to sell adulterated foods and make false claims about a food or drug and also carried these bans into interstate commerce. An existing Department of Chemistry was designated to carry out tests and enforce the law. The primary concern was to use scientific methods to analyze the risk to human health and safety.

In 1933 the FDA recommended an overhaul of the obsolete 1906 Food and Drugs Act, launching a prolonged legislative battle that came to a conclusion due to a crisis in 1937. In that year, the popularity of S.E. Massengill Co.'s elixir, sulfanilamide, for the treatment of sore throats in people with streptococcal infections, drove demand for a palatable form for children. The company chemist added diethylene glycol to sweeten the elixir, without going through testing of the new additive. Diethylene glycol is known today as antifreeze and ingesting it carries lethal consequences. In 1937, 107 people, mostly children, died before the U.S. Food and Drug Administration was able to gather all of the supply.

The incident led to the Food and Drug Act of 1938, which, among other things, required drug makers to show their products were safe before they went on the market. That gave rise to the regulatory pathway that drugs in development go through prior to approval, beginning with animal testing prior to human testing.

Box 6.1 Laws and regulations

Laws are written by elected representatives in the U.S. Congress. The United States Code (U.S.C.) is the codification by subject matter of the general and permanent laws of the United States. It is divided by broad subjects into 50 titles and published by the Office of the Law Revision Counsel of the U.S. House of Representatives. An electronic version is available at www.gpoaccess.gov/uscode/index.html.

Regulations are written by government employees to explain the practical application of the laws. The Code of Federal Regulations or CFR is the book in which all regulations are codified. Title (volume) 21 contains regulations pertaining to food and drugs. These regulations represent how the FDA interprets the Acts or laws which Congress passes. An electronic version of the CFR is accessible at www.ecfr.gov/.

Citation example: [21 CFR 302] refers to title 21 of the Code of Federal Regulations, section 302.

Citation example: [21 U.S.C. 321 (g)1] refers to title 21, United States Code, section 321, subsection (g) part 1.

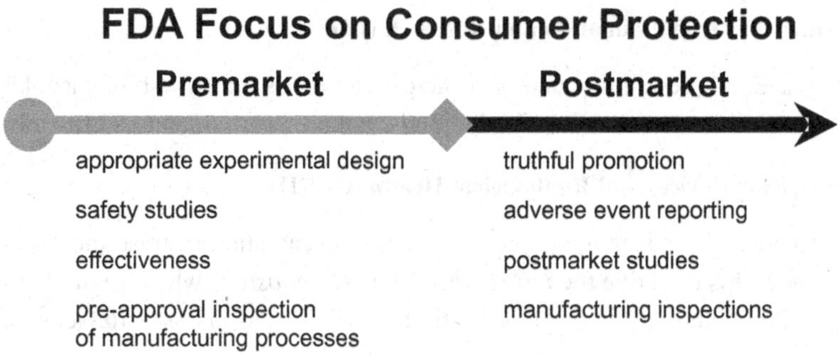

FDA Focus on Consumer Protection

Premarket   Postmarket

appropriate experimental design   truthful promotion

safety studies   adverse event reporting

effectiveness   postmarket studies

pre-approval inspection   manufacturing inspections
of manufacturing processes

Figure 6.1 The FDA focuses on consumer protection. (Slide courtesy from presentation made by Dr Jan Stegemann, Rensselaer Polytechnic Institute.)

6.1.2 Role of the FDA and significance for product development

The FDA is thus a federal organization that enacts the laws with the goal of protecting consumer safety in the development and sale of food and medical products (drugs/devices/diagnostics). The FDA ensures that the claims made by a medical product accurately reflect its risks and benefits so that users and purchasers of the product can make sound judgment on the balance they want to strike between benefits and risks of taking the product. Figure 6.1 highlights the key issues that the FDA reviews in the preclinical and clinical areas.

In effect, from a manufacturer's viewpoint, the FDA is not only the gatekeeper to the market, but the FDA's final marketing approval of a product will also specify the claims that the manufacturer can make. *The specific wording on the approved label strongly influences and defines the patient population to whom the product can be sold.* The label is generated from the data that the manufacturer provides to the FDA. *Thus, it is clear that the product development process has to be designed with the end label (indication) in mind* (see Section 5.7 for details).

6.2 Organization and scope of the FDA

6.2.1 Divisions of the FDA

Key divisions of interest for biomedical product development and review are:

Office of the Commissioner

Policy, planning, administration, ombudsman, *office of combination products, office of orphan products*, financial management and other functions.

Center for Biologics Evaluation and Research (CBER)

Regulates biological products such as vaccines, biologics, cellular, tissue and gene therapies, blood, blood-derived products, devices, and tests used to safeguard blood from infectious agents, and xeno-transplantation products.

Center for Drug Evaluation and Research (CDER)

Regulates all prescription and over-the-counter drugs (includes biological large molecule drugs like monoclonal antibodies and cytokines, etc.), monitors drug advertising.

Center for Devices and Radiological Health (CDRH)

Regulates all devices including those emitting radiation (ultrasound, electronic). Has under this directive the *Office of In Vitro Diagnostics*, which regulates all aspects of in-home and laboratory diagnostic tests (in vitro diagnostic devices, or IVDs).

Center for Food Safety and Applied Nutrition (CFSAN)

Responsible for the safety of 80 percent of all food consumed in the United States – the entire food supply except for meat, poultry, and some egg products.

Center for Veterinary Medicine (CVM)

Assures that animal food products, and drugs used to treat animals are safe.

National Center for Toxicological Research (NCTR)

Conduct peer-reviewed scientific research that supports and anticipates the FDA's current and future regulatory needs.

6.2.2   What the FDA does not regulate

The categories of products that are consumed by humans or that could affect human health and the regulatory government agencies are shown in Table 6.1

Table 6.1 Products regulated by government agencies other than the FDA

Item	Regulated by
Advertising – make sure it is not misleading to consumer	Federal Trade Commission
Alcohol – labeling and quality	Bureau of Alcohol, Tobacco, and Firearms, Treasury Department
Consumer products like household goods, appliances, toys, paint, packages.	Consumer Product Safety Commission
Illegal drugs with no approved medical use	Drug Enforcement Administration, U.S. Department of Justice
Health Insurance	Questions about Medicare should be directed to the Centers for Medicare & Medicaid Services (CMS)
Meat and Poultry	Food Safety and Inspection Service, U.S. Department of Agriculture (USDA)
Pesticides	FDA, USDA, and the Environmental Protection Agency (EPA) share the responsibility for regulating pesticides
Restaurants and Grocery Stores	Local county health departments
Water	EPA has the responsibility for developing national standards for drinking water from municipal water supplies. FDA regulates the labeling and safety of bottled water.

Table 6.2 Products regulated by the FDA

Blood-related biologics – blood substitutes, etc.	Product and manufacturing establishment licensing; safety of the nation's blood supply; research to establish product standards and develop improved testing methods
Cosmetics	Safety and labeling
Drugs (includes biological large molecule drugs)	Product approvals; Over-the-counter (OTC) and prescription drug labeling; drug manufacturing standards
Foods	Labeling; safety of all food products (except meat and poultry); bottled water
Medical devices	Premarket approval of new devices; manufacturing and performance standards; tracking reports of device malfunctioning and serious adverse reactions
Radiation-emitting electronic products	Radiation safety performance standards for microwave ovens, television receivers, diagnostic X-ray equipment, cabinet X-ray systems (such as baggage X-rays at airports), laser products; ultrasonic therapy equipment, mercury vapor lamps, and sunlamps; accrediting and inspecting mammography facilities
Veterinary *products*	Livestock feeds; pet foods; veterinary drugs and devices

6.2.3   What does the FDA regulate?

The FDA regulates the products listed in Table 6.2.

There are three main divisions (see Section 6.2.1) at the FDA that receive and review human health product applications for market approval, and that are discussed further in this chapter:

CDER – regulates drugs (includes large molecule biologic drugs)

CBER – regulates tissue, cellular gene therapies, vaccines, the collection of blood and blood components used for transfusion or for the manufacture of pharmaceuticals derived from blood and blood components.

CDRH – regulates medical device manufacturers

OIVD (Office of In Vitro Diagnostics; part of CDRH) – Regulates diagnostic tests

6.2.4   Friends not foe

The FDA's goal is to get safe and effective biomedical products to the public. This is conceptually the same goal that the manufacturers have. Conflicts usually arise due to the uncertainty of the science and a different view of risk versus benefit between the company/sponsor and the FDA reviewers. In general, while the FDA is obliged to fulfil its duty to get biomedical products to the public as quickly as possible, the FDA is not driven by commercial concerns and will not change its process or make exemptions to reduce costs to the sponsor. However, the PDUFA (Prescription Drug User Fee Act, 1992) has made the FDA more answerable as a service organization to the public and to industry. This act increased the application fees paid by industry sponsors to the FDA to help fund the review of new drugs; in turn, the FDA has made several performance promises (e.g. set a maximum time for review of applications). A similar Act called The Medical Device User Fee and Modernization Act of 2002 (MDUFMA) charges user fees for device premarket reviews. (Note: Other significant provisions of the MDUFMA allow for establishment inspections to take place by accredited third parties and also put in place new regulatory requirements for reprocessed single-use devices.)

The FDA has also made great efforts in recent years to work with industry to help sort through complex new product applications [such as combination products (Section 6.8) or artificial intelligence software (Section 6.9)] and to give them guidance and clarity on internal processes and reviews at the FDA. The FDA website is a great example of this effort and is a handy reference on most policy and process questions. The FDA also presents at industry conferences and gives talks to convey their current thinking and opinions to industry.

6.2.5   Science rules – most of the time!

The FDA is an organization driven by the scientific method and scientific principles of analysis. The product development plan and in particular, the clinical study, thus have to be designed to collect data using accepted scientific and experimental methods of inquiry. Statistically valid analysis of the results must be used to support the claim for the indication. However, the FDA usually errs on the side of caution in the interpretation of the data and evaluation of risk. They are always involved in a delicate balancing act in considering the final approval of new drugs, where political pressure (take the case of the application to make the "morning after pill" available without prescription; see Box 6.2), internal scientific review, and public pressure (e.g. in creating a process to make AIDS drugs and now other drugs available to patients before final approval) all play a significant role. In general, despite existing public opinion or controversy, the sponsor organization must interact with the FDA solely on the basis of the data and scientific evidence, wherein the dialogue with the FDA is a formal and highly specific interaction.

6.2.6   International harmonization

The FDA also works outside U.S. national borders to protect U.S. consumer health, as over 80% of seafood, 20% of all fresh produce, and millions of other FDA-regulated products come from other countries. On another perspective, most U.S. manufacturers that want to sell their FDA-regulated products outside the United States need to be aware of international differences in regulation. In recognition of the increasing global trade in medical products and the myriad complex regulations in place in various countries, there is an ongoing effort to harmonize the regulatory regimes in the three largest markets – European Union, Japan, and the United States, eventually hoping to generate a common filing process and format for these regulatory agencies. The U.S. FDA is an active participant in this International Conference for Harmonization (ICH).

Box 6.2 Plan B: science, religion, and society – a debate over approval

An emergency contraceptive (called morning after pills or Plan B, which needs to be taken within 72 hours after unprotected sex to prevent ovulation or in some cases, implantation of a fertilized egg) containing two levonorgestrel pills (a synthetic hormone used for over 35 years in birth control pills) were approved in 1999 for prescription use. An application for converting plan B to an over-the-counter (OTC) drug (nonprescription purchase) was not approved despite strong internal (CDER) and external (external scientific advisory board voted 23 to 4 in favor of approval) support for the scientific evidence. There was a tremendous amount of pressure from President George W. Bush's administration, conservative groups (who felt that this would condone sexual activity for pre-teens and teenagers and let them avoid medical care) and anti-abortion-rights groups (who see this as an abortion pill). Despite overwhelming evidence that the drug is safe and effective, the FDA did not approve it. Dr. Susan Wood, who headed the FDA's Office of Women's Health until August 2005, supported approval of the OTC label. She resigned in August 2005, when the then-head of the FDA, Lester Crawford, announced that the agency had postponed decision on plan B for months or years. The debate continued at the state levels and at many pharmacies, where individual pharmacists refused to fill prescriptions for plan B. In mid 2006, the FDA finally approved plan B for over the counter sales to women over the age of 18 only.

COVID-19 vaccine approval and the presidential election pressure in November 2020

In October 2020, President Trump was campaigning for reelection under significant headwinds of the economy and a spiraling COVID case count, largely resulting from a poorly coordinated COVID response by his administration. In order to address the polling deficit and the struggling economy, the president put tremendous pressure on the FDA to approve vaccines before election day. The president put public pressure through tweets and public comments on officials to speed up their timeline for developing and approving a vaccine, to the point where the FDA Commissioner Dr. Hahn issued an opinion-editorial in the *Washington Post* (August 5, 2020; www.washingtonpost.com/opinions/fda-commissioner-no-matter-what-only-a-safe-effective-vaccine-will-get-our-approval/2020/08/05/e897d920-d74e-11ea-aff6-220dd3a14741_story.html) assuring the public, "I have been asked repeatedly whether there has been any inappropriate pressure on the FDA to make decisions that are not based on good data and good science. I have repeatedly said that all FDA decisions have been, and will continue to be, based solely on good science and data. The public can count on that commitment." News reports around this time confirmed that the White House had tried to block the issuance of the two-month follow up safety monitoring period that would be required for any expedited review for new vaccines against COVID-19. Specifically, the guidelines for emergency use authorization of a COVID-19

Box 6.2 (*cont.*)

vaccine, released by the FDA on October 6, 2020, required that vaccine makers should follow trial participants for at least two months to rule out any major side effects before seeking emergency approval, dashing any White House attempts help fulfil President Trump's unsubstantiated promises to have a vaccine ready before election day November 3, 2020 without seeming concern to the potential safety of millions. The FDA Commissioner Stephen Hahn said in a statement on October 6, 2020, that he hoped the guidelines would help "the public understand our science-based decision-making process that assures vaccine quality, safety and efficacy."

However, a few months ago, under intense pressure from President Trump and trade advisor Peter Navarro, the FDA had approved an emergency use authorization for hydroxychloroquine, touted by the president as a miracle cure for COVID-19, only to rescind it 2 months later when a larger study showed it had no benefit in COVID-19 patients. Around the same time, the emergency use authorization by the FDA commissioner's office of convalescent plasma treatment was similarly controversial, with several health officials rejecting the available data as too weak. The FDA commissioner later offered an apology after publicly mis-stating the life-saving potential of the treatment.

Should agencies like the FDA be divorced from the debates and current opinions that are prevalent in society?

The goals of the ICH are to make the international regulatory processes for medical products more efficient and uniform. A clear example of inefficiencies is seen in the fact that despite similar concerns among these three regions about safety, efficacy and quality, time-consuming and expensive clinical trials need to be repeated in each region. For more details visit the website www.ich.org. Guidelines created in joint discussion at the ICH are then adopted by individual regulatory bodies (EU, Japan, Canada, and the United States as founding members, and others such as South Korea, India, China, Singapore, Switzerland). Guidelines on various topics have been published by the ICH, including safety and toxicology studies and common format for submission of data prior to first in human testing (Common Technical Document), and have been integrated and published as FDA guidance to industry (refer to websites listed in Table 6.3 for more information).

6.3   Regulatory pathways for drugs (biologicals or synthetic chemicals)

6.3.1   Classify your product

The first step is to decide whether the product is regarded as a drug, device, or diagnostic (or combination product) and then further classify it according to perceived risk factors. See following sections to review definitions and see examples in Box 6.3.

Table 6.3 Some useful FDA websites for drugs/biologics

Biologics and small molecule drugs (CDER and CBER)	
Starting point – drugs/CDER home page	www.fda.gov/drugs
Starting point at EMA website	www.ema.europa.eu/en/human-medicines-regulatory-information
Guidance documents for drugs	www.fda.gov/drugs/guidance-compliance-regulatory-information
	www.ema.europa.eu/en/human-medicines-regulatory-information
Drug Development and Approval Review Processes At the FDA and EMA	www.fda.gov/drugs/development-approval-process-drugs
	www.ema.europa.eu/en/from-lab-to-patient-timeline (interactive site)
Home page for Vaccines Blood and Biologics (CBER at FDA)	www.fda.gov/vaccines-blood-biologics
Biologicals (Guidance docs at EMA)	www.ema.europa.eu/en/human-regulatory/research-development/scientific-guidelines/biologicals/biologicals-active-substance
Drug Master Files and Guidance for Industry At FDA and EMA	www.fda.gov/drugs/forms-submission-requirements/drug-master-files-dmfs
	www.ema.europa.eu/en/active-substance-master-file-procedure
Institutional Review Boards (IRBs)	www.fda.gov/science-research/guidance-documents-including-information-sheets-and-notices/information-sheet-guidance-institutional-review-boards-irbs-clinical-investigators-and-sponsors
Office of Generic Drugs at FDA	www.fda.gov/about-fda/center-drug-evaluation-and-research-cder/office-generic-drugs
Office of Biosimilars at EMA	
Generic Drug review at EMA	www.ema.europa.eu/en/human-regulatory/overview/biosimilar-medicines-overview
	www.ema.europa.eu/en/human-regulatory/marketing-authorisation/generic-medicines/generic-hybrid-applications
Orphan Products Development. At FDA and EMA	www.fda.gov/industry/developing-products-rare-diseases-conditions
	www.ema.europa.eu/en/human-regulatory/research-development/orphan-designation/applying-orphan-designation
Good Laboratory Procedures (GLP) in the Code of Federal Regulations	21 CFR Part 58 www.accessdata.fda.gov/scripts/cdrh/cfdocs/cfcfr/cfrsearch.cfm?cfrpart=58
IND details and guidance documents	www.fda.gov/drugs/types-applications/investigational-new-drug-ind-application
NDA details and guidance documents at the FDA Marketing Authorization Applications (MAA) at EMA	www.fda.gov/drugs/types-applications/new-drug-application-nda
	www.ema.europa.eu/en/human-regulatory/marketing-authorisation
ANDA details and guidance documents	www.fda.gov/drugs/types-applications/abbreviated-new-drug-application-anda
Office of Combination Products	www.fda.gov/combination-products
International Conference on Harmonization of Technical Requirements for Registration of Pharmaceuticals for Human Use (ICH)	www.ich.org

Box 6.3 Practical tips for successful FDA meetings

There are three types of FDA meetings corresponding to major milestones in drug development:

1. Pre-IND meeting – Goal is to confirm that the nonclinical studies, drug product formulation, and the chemistry, manufacturing, and controls (CMC) are sufficient to support FDA approval to move into human testing.
2. End-of Phase II meeting – Update FDA on results of Phase I and II and gain clarity on potential Phase III study endpoints and size etc.
3. Pre-NDA/BLA meetings – Goal is to review marketing application contents and discuss any issues that have emerged in the development process, ranging from results of pivotal studies, to CMC operational updates and how summaries need to formatted, etc.

A formal meeting (also called a Type C meeting) can also be requested by the sponsor at any time in between with details on the reason for the request (e.g. review of changes in Phase II study endpoints)

Points for a successful meeting with FDA:

1. Understand the purpose of the meeting clearly (as above).
2. Have the right people at the table. Sponsors should select representatives who are leaders of functional areas of potential queries by the FDA, e.g. head of clinical development or head of CMC. Do not take senior executives who may not be familiar with the development details but instead take people who are leading those areas of development with the best working knowledge of the product. The sponsors representatives should look at the FDA as a partner not an adversary.
3. Practice and prepare. The FDA will usually provide (the day before) written responses to the questions sent in by your team when requesting the meeting. Go over these responses and their implications in a practice meeting and appoint one lead person to discuss each question/point in the meeting. Practice responses and follow-up questions and be mindful of the scheduled length of the meeting so that all critical points can be covered. This may mean moving on if one point is taking too long.
4. Word your discussion and questions in a way to get clarity on issues and to gain binding agreement. Most importantly, put the evidence and data that have prompted your team to make a particular decision on the study design and then inquire if the FDA agrees with this decision or alternative. This is an art and it is helpful to have someone who has gone through these before guiding your wording. Some example points to query are below by section:

   Pre-IND/Phase I – Polymorphs, enantiomers, or other unique physico-chemical properties. Reasons for selection of specific form of compound for drug product. Qualification of impurities.

> **Box 6.3** (*cont.*)
>
>> End-of-Phase II meeting – Agreement on final drug product synthesis
>> scheme, specifications, impurities, etc. Assays and validation testing;
>> CMC points including stability; discuss approvable endpoints and
>> request opinion on study design and statistical significance of results of
>> Phase III study.
>
> It is key to remember that FDA staff are similar to you and your team, as scientists,
> and are collaborators, not adversaries in the process. Being well prepared,
> providing adequate briefing information, and keeping to scheduled times are all
> important factors that can help smooth out interactions and make these FDA
> meetings successful.

6.3.2   Definition of drug product

A drug is defined (at www.fda.gov/cder/drugsatfda/Glossary.htm) as:

- A substance recognized by an official pharmacopoeia or formulary.
- A substance intended for use in the diagnosis, cure, mitigation, treatment, or
  prevention of disease.
- A substance (other than food) intended to affect the structure or any function of
  the body.
- A substance intended for use as a component of a medicine but not a device or a
  component, part or accessory of a device.

Biologic drug products are included within this definition and are generally covered
by the same laws and regulations, but differences exist regarding their manufacturing
processes (chemical process versus biological process.) Note: the definition of a drug
is originally found in 21 U.S.C. 321(g)1.

A *biologic product* is defined (in www.fda.gov/cder/drugsatfda/Glossary.htm) as
"any virus, serum, toxin, antitoxin, vaccine, blood, blood component or derivative,
allergenic product, or analogous product applicable to the prevention, treatment, or
cure of diseases or injuries." Biologic products are a subset of 'drug products'
distinguished by their manufacturing processes (biological process vs. chemical
process). *In general, the term "drugs" also includes therapeutic biological products
(the meaning used in this book).* While many biologics are treated as drugs, others
may be treated as devices and have to pass through the CDRH (Section 6.5).

This section gives a general overview of how a new drug passes through the FDA
and describes the impact of the regulatory process on the product development plan.
Figure 6.2 shows a schematic of the regulatory paths for marketing approval for
generics (copies of already approved drug molecules) and new innovative drug
molecules. While planning the regulatory path for drugs some of the following points
are useful to keep in mind:

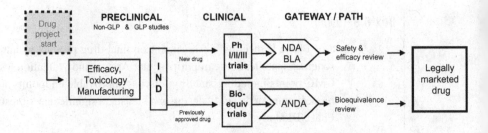

Figure 6.2 Regulatory paths to market for a drug (biological or chemical). IND = investigational new drug; NDA = new drug application; BLA = biologics license application (equivalent to an NDA for a biologic drug); ANDA = abbreviated NDA. The bioequivalence trials and ANDA path for generic medicines requires much less data and is much shorter compared to the NDA.

- Define the exact indication and clinical trial endpoints (see Chapter 5) in order to obtain the desired marketing claims for the product.
- Decide if this is an orphan drug product – see Section 6.4.
- Complete required toxicity, manufacturing and other preclinical packages for the IND application.
- Meet with the FDA before filing the IND (see Section 6.3.3) and then at designated times after to keep open discussion on the design of the clinical trials and to gain agreement on the endpoints. Key issues in these meetings are clarification of the size, scope and characteristics of the clinical trial and the manufacturing process.
- Follow guidance documents and regulatory requirements to file INDs, NDAs, biologics license applications (BLAs), etc. Smaller companies may want to work with an outside regulatory affairs consultant or legal counsel who can help make sure they are fully compliant with the requirements and ensure a smooth path to approval.
- Understand that the review team from the FDA will consist of the following:
  - Project manager
  - Medical officer
  - Chemist
  - Microbiologist
  - Statistician
  - Pharmacologist
  - Establishment/facility reviewer
  - Support Personnel

In brief, there are three main regulatory gateways and paths for a drug product to reach the market: (1) approval of an NDA, (2) abbreviated pathway with an ANDA, or 505(b)2 submission, and (3) OTC.

Note: The regulatory pathway for a new drug product is discussed first in the sections below, followed by generics, re-purposed drugs and OTC pathways.

6.3.3   Preclinical studies regulated by the FDA

Preclinical studies that are submitted to the FDA should be well described and summarized and the data should be completely transparent with detailed records. Formal preclinical studies should be carried out following the guidelines in 21 CFR 58, which are generally described as current good laboratory practices (cGLP). Early efficacy studies and other in vitro studies can be non-GLP as long as the experiments are scientifically well-designed with controls and the data are analyzed using statistically valid methods.

Before market approval of a drug, the FDA *will* inspect and audit one or more of the following: development facilities, planned production facilities, clinical trial sites, institutional review boards, and laboratory facilities in which the drug was tested in animals.

Specific preclinical studies that are evaluated by the FDA include
Efficacy data establishing the utility of the drug in treating the specific indication
Pharmacology data on drug behavior
Toxicology data on drug safety
Full characterization of the drug molecule using current knowledge and technologies
Manufacturing process for the drug

A pre-IND meeting (as described in Section 6.3.5) can be held to discuss the planned GLP toxicology studies and the plans for Phase I and II clinical studies.

Toxicology data submitted prior to the start of clinical studies should support the length of exposure to the drug planned in the Phase I studies. The primary goals of preclinical safety evaluation are: (1) to identify an initial safe dose and subsequent dose escalation schemes in humans; (2) to identify potential target organs for toxicity and for the study of whether such toxicity is reversible; and (3) to identify safety parameters for clinical monitoring. Some chronic toxicity testing studies may continue in parallel with Phase I clinical testing. Preclinical toxicology or safety studies must be carried out under strict GLP guidelines and include the following:

- Safety pharmacology studies
- Single and repeat dose toxicology studies in two species of mammals (to exceed or equal duration of human trials)
- Genotoxicity studies (in vitro studies evaluating mutations and chromosomal damage)
- Reproduction toxicity studies (animal studies)
- Other supplementary studies if safety concerns are identified (e.g. neurotoxicity studies, cardiotoxicity studies)

Regulatory guidelines for these studies can be accessed at www.ich.org/cache/compo/276-254-1.html.

Pharmacology studies are reviewed by the FDA to understand the effect and behavior of the drug molecule in the body. Pharmacokinetic (PK) studies (the half-life of a drug, exposure, etc.) and an examination of the absorption, distribution, metabolism, and excretion (ADME) behavior of a drug molecule must also be submitted to the FDA. Additionally, data on the effect of the drug on various key enzymes or on the targeted protein and physiology (pharmacodynamics; PD) are usually collected.

Chemical and manufacturing details (known as the CMC section: chemistry, manufacturing, and controls) are also provided, including a detailed description of the drugs' manufacturing process and adherence to established standards and FDA guidelines. Specifically, the drug must be manufactured in facilities that are following general cGMP and adhering to specific FDA guidance that pertains to the product at hand. Manufacturing practices are further discussed in Chapter 7.

Specific issues for preclinical testing of biological drugs (proteins, glycoproteins, and other biological macromolecules) are addressed in guidance documents published by the FDA. These concerns generally review issues related to local tolerance (most biologics are administered by injection) and immunogenicity studies. Other genetic and reproductive toxicology studies may be requested by the FDA for certain classes of drugs. Other tests that may be required for biologics include mutagenicity (e.g. growth hormones or other receptor agonists may induce or stimulate growth of malignant cells expressing the receptor). Gene therapy products may be evaluated for tissue distribution and activation, while monoclonal antibodies may be studied for the formation of neutralizing antibodies, and vaccines may be studied for the formation of non-specific antibodies that may cause immunogenicity or adverse reactions to the vaccine. However, the specific studies required for safety pharmacology endpoints are determined on a case-by-case basis with guidance documents issued for specific classes of therapeutics.

As an example, ICH guidelines on preclinical safety testing of biological drugs illustrate one such specific concern: "With some biopharmaceuticals there is a potential concern about accumulation of spontaneously mutated cells (e.g., via facilitating a selective advantage of proliferation) leading to carcinogenicity. The standard battery of genotoxicity tests is not designed to detect these conditions. Alternative in vitro or in vivo models to address such concerns may have to be developed and evaluated" (*ICH Harmonized Tripartite Guideline*, Preclinical safety evaluation of biotechnology-derived pharmaceuticals, S6, 1997).

Other points of detailed review for biologicals include an analysis of the ingredients used in the biological manufacturing process, with risks coming from host cell contaminants (cells used to make the biological drug molecule), and the detailed physical characterization of the product and its formulation and stability. In all cases, the product should be sufficiently characterized to allow an appropriate design of preclinical safety studies.

6.3.4   Filing an investigational new drug application (IND; or form FDA 1571)

An IND must be filed to conduct a clinical study on an unapproved drug (a full IND required as described below) or on an already-approved drug that is being tested for new unapproved indications, at new doses, on in novel combinations with any other drug (a much shorter IND is filed).

An IND is submitted by a sponsor-investigator, defined as one who initiates and conducts a clinical trial (if submitted by a physician to whom the manufacturer will supply the drug, then the physician is the sponsor for that IND submission). Therefore, INDs are usually of two types – a commercial IND and a research IND (commercial sponsor vs. physician or researcher driven). If an academic researcher wants to carry out a study using an approved or generic drug, they have to file an IND that describes the new studies as discussed below. In general, there are far more research INDs submitted to the FDA per year than commercial INDs. A new IND application comprises several volumes and hundreds of pages, whereas a research IND may only be tens of pages, as a research IND mainly describes rationale and the clinical protocol for the proposed study and references the content of the original IND.

The IND application can reference a prior IND for the same drug and indication if accompanied by a letter of permission from the sponsor of the original IND. Clinical studies can be initiated 30 days after the date of receipt of the IND by FDA, unless the sponsor receives requests for more information by the FDA within those 30 days. No specific approval will be sent at the expiry of the 30-day period if no concerns or questions are found by the FDA, and tacit approval to initiate clinical studies is assumed (but a quick confirming phone call or communication with the FDA is advisable).

Exemption for research use for cancer: Recent guidelines for clinical testing of cancer drugs exempt them from IND filings, provided the clinical studies (i) will not be filed with the FDA, (ii) will not be used to promote use in any indications other than approved indications and (iii) will not be used to file for expansion of indications. Then it falls on the institutional review board (IRB) to determine the risk for the study.

An IND submission must include the following (see 21 CFR 312 and guidelines for more detail):

- *Introductory statement and general investigational plan*: a summary overview of the investigational drug (formulation, dosage, administration) and the sponsor's plan for development.
- *Investigator's brochure*: contains all key nonclinical, clinical, and CMC data that support the clinical trial, providing the investigator and IRB (institutional review board) with scientific rationale for the proposed trial and should provide enough data to allow them to make an unbiased risk–benefit assessment.
- *Clinical protocol*: (described in detail in 21 CFR 312.23 and other documents) describes how clinical trials will be carried out, with an estimate of number of subjects, inclusion criteria, dosing plan, etc.; Phase II studies should have even more detail, including endpoint measurements and detailed statistical analysis methodology.
- *Chemistry manufacturing and controls (CMC) section*: must provide enough information to demonstrate the identity, quality, purity, potency, and formulation of the drug product. This includes description of substance and evidence to support its structure, overview of the manufacturing and packaging process, analytical methods

used to measure identity, potency, purity, stability, etc. Assurance must be given that the drug product is made under current cGMP.

- *Pharmacology and toxicology information*: includes characterization of toxic effect with respect to target organs, dose dependent effects etc. Single and repeat dose toxicity studies

6.3.5   Working with the FDA in formally arranged meetings

Pre-IND meeting

An IND (investigational new drug) submission is filed before beginning clinical testing. The purpose of a pre-IND meeting is to discuss the design and scope of preclinical/animal studies required to initiate clinical trials and also to discuss the design and scope of the clinical trials. This is a formal meeting requested in writing. A record of the minutes will be kept by the FDA and made available to the sponsor after the meeting. A list of objectives, outcomes, and specific questions, along with a briefing document should be sent to the FDA before the meeting. The discussion is focused on data and scientific methodology without any commercial or emotional issues being raised with the FDA. The sponsor typically would propose a study design (s) with explanations based on scientific analysis or statistical methodology and seek agreement from the FDA rather than asking the FDA to suggest protocols. This is also typically a multidisciplinary meeting, including FDA representatives from clinical, chemistry, and statistical disciplines.

End-of Phase-II meeting

This is typically the next meeting with the FDA, although an end-of-Phase I meeting can also be requested. In particular, if the program is fast-tracked, then an end-of-Phase I meeting is usually appropriate. The end-of-Phase II meeting is a critical meeting in the development process, as the study protocol, endpoints and statistical methodologies of the Phase III studies and other details on manufacturing, formulation and dosing for a successful NDA (for small molecule or synthetic drugs) or BLA (for large molecule or biological drugs) are agreed on during this meeting. The sponsor is expected to provide proof of efficacy and other data to support the Phase III design and endpoints and to show that the drug is performing a desired function. Figure 6.3 (from the FDA website) shows the various points of interaction with the FDA during the drug development process and Box 6.3 has some practical tips on preparing for these meetings.

6.3.6   New drug application (NDA) submission

The submission of the NDA is the key component in the regulatory approval process. The NDA contains clinical and non-clinical test data and analyses, drug chemistry information and description of manufacturing procedures. The NDA consists of thousands of pages of information to be reviewed by highly qualified individuals/teams from clinical, pharmacology/toxicology, chemistry, statistics,

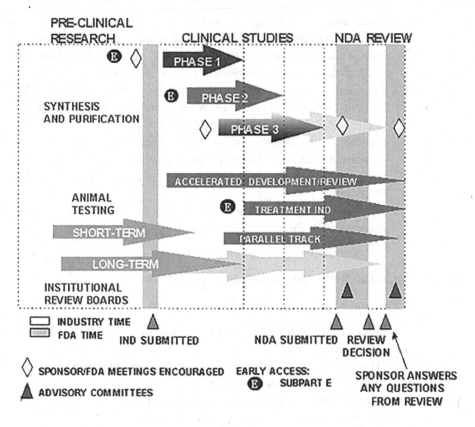

PRE-CLINICAL RESEARCH

CLINICAL STUDIES

NDA REVIEW

PHASE 1

PHASE 2

PHASE 3

SYNTHESIS AND PURIFICATION

ACCELERATED DEVELOPMENT/REVIEW

ANIMAL TESTING

TREATMENT IND

SHORT-TERM

PARALLEL TRACK

LONG-TERM

INSTITUTIONAL REVIEW BOARDS

☐ INDUSTRY TIME
☐ FDA TIME

IND SUBMITTED

NDA SUBMITTED   REVIEW
                        DECISION

◇ SPONSOR/FDA MEETINGS ENCOURAGED
▲ ADVISORY COMMITTEES

EARLY ACCESS:
Ⓔ   SUBPART E

SPONSOR ANSWERS
ANY QUESTIONS
FROM REVIEW

Figure 6.3 Chart showing FDA–sponsor interaction points (from the FDA website). Interaction points are marked in the drug development process chart above as yellow diamonds.

biopharmaceutical and microbiology disciplines. The FDA has specific and highly detailed guidelines on formatting, assembling and submitting the NDA. The NDA has over 20 sections which include the following key sections:

- *Application summary*: all review groups read this summary (50–200 pages) of the entire application and it must give a clear idea of the drug and its application. This must include the proposed package insert (label), pharmacologic class, scientific rationale, indication or intended use, potential clinical benefits, any foreign marketing history, CMC summary, nonclinical studies, human pharmacology studies, microbiology summary (for antibiotics only), data and summary of clinical results, statistical analysis, and discussion of risk–benefit relationship.
- *CMC chemistry, manufacturing, and controls section*: describe physical and chemical characteristics, methods of analysis, stability, etc. manufacturing process; drug master file (Section 6.3.5) authorization letters, drug product packaging details, etc.

- *Nonclinical pharmacology and toxicology*: summary or description of all animal and in vitro studies with the drug, including individual study reports.
- *Human pharmacokinetics and bioavailability*: data from Phase I and summary of all pharmacokinetic studies performed with overall conclusions.
- *Clinical data*: includes clinical pharmacology, controlled clinical trials, with an integrated summary of effectiveness data demonstrating substantial evidence of effectiveness for each indication claimed. Clinical trials done in foreign countries can be submitted as long as certain provisions, which are described in Section 6.3.7, are met. This section also contains statistical analysis of the data, with clinical trial reports, summary of effectiveness and safety, and a summary of risks and benefits.
- *Safety update reports*: this section is used to update new safety information that is collected while the application is in review.

6.3.7   Clinical trials done in foreign countries

Data can be gathered in foreign countries, as long as the studies are conducted:

- In accordance with the Declaration of Helsinki or local laws and protocols, whichever is stricter
- On a patient population representative of the U.S. population
- Under a standard of care similar to the United States
- With data being available for audit by the FDA
- Individual patient data is presented to the FDA

See 21 CFR 312.120 for more details on conditions under which the FDA accepts data from foreign clinical trials done under an IND and those done without an IND: www.accessdata.fda.gov/scripts/cdrh/cfdocs/cfcfr/CFRSearch.cfm?fr=312.120.

6.3.8   Drug master files

Master files are used by CDER/CBER in a similar manner to the device master files (Section 6.5.7). A drug master file (DMF) is a voluntary submission to the Food and Drug Administration (FDA) that may be used to provide confidential detailed information about facilities, processes, or articles used in the manufacturing, processing, packaging, and storing of one or more human drugs. The information contained in the DMF may be used to support an investigational new drug application (IND), a new drug application (NDA), or an abbreviated new drug application (ANDA) or another abbreviated approval pathway, the 505(b)2 application. DMFs are generally created to allow a party other than the holder of the DMF to reference material without disclosing to that party the contents of the file. For example, an IND or NDA sponsor can refer to a DMF submitted by a contracted manufacturing facility to support their application. Further details and guidance are available at www.fda.gov/cder/guidance/dmf.htm.

Drug master files are classified by content into five types by the FDA:

- Type I: Manufacturing Site, Facilities, Operating Procedures, and Personnel
- Type II: Drug Substance, Drug Substance Intermediate, and Material Used in Their Preparation, or Drug Product
- Type III: Packaging Material
- Type IV: Excipient, Colorant, Flavor, Essence, or Material Used in Their Preparation
- Type V: FDA Accepted Reference Information

6.3.9 Abbreviated pathway for duplicate drugs (generic or biosimilar drugs) – the ANDA

When a company wishes to market a copy of a drug that has been on the U.S. market after its patent has expired, an abbreviated NDA (ANDA; or ABLA for biologics) is filed. The ANDA must be filed for indications that were already approved for the original drug molecule with no adverse findings since approval. This ANDA filing requires the manufacturer of the copy drug to certify the original patent(s) is expired and to demonstrate biological and pharmaceutical equivalence to the original drug. The label granted is the same as the approved reference drug. This filing is only approved after the generic manufacturing company has shown that the key patents of the reference drug are invalid or expired.

The FDA maintains the key patents and a description of therapeutic equivalence standards for each approved drug in a regularly updated book called the *Orange Book* (for the color of its cover). Therapeutic equivalence is demonstrated by showing bioequivalence (same rate and extent of absorption) and by pharmaceutical equivalence (same dose, dosage form, strength).

For *biosimilars*, or copies of approved biological products (e.g. proteins such as monoclonal antibodies), the 351(k) pathway for approval (section 351[k] of the PHS Act 28) provides a similar path as the ANDA for small molecule drugs, with certain specific checkpoints. These biological drugs are complex molecules that are much more difficult to characterize compared to small molecule synthetic chemical drugs. Hence, the requirements from the FDA and European Medicines Agency (EMA) require additional preclinical and quality studies to be done to ensure the similarity between the biological drugs and pharmaceutical drugs, replicating the original molecule manufacturer's data and adding to it. The biosimilar manufacturer usually has to replicate the Phase I clinical studies carried out by the original drug in entirety but has the benefit of a much smaller pivotal clinical study to show efficacy. However, the extent of further efficacy study in patients is dependent on the high sensitivity risks determined by the FDA and analysis of the structural and functional differences from the reference molecule. [For further information on biosimilars, see Royzman and Shah (2020)].

6.3.10   An abbreviated approval pathway for reformulated versions of drugs 505(b)(2)

A 505(b)(2) application is an NDA submitted under section 505(b) and approved under section 505(c) of the FD&C Act that contains full reports of investigations of safety and effectiveness, where at least some of the information required for approval comes from studies not conducted by or for the applicant. In contrast to an ANDA, a 505(b)(2) application allows greater flexibility as to the characteristics of the proposed product and may not necessarily be rated therapeutically equivalent to the listed drug it references.

505(b)(2) applications are usually made for re-purposed drugs where a change in formulation or method of dosage provides some significant benefit to the patients, usually by reducing side effects or providing a more convenient dosage. This pathway results in significant costs savings in the nonclinical development area and also may allow a streamlined clinical trial process for approval. [For additional information, see Salminen et al. (2019).]

6.3.11   Regulatory pathway for over-the-counter drugs

Over-the-counter (OTC) drug products are those drugs that are available to consumers without a prescription. There are more than 80 classes (therapeutic categories) of OTC drugs, ranging from acne drug products to weight control drug products. As with prescription drugs, CDER oversees OTC drugs to ensure that they are properly labeled and that their benefits outweigh their risks. There are more than 100,000 OTC drug products marketed, encompassing about 800 significant active ingredients. The FDA maintains a list of OTC drug monographs. These monographs are a kind of "recipe book" covering acceptable ingredients, doses, formulations, and labeling. Products conforming to the monograph do not need FDA clearance to be marketed. New ingredients entering the OTC marketplace for the first time, especially those that had previously been approved through NDAs must apply for the switch to OTC through the NDA process.

6.3.12   Post-market clinical studies (Phase IV) and safety surveillance by the FDA

Phase IV studies are post-marketing studies that may be imposed upon a pharmaceutical firm as a condition for drug approval. Phase IV investigations that are voluntarily conducted by the industry are typically used to support additional indications or to extend the life cycle of the product.

The FDA has an active surveillance program that requires sponsors to report any serious side effects not listed on the label within 15 days of learning of such events, and to submit quarterly safety reports for 3 years post-approval. The FDA has a separate office that specifically tracks adverse events in the post-approval marketplace through various sources, the manufacturer being one of them. The reach of this program and the sources it uses to monitor for safety in the post-approval phase are shown in Figure 6.4.

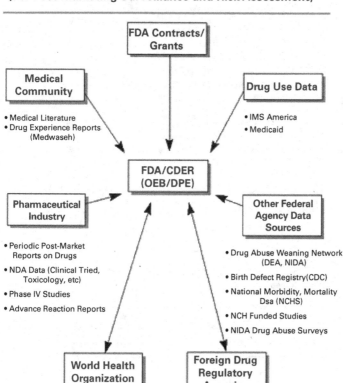

Drug Experience/Epidemiologic Sources Available to FDA (For Post-Marketing Surveillance and Risk Assessment)

Figure 6.4 Safety surveillance program sources. (Chart from the FDA CBER website.)

6.3.13 Schematics of IND, NDA, and ANDA review processes

Figures 6.5–6.7 are excerpted from the FDA website and schematically show the CDER review process for IND, NDA and ANDA submissions. The flow charts in Figures 6.5–6.7 are largely self-explanatory.

6.3.14 Speeding up access to drugs

Fast Track

The following information concerning the Fast Track process is from the FDA website, www.fda.gov/oashi/fast.html:

The Fast Track Drug Development Program facilitates the development and expedites the review process of drugs intended for the treatment of a serious or life-threatening condition. The Fast Track drug program was first initiated in section 112 of the FDA Modernization Act of 1997.

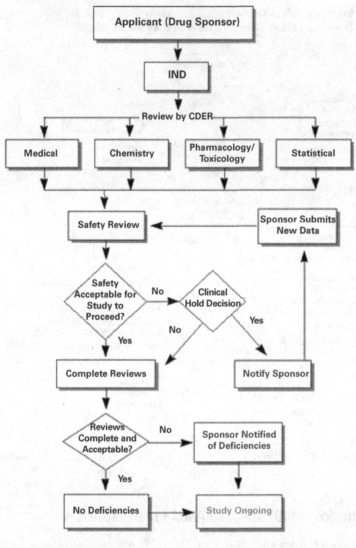

Figure 6.5 IND review process.
(From the FDA CDER website.)

A drug can receive Fast Track designation if it meets three criteria:

(i)   the drug must be targeted for use by a person that has a serious or life-threatening condition

(ii)   the drug must be intended to treat a serious condition, and

(iii) the drug must have the potential to address unmet medical needs.

A drug that receives Fast Track designation is eligible for some or all of the following:

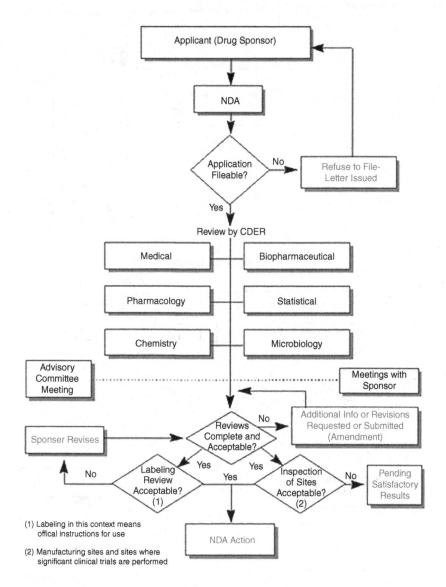

Figure 6.6 NDA review process.
(Chart from the FDA CDER website.)

- More frequent meetings with FDA to discuss the drug's development plan and ensure collection of appropriate data needed to support drug approval
- More frequent written correspondence from FDA about issues such as the design of the proposed clinical trials
- Eligibility for Priority Review or Accelerated Approval, i.e., approval on an effect on a surrogate, or substitute endpoint reasonably likely to predict clinical benefit

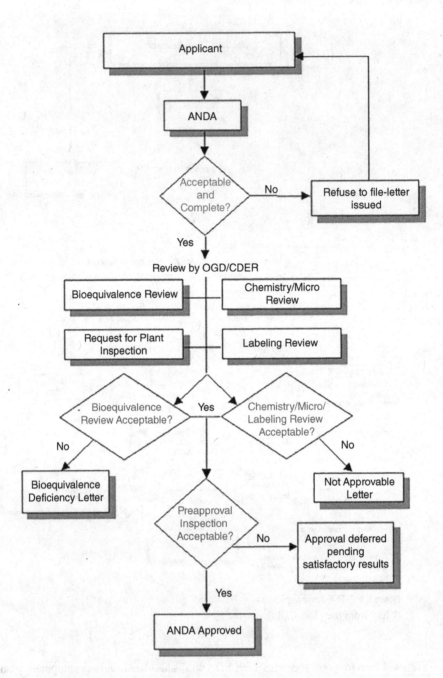

Figure 6.7 Generics review process.
(From the FDA CDER website.)

- Eligible for Rolling Review, which means that a drug company can submit completed sections of its New Drug Application (NDA) for review by FDA, rather than waiting until every section of the application is completed before the entire application can be reviewed. NDA review usually does not begin until the drug company has submitted the entire application to the FDA, and
- Dispute resolution if the drug company is not satisfied with an FDA decision not to grant Fast Track status.

Fast Track designation must be requested by the drug company. The request can be initiated at any time during the drug development process. FDA will review the request and make a decision within sixty days based on whether the drug fills an unmet medical need in a serious disease.

Once a drug is designated for Fast Track review, the FDA works closely with the sponsor, with frequent reviews of the data and protocols, often allowing parallel clinical and continuing preclinical development. In addition, most drugs that are eligible for Fast Track designation are likely to be considered appropriate to receive a Priority Review (see below).

Example of fast track drug development: In November 2004 a drug called Tarceva was approved by the FDA though the Fast Track program. Tarceva is used for the treatment of people with locally advanced or metastatic non-small cell lung cancer. In early data from clinical trials, Tarceva showed that people taking the drug had two months additional survival (from five months to seven months). Although this increase in improvement may seem rather small, the potential of the drug to bring additional benefits and improvements was enough for acceptance into the Fast Track program.

Accelerated approval

The following information concerning the accelerated approval process is from the FDA website, www.fda.gov/oashi/fast.html:

The accelerated approval process allows earlier approval of drugs that treat serious diseases, and that fill an unmet medical need, based on a surrogate endpoint [also discussed in Section 5.8.3]. The FDA bases its decision on whether to accept the proposed surrogate endpoint on "adequate and well-controlled" studies that demonstrate the effect of the drug and may usually require longer term post-approval confirmatory studies. For example, instead of having to wait to learn if a drug actually can extend the survival of cancer patients, the FDA might now approve a drug based on evidence that the drug shrinks tumors because tumor shrinkage is considered reasonably likely to predict a real clinical benefit. In this example, an approval based upon tumor shrinkage can occur far sooner than waiting to learn whether patients actually lived longer. The drug company will still need to conduct studies to confirm that tumor shrinkage actually does predict that patients will live longer. These studies would be carried out as Phase IV confirmatory trials. The Fast Track Program is most useful in clinical product development up to the submission of the final NDA application.

Priority review

The following information concerning priority review is from the FDA website, www .fda.gov/patients/learn-about-drug-and-device-approvals/fast-track-breakthrough-ther apy-accelerated-approval-priority-review:

Prior to approval, each drug marketed in the United States must go through a detailed FDA review process. In 1992, under the Prescription Drug User Act (PDUFA), FDA agreed to specific goals for improving the drug review time and created a two-tiered system of review times – Standard Review and Priority Review.

Standard Review is applied to a drug that offers at most, only minor improvement over existing marketed therapies. The 2002 amendments to PDUFA set a goal that a Standard Review of a new drug application be accomplished within a ten-month time frame.

A Priority Review designation is given to drugs that offer major advances in treatment, or provide a treatment where no adequate therapy exists, as seen with drugs designated Breakthrough Therapy. Priority Review status can apply both to drugs that are used to treat serious diseases and to drugs for less serious illnesses. A Priority Review designation means that the time it takes FDA to review a new drug application is reduced. The goal for completing a Priority Review is six months.

The distinction between priority and standard review times is that additional FDA attention and resources will be directed to drugs that have the potential to provide significant advances in treatment.

Such advances can be demonstrated by, for example:

- evidence of increased effectiveness in treatment, prevention, or diagnosis of disease;
- elimination or substantial reduction of a treatment-limiting drug reaction;
- documented enhancement of patient willingness or ability to take the drug according to the required schedule and dose; or
- evidence of safety and effectiveness in a new subpopulation, such as children.

A request for Priority Review must be made by the drug company. It does not affect the length of the clinical trial period. FDA determines within 45 days of the drug company's request whether a Priority or Standard Review designation will be assigned. Designation of a drug as "Priority" does not alter the scientific/medical standard for approval or the quality of evidence necessary. For a drug that has a potential hundreds of millions of dollars in annual revenue, four months of additional revenue can be worth tens or hundreds of millions of dollars.

6.3.15   Market exclusivity for new drugs and the Hatch–Waxman Act 1984

A 5-year period of exclusivity is provided by the Federal Food, Drug, and Cosmetic Act under section 505(c)(3)(E) and 505(j)(5)(F). Exclusivity is available for new chemical entities (NCEs), which by definition are innovative, and for significant changes in already approved drug products, such as a new use. A 5-year period of exclusivity provides the holder of an approved new drug application limited protection from new competition in the marketplace for the innovation represented by its approved drug product. This exclusivity was specifically provided in the 1984 Hatch–Waxman amendments to the Food, Drug, and Cosmetic Act, to encourage R&D investment and innovation and also *to speed entry of generic drugs* into the market. As a balance against that exclusivity, ANDAs, filed by generic manufacturers, could now reference NDAs containing data that had not been developed by the generic manufacturers. No ANDA may be submitted during the 5-year exclusivity period, unless they contain a certification of patent invalidity or noninfringement, in which case they can be submitted after 4 years. Additionally, the FDA grants a specific 6-month exclusivity after expiry of the patent term, if the drug manufacturer carries out studies to test the drug in pediatric populations.

6.3.16   Drugs: helpful FDA websites and the Electronic Orange Book

The list of websites in Table 6.3 provides a starting point for exploration of specific regulatory processes or issues.

The *Orange Book*: Information on drugs that have been approved is listed in the *Orange Book & Supplements/Electronic Orange Book* (actual title: *Approved Drug Products with Therapeutic Equivalence Evaluations*).

[Reference: www.accessdata.fda.gov/scripts/cder/ob/index.cfm.]

The *Orange Book* is composed of four lists: (1) approved prescription drug products with therapeutic equivalence evaluations; (2) approved over-the-counter (OTC) drug products for those drugs that may not be marketed without NDAs or ANDAs because they are not covered under existing OTC monographs; (3) drug products with approval under Section 505 of the Act administered by the Center for Biologics Evaluation and Research; and (4) a cumulative list of approved products that have been discontinued from marketing, have had their approvals withdrawn for other than safety or efficacy reasons subsequent to being discontinued, have never been marketed, are for exportation, or are for military use. This publication includes indices of prescription and OTC drug products by trade or established name (if no trade name exists) and by applicant name (holder of the approved application). The *Orange Book* also has an addendum with patent and exclusivity information on each drug.

6.4   Orphan drugs

The FDA's Office of Orphan Products only designates orphan drugs (that treat rare diseases affecting U.S. patient populations less than 200,000) and provides grants and guidance, but does not influence the approval and review process of CDER or CBER. An orphan drug designation is given to a specific drug molecule for a specific indication (disease). The Orphan Drug Act (1983) gives 7 years of market exclusivity for the first such orphan (designated) drug to gain market approval for a specific indication. This market exclusivity is given to the sponsor as an incentive to develop treatments for rare diseases. Additional incentives include tax credits for clinical research undertaken by a sponsor in support of their application. The success of this Act is seen by the fact that fewer than 10 such products were approved in the period 1973–1983 while over 200 orphan drugs were approved from 1983 to the present. This Act has provided a commercialization pathway for drug molecules whose patents have expired and had promise to treat these rare diseases but had no financial incentive for entrepreneurs to take risks and invest in development. On the other hand, the exclusivity period granted to such drugs and the very low patient count for these diseases has also led to very high pricing strategies.

The European Union has since also put in place an Orphan Drug program with a 10-year market exclusivity. Countries such as Australia and Japan have similar programs, with slight variations, and others are considering such legislation.

6.5   Devices: regulatory pathways and development considerations

This section presents a general understanding and overview of the device review processes focusing largely on the U.S. FDA (Figure 6.8). Specific mention is made of European or other countries' regulatory body practices in certain sections, especially where they deviate significantly from general process with the FDA. In vitro diagnostics (most of which are regulated as devices) are also covered in this section where relevant but discussed separately in more detail in Section 6.6. Also discussed here are specific regulatory controls that the FDA will use to review the data generated during preclinical and clinical testing of the devices. In particular, it is extremely important to read the general guidance documents or special

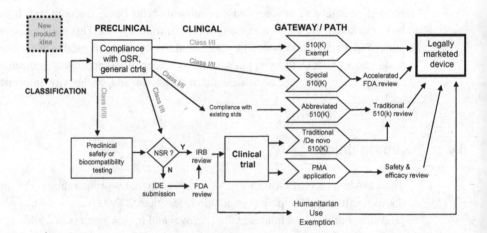

Figure 6.8 Regulatory paths to market for a medical device. QSR = quality system regulation; NSR = non-significant risk study; IRB = institutional review board.

controls for devices and to address those issues thoroughly during testing of the devices.

In brief there are four possible regulatory pathways for medical devices and an additional one for diagnostics to get to market : (1) exempt from FDA review; (2) 510 (k) clearance process for devices that are similar in action to another approved device; (3) de novo process for new to the market devices (without a predicate) whose safety profile and technology is well understood; and (4) PMA review for new to the market devices deemed high risk. For diagnostics – if they are exempt from (2) and (3), they have to go through CLIA categorization. These are explained in greater detail in the following sections.

A 510(k) (described in greater detail in Section 6.5.9) is a premarketing submission made to FDA to demonstrate that the device to be marketed is as safe and effective, that is, substantially equivalent (SE), to a legally marketed device that is not subject to premarket approval (PMA). A PMA (described in Section 6.5.10) is an application submitted to FDA to request approval to market, or continue marketing, a Class II or more typically, a Class III medical device. PMA approval is based on scientific evidence providing a reasonable assurance that the device is safe and effective for its intended use or uses.

6.5.1 Step1: determine the jurisdiction of the FDA center – is it a device?

Medical devices range from simple tongue depressors to complex programmable pacemakers with micro-chip technology, X-ray machines, and include in vitro diagnostic products, such as general-purpose lab equipment, reagents, and test kits, which may include monoclonal antibody technology. Diagnostics are discussed in a separate section below, but are usually classified as a type of medcial device, as are stand-alone software products (software as medical device – SaMD).

A device is defined by its primary mode of action in the indication specified by the company. For example, a wound dressing containing anti-bacterials has as its primary mode of action the creation of a physical barrier for the wound and hence is not regulated from the standpoint of the anti-bacterials which only enhance the primary mode of action. On the other hand, if the wound dressing is made of a biodegradable matrix whose sole purpose is to deliver the anti-bacterial compound, the dressing is then reviewed as a drug.

The specific definition of a device is detailed in section 321(h) of the Federal Food and Drug Control Act [21 U.S.C. 321(h)]:

an instrument, apparatus, implement, machine, contrivance, implant, in vitro reagent, or other similar or related article, including a component part, or accessory which is:

* recognized in the official National Formulary, or the United States Pharmacopoeia, or any supplement to them,
* intended for use in the diagnosis of disease or other conditions, or in the cure, mitigation, treatment, or prevention of disease, in man or other animals, or

> **Box 6.4** Drug or device?
>
> Here is an example of why it is important to be very careful in approaching this process of classification: A company (unnamed by request) that was already marketing several catheters, designed a new system with which a bolus of CO_2 gas could be injected into a blood vessel. This bubble of gas displaced the blood and gave good images of large vessels and higher resolution images of the small vessel during imaging. The submission for requesting classification of the device under a Class II inadvertently described the intention of use of the gas as a "contrast media."
>
> The FDA wrote back that this system would have to be reviewed as a drug as all "contrast media" were automatically reviewed as drugs. This reclassification would have led to an increase of $5\times$ for the project costs and $3\times$ for the time to market.

* intended to affect the structure or any function of the body of man or other animals, and which does not achieve any of its primary intended purposes through chemical action within or on the body of man or other animals and which is not dependent upon being metabolized for the achievement of any of its primary intended purposes.

If the product fits this definition, it will be regulated as a medical device and is subject to premarketing and post-marketing regulatory controls.

6.5.2   Step 2: classify the medical device – what controls and regulations apply?

In order to understand the impact of the regulatory process on your product's commercialization path, the next step is to classify the device by its indication and intended use and the risk it poses to patient/user if it malfunctions or fails. This classification of devices will indicate the type of submission to be made to the FDA to commercialize the device.

Class I   General Controls apply
           Exemptions / Without Exemptions
Class II   General Controls and Special Controls apply
           Exemptions / Without Exemptions
Class III General Controls apply and Premarket Approval required

Even the lowest risk devices (Class I) are subject to certain general controls. The regulatory controls increase with perceived risk (Class III being highest risk). Enter the keyword or device name into the classification database at the FDA website (www .accessdata.fda.gov/scripts/cdrh/cfdocs/cfpcd/classification.cfm) to see how other similar devices have been classified. If the device is completely novel, you will still look for descriptions of other devices with similar indications, methods of use or other such comparison within the medical specialty of intended use to gain a better idea of how the FDA will classify the device – both for its regulation number and the risk class.

Box 6.5 Product classification database example

Device	Blade, scalpel
Regulation description	Manual surgical instrument for general use.
Regulation medical specialty	General & Plastic Surgery
Review panel	General & Plastic Surgery
Product code	GES
Submission type	510(k) Exempt
Regulation number	878.4800
Device class	1
GMP exempt?	No

This result of a search of the classification database at the CDRH shows that a scalpel blade has a device regulation number 21 CFR 878.4800, three- letter product code GES, and is a Type I device exempt from reporting to the FDA before marketing, but not exempt from showing that GMP processes for manufacturing are in place.

The example in the sidebar (Box 6.5) shows the detailed classification of a scalpel blade (a scalpel handle has a separate classification) in the FDA product database.

6.5.3 Step 3: determine the specific path and marketing application required to be submitted

Class I *exempt* devices do not need to get approval before marketing but still need to adhere to some general controls (guidelines for making, storing, packaging, and selling). Class I and Class II devices that are not exempt can go through a premarket notification 510(k) submission to get *clearance* and most Class III devices will go through a premarket approval (PMA) process to get *approval* and will be subject to special controls, such as FDA guidance documents, FDA accepted international standards, and the Quality System Regulation (QSR).

The 510(k) submission is dependent on first locating a substantially equivalent device (predicate device) already on the market and then filing data that demonstrates such equivalence of the new device. Substantial equivalence has to be established both in the technological characteristics and intended use of the new device in comparison to a chosen predicate device.

To locate predicate 510(k) devices, start with a known device currently on the market, and enter the manufacturer name in the product classification database (see Section 6.5.2). Note the three-letter code (product code) for the classified device. Enter this product code again in the 510(k) products database (www.fda.gov/search/databases.html) search function and you will see a list of all possible predicate devices.

Table 6.4 Comparison of the PMA and 510(k) processes

Characteristic	PMA submission	510(k) submission
Time to collect data	Several years	Several months
Submission size	Several thousand pages	Much less
Manufacturing details	Process, methods details required	Typically not required
Preapproval inspection of device manufacturing facility	Required	Not required
Clinical trial site review	Often required	Not required
Review time	1 year	90 days
Post approval annual reports	Required	Not required
Submission availability through freedom of information (FOI)	Not available by FOI	Available
Scientific advisory panels convened to assist FDA in review	Sometimes	Rarely

If a clinical trial is needed to demonstrate substantial equivalence, an investigational device exemption (IDE) must be applied for and approval obtained from the FDA before starting such trials. On submission of the appropriate data in a 510(k) submission, the FDA will review the data (on average a 90–120-day process) and if they agree, the new device will receive *clearance* (terminology and process distinct from PMA "approval") for marketing under regulation 510(k). If the FDA determines, in response to a 510(k) application, that the device is NOT substantially equivalent (NSE) to a previous legally marketed device, then the device is classified as Class III and will have to go through the PMA process. The sponsor can submit a request to reclassify as Class I or II and go through the de novo 510(k) application. The sponsor can also directly submit a De Novo classification request to the FDA without having to submit a 510(k) application. Table 6.4 highlights the differences between a 510(k) and PMA pathway for regulatory approval/clearance.

A premarket application (PMA) process is necessary when the new device is not substantially equivalent to any other device that has been cleared through 510(k) process. The PMA process is more complex and will usually take two formal submission steps (refer to Figure 6.8) depending on the perception of significant risk (SR) or non-significant risk (NSR) prior to clinical testing of the device. The first step (if SR established) is the submission and approval of an investigational device exemption (IDE) to allow clinical trials to be conducted. The data in an IDE must assure the reader that the device, all preclinical testing, and the investigational plan are described and provide adequate justification for the initiation of the clinical trial. The FDA website also highlights the most common problems that lead to rejection of an IDE (www.fda.gov/medical-devices/investigational-device-exemption-ide/ide-applica tion), some of which are: inadequate support for conclusions or clinical trial designs, failure to identify all risks, and failure to develop adequate monitoring methods in the clinical trial investigational plan. The clinical trials should ascertain the safety and efficacy of the device in its intended use. The data will be analyzed and submitted in a PMA to the FDA. On review of the safety and efficacy data, the FDA *approval* will establish market indication and label along with details on post-approval studies or follow up. Typical review time is 1 year.

6.5.4 Device regulatory pathways in the European Union (EU)

In the EU, every marketed medical device must carry a Conformité Européenne (CE) mark indicating that it conforms to relevant directives set forth in the (2017) *EU Medical Device Regul*ation (MDR), which is being phased in through the year 2020.

A device with a CE mark can be marketed in any EU member state. The manufacturer can also certify compliance and apply a CE mark if they consider their device to be non-implantable and low risk (Class I). High-risk devices (Class IIa, IIb, or III, similar to the FDA risk-level classifications) must undergo a more extensive review.

The EU has a complex set of regulations that guide high-risk medical device approvals. An application can be filed in any EU member state and is then reviewed by a "notified body" (NB) in that state. These notified bodies are third-party reviewing organizations accredited by European Commission (EC) authorities or the state's Competent Authority, or health agency, to assess and assure conformity with requirements of the relevant MDR directives. NBs are private companies that contract with manufacturers to supply these certifications for a fee. Once the NB agrees that the device meets requirements for conformity, the NB issues a CE mark, and the device can then be marketed in EU member states.

Until recently, the CE mark authorized marketing "without further controls and no further evaluation," but the 2017 MDR has tightened requirements for approval of devices based on their similarity to previous predicate devices. Most devices will need clinical data to be presented for approval. A major impact of the new MDR regulations is that sponsors are required to carry out "proactive post-market surveillance" of their devices.

Note: Since the EMA mainly assesses devices for safety rather than efficacy, the path to market in the EU is seen as slightly less onerous than through the U.S. FDA. However, the new MDR regulations and decentralized review processes that vary with each member country seem to make it easier for a medical device to get to 510(k) FDA approval and revenues rather than the CE Mark.

6.5.5 Working with the FDA in formal meetings

Devices

A pre-submission (of the IDE) informal meeting is welcomed. This meeting should be used to inform the FDA of new technologies, verify the approach to approval, and to evaluate the clinical protocol and any preclinical studies that may need to be done.

Two formal meetings, the proceedings of which are binding and recorded, are encouraged before launching clinical trials for devices [also see FDA Guidance Document: Early Collaboration Meetings, 2001, at www.fda.gov/regulatory-informa tion/search-fda-guidance-documents/early-collaboration-meetings-under-fda-modern ization-act-fdama-final-guidance-industry-and-cdrh]. Depending on FDA's familiarity with the device and the degree to which relevant precedents exist, one or more informal meetings may be useful before a Determination or Agreement meeting can be productive.

(1) Only for devices requiring a PMA (Pre-Market Application; requiring submission of detailed clinical tests for approval):

A Determination Meeting, described in 21U.S.C.360c(a)(3)(D), is intended to provide the applicant with the Agency's determination of the type of valid scientific evidence that will be necessary to demonstrate that the device is effective for its intended use. As a result of this meeting, FDA will determine whether clinical studies are needed to establish effectiveness and, in consultation with the applicant, determine the least burdensome way of evaluating device effectiveness with a reasonable likelihood of success. The applicant can expect that the FDA will determine if concurrent randomized controls, concurrent non-randomized controls, historical controls, or other types of evidence will be acceptable. (quoted from the FDA guidance document)

(2) For devices requiring PMA or 510(k):

The Agreement Meeting is open to any person planning to investigate the safety or effectiveness of a class III product or any implant. The purpose of this meeting is to reach agreement on the key parameters of the investigational plan (see 21 CFR 812.25), including the clinical protocol. The Agreement Meeting is available to submitters of 510(k)s for eligible devices.

Diagnostics

Interactions with the FDA, CLIA, and OIVD offices are similar to those described above, as most diagnostics are classified as devices. The timing of interactions will differ based on development path and type of diagnostic. See Office of IVD (OIVD) website (Section 6.5.13, Table 6.5) for more details.

6.5.6   General controls and exempt devices (FDA and EMA)

Some Class I devices may also be exempt from premarket notification [510(k)] which means that there is no need to inform the FDA before launching onto market. However, these devices are not exempt from other *general controls*, meaning that they must be

- Manufactured under a quality assurance program
- Be suitable for the intended use
- Be adequately packaged and properly labeled

Additionally, as part of the general controls,

- The production and distribution company must be registered with the FDA and
- Device listing forms must be on file with the FDA.

The European Commission regulations for medical devices (MDR) make it imperative for every company to have a quality system in place even if the device is a Class I device which does not require any notified body review. In addition, the companies need to have a designated qualified person in charge of regulatory activities.

6.5.7 Preclinical considerations – special controls and QSR for Class II and III devices

Special controls are specific issues that the FDA has identified for a type of device (e.g. all scalpel blades) and would like to see addressed by the sponsor when submitting either a 510(k) or a PMA for any product that falls under a similar classification. The FDA issues guidance documents detailing these concerns. These special controls guidance documents must be read and thoroughly addressed in preclinical design and testing of the product and in clinical trial designs. An example of special controls guidance is given in Box 6.7.

The QSR (Quality Systems Regulation) (21 CFR Part 820) is the device equivalent of the pharmaceutical Good Manufacturing Process (GMP) regulations. However, unlike the GMP regulations which primarily affect product manufacturing processes, QSR impacts the preclinical stage significantly as it includes Design Controls (21 CFR 820.30) which apply to preclinical studies for all Class II and III (and some Class I) devices as soon as they move beyond initial concept and feasibility testing studies. The Design Controls section of the Quality System Regulation outlines the requirements that each manufacturer of any Class II or Class III device, and certain Class I devices, must meet when designing such products or related processes, and when changing existing designs and processes.

The Design Control process has multiple components and steps (e.g. list of design inputs; design verification process) that require the company to formally manage and document the entire design and development process, as standard operating procedures (SOPs) that comply with and specifically addresses regulatory requirements. All relevant activities must be documented in the design history file (DHF) which must be regularly updated during the development process and filed with the FDA. Most importantly for regulatory purposes, *design review activities and results* must be documented and personnel participating in the review must be identified. This NPD plan must be in place (with modifications recorded in the DHF) before development begins. Internal audits must be performed to continuously improve and monitor quality systems.

It is important to note that despite the emphasis on regulatory requirements and compliance in the section above, the importance of having a quality system in place with appropriate documentation of design steps, formalized review processes, and so on can help a young company with a first-time product to develop a more reliable product with a consistent production output. This reliability and consistency established with the help of design and production controls will eventually provide safer products for patients and also save money in the long run for the manufacturer by improving yields and reducing the risk of product returns or lawsuits.

The QSR also requires the firms to demonstrate that specific management systems are in place for controlling the device development, production, and marketing processes. These control systems are discussed further in Chapter 7 and in Appendix 7.2. The final stages of development (process and production controls) also require documentation of the production history of a particular batch in the device history record (DHR) and the documentation of the device specifications and manufacturing procedures in the device master record (DMR). The DMR is the term used in

---

Box 6.6 Design history file (DHF)

What type of records are maintained in a DHF?

A DHF must be created and maintained for each device made under design controls. The firm should first put into place formal procedures that detail how the DHF will be established and maintained. The DHF should contain the following records (not a comprehensive list) or references to specific records and their location:

- The design and development plan along with evidence showing that the device was designed in accordance with this plan and in compliance with Subpart C of the Regulations.
- Design input documents (see Section 5.10.1)
- Engineering notebooks, which contain relevant information recorded after Design Control
- Risk analyses documents
- A design history file (DHF) index and copies of controlled documents used during the design process, including records of product builds and testing
- Design output
- Pre-production design change control records
- Design review and transfer records
- The initial device master record (DMR)
- Issues tracking matrix

---

the QSR for all of the routine documentation required to manufacture devices that will consistently meet company and regulatory requirements through the life of the product. The quality system must address process risks and ensure that adequate controls are in place around identified risks, including process validation, process controls, and quality controls. Adequate record-keeping and documentation will demonstrate process and product are operating as per expectations. The quality system covers the design process as well and supporting functions of product realization. Thus, the quality system, when implemented, will usually encompass the entire organizational structure in particular with an emphasis on management responsibilities for review of the quality system outputs.

The value of putting such a system in place from the beginning stages of device concept development will help the company in compliance to regulations, but more importantly, it is worth noting that this quality system and the attention and documentation to production process validation, quality inspections, and so on will pay off well when production is scaled up. The consistent application of the quality system will likely result in significant savings from high yields and low rate of field returns due to better device quality designed in from the start.

The recording requirements of the QSR (DHF, DHR, and DMR) thus regulate the device from concept to development. The goal of the QSR is to create a set of self-correcting systems that will reliably produce a high-quality design, with a controlled and predictable production process.

Box 6.7. Special Controls example guidance for industry and FDA staff

Class II Special Controls Guidance Document: Dental Bone Grafting Material Devices

(The following content is an extract from the original document.)

Table of Contents

1. Introduction	6. Risks to Health
2. Background	7. Material Characterization
3. The Content and Format of an Abbreviated 510(k) Submission	8. Biocompatibility
4. Scope	9. Sterilization
5. Device Description	10. Labeling

Risks to Health

In the table below, FDA has identified the risks to health generally associated with the use of the bone grafting material device addressed in this document. The measures recommended to mitigate these identified risks are given in this guidance document, as shown in the table below. You should also conduct a risk analysis, before submitting your 510(k), to identify any other risks specific to your device. The 510(k) should describe the risk analysis method.

Identified Risks	Recommended Mitigation Measures
Ineffective Bone Formation	Section 7 – Material Characterization
Adverse Tissue Reaction	Section 8 – Biocompatibility
Infection	Section 9 – Sterilization
Improper Use	Section 10 – Labeling

Sections refer to additional parts of the document as above.

Note: The QSR manual on the FDA website is recommended reading.

Similarly, the European Medicine Agency (EMA) requires each manufacturer of medical devices and in vitro diagnostics to establish a quality management system (QMS) and have a corresponding responsible person designated within the company. The directives in the MDR (Medical Device Regulation) and IVDR (In Vitro Diagnostic Regulation) lay out the QMS requirements in detail, which, in principle, are not much different in scope and intent from the FDA's QSR requirements.

6.5.8   The use of master files

Master files are data submitted to the FDA by non–sponsor organizations to allow the FDA to review technical or scientific data from a process or some technical test related to a sponsor's IDE or PMA application. The sponsor (academic or other institution) can reference the master file in their submission (with the permission of the other party) while still maintaining secrecy between the sponsor and the other party that has filed the master file.

Information in a master file may be incorporated by reference in a client's PMA, 510(k), or IDE, or other submissions to FDA. The use of information in a master file can only be authorized by the master file holder or by a designated agent.

6.5.9   510(k) submission type and content and CE technical documentation

There is no specific 510(k) form but instead a format for the submission is described in 21 CFR 807 and in multiple guidance documents published by the FDA on their website. The term "510(k)" refers to the original section of the CFR that described the process of clearing a device that demonstrates *substantial equivalence* to a device already on the market before 1976 or another 510(k) cleared device on the market currently (a *predicate device*). As the 510(k) process is much shorter than the PMA route, most companies try to qualify their devices into the 510(k) pathway. *However, as this process may not generate adequate clinical data to satisfy insurance companies or other payers, often manufacturers will benefit by performing more extensive clinical studies to show the specific clinical benefits of the new device or technology* (see Chapter 8 for detailed discussion).

The first step is to find a predicate device that is most similar (substantially equivalent) to the device intended for submission through the 510(k) program. The next step is to read all guidance documents published on this type of device by the FDA. The FDA has also published detailed guidance on the content of the application on their website (see References and Additional Readings at the end of this chapter and Table 6.5 for website links).

In order to show substantial equivalence to a predicate device, the sponsor has to show

- that the new device has an intended use that is substantially equivalent to a predicate device

AND

- that the technological characteristics of the new device are substantially equivalent to the predicate device.

Typically, a device may be used for multiple indications. Each indication for use must be cleared with its own 510(k). Design inputs, required as part of the design controls, typically contain the information that establishes the technological characteristics in comparison to predicate devices.

The four types of 510(k) submissions are

1. Traditional 510(k) – if the new device is not a modification of one of the sponsor's own previously cleared devices and does not need to conform to any special control

or guidance document from the FDA. The FDA has 90 days to review this submission.

2. Special 510(k) – if the sponsor has modified their own 510(k) cleared device, but has not added a new indication or altered fundamental scientific technology of the device. The FDA has 30 days to review this submission.

3. Abbreviated 510(k) – if the new device has to conform to a special control or guidance, a declaration of conformance has to be included in this abbreviated 510 (k), stating that the device meets the referenced standards. Detailed test reports, like those required in a traditional 510(k), are not required. The FDA has 90 days to review this submission.

4. De novo 510(k) – if this is a 510(k) without a predicate device, where the sponsor can demonstrate that the device has few risks and that the detailed PMA reviews for safety and effectiveness are not required. This approach should be confirmed in discussion with the FDA.

A 510(k) submission typically contains the following sections:

- Cover letter summarizing classification of device, reason for 510(k), identification of predicate device with details of that product trade name, 510(k) number, product code.
- (optional) Statement of substantial equivalence: narrative description or rationale with list of predicate devices
- Labeling: all printed material associated with device and patient information brochures)
- (optional) Advertising and promotional material
- Comparative information: this is the most important section, containing data with critical choice of comparison parameters and including comparison table with new device and predicate device data or characteristics, bench and clinical testing data and any relevant supporting materials; clinical data that may be requested by reviewers.
- (if necessary) Biocompatibility assessment: FDA version of the international standard ISO 10993 at www.fda.gov/media/85865/download (accessed Nov 2020) is used to determine testing. Recommended to discuss with FDA before conducting testing.
- (if necessary) Shelf life: stability data.
- Indication for use form: list of indications for use; new indications in an abbreviated 510(k)
- 510(k) summary: Publicly available information on FDA website once the 510(k) is cleared

The *CE technical documentation* is the equivalent of a 510(k) premarket submission for the EMA CE mark application. In Europe, technical documents will be reviewed by a notified body to assess conformity of the device with requirements of the MDR regulation. Annexes II and III of the *Medical Device Regulations* contain a table of contents as guidance of the structure for the technical documentation submission. Figure 6.9 summarizes content from the design history file (see Box 6.6) that is

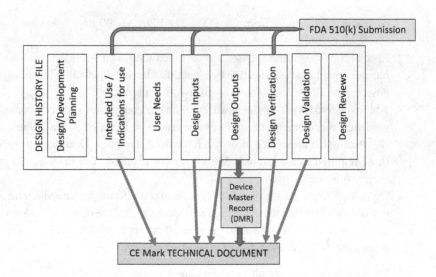

Figure 6.9 The elements of the design history file used for the 510(k) submission and as inputs for the CE Technical Documentation needed.
(Adapted from descriptions at www.consultys.ch/technical-dossier.)

used for preparation of the 510(k) market submission and the CE mark Technical Documentation.

6.5.10   PMA submission content

Typically, PMA submission documents take up 1,500 pages or more. Many years of effort go into compiling and collecting data necessary for the PMA. A PMA must provide a "reasonable assurance" that the new device is both safe and effective. The evidence is typically a compilation and statistical analysis of data from clinical trials, the majority of which are designed as controlled studies with patients randomly assigned into the treatment or control groups. There are administrative elements of a PMA application, but good science and scientific writing is a key to the approval of PMA application.

A PMA submission includes the following sections (see Section 6.5.13, Table 6.5 for website link):

- Summary of safety and effectiveness
- Device description, intended use, and manufacturing data
- Performance standards referenced
- Technical data (nonclinical): Nonclinical laboratory studies' section includes information on microbiology, toxicology, immunology, biocompatibility, stress, wear, shelf life, and other laboratory or animal tests. Nonclinical studies for safety evaluation must be conducted in compliance with 21 CFR Part 58 (Good Laboratory Practice for Nonclinical Laboratory Studies).
- Technical data (clinical): The Clinical Investigations Section includes study protocols, safety and effectiveness data, adverse reactions and complications, device failures and replacements, patient information, patient complaints, tabulations of

data from all individual subjects, results of statistical analyses, and any other information from the clinical investigations. Any investigation conducted under an investigational device exemption (IDE) must be identified as such. Data collected in clinical trials done in foreign countries can be submitted as long as the trials meet certain provisions (described in Section 6.3.7).
- Labeling: Any printed matter associated with the device, including directions for use, warnings or other labels.
- Environmental assessment

6.5.11   Types of PMA submissions

Traditional PMA submission: This is a one-time complete submission, usually made when the device is approved in another country and has already completed clinical testing.

Modular PMA submission: Each pre-agreed component of the PMA submission is reviewed and locked in until the final module is received, at which point the entire PMA is complete. The review period is thus shorter, with rolling submissions based on the sponsor's timetable.

Streamlined PMA submissions: In this program being piloted by the Division of Clinical Laboratory Devices, the clinical trial protocol is discussed and an agreement reached between the sponsor and the FDA, typically addressing a guidance document prepared by the FDA. This up-front agreement and familiarity with the device and protocol allows for a faster review.

6.5.12   Humanitarian use devices (HUDs)

A humanitarian use device (HUD) treats or diagnoses a disease or condition that affects fewer than 4,000 individuals in the United States per year and can be approved by the FDA under the Humanitarian Device Exemption (HDE). The HDE is similar in both form and content to a premarket approval (PMA) application, but is exempt from the effectiveness requirements of a PMA. An HDE application is not required to contain the results of scientifically valid clinical investigations demonstrating that the device is effective for its intended purpose.

The FDA recognizes that sometimes a condition is so unusual that it would be difficult for a company to scientifically demonstrate effectiveness of their device in the large number of patients that usually must be tested. Additionally, the applicant must demonstrate that no comparable devices are available to treat or diagnose the disease or condition, and that they could not otherwise bring the device to market. In these special situations, the FDA may grant a HDE to market the device, provided that: (1) The device does not pose an unreasonable or significant risk of illness or injury and (2) The probable benefit to health outweighs the risk of injury or illness from its use, taking into account currently available devices or alternative forms of treatment.

6.5.13   Devices: helpful websites

Table 6.5 contains helpful FDA and European websites.

Table 6.5 Some useful U.S. FDA and European websites for devices

All devices (CDRH)	
Starting point	www.fda.gov/medical-devices
Device Advice (another good starting point)	www.fda.gov/medical-devices/device-advice-comprehensive-regulatory-assistance
How-to guide	www.fda.gov/medical-devices/device-advice-comprehensive-regulatory-assistance/how-study-and-market-your-device
Classifying a device	www.fda.gov/medical-devices/classify-your-medical-device/how-determine-if-your-product-medical-device
QSR Manual and design controls details – Note: **excellent advice and guidelines for product development**	www.fda.gov/medical-devices/device-advice-comprehensive-regulatory-assistance/quality-and-compliance-medical-devices
GMP manufacturing/quality systems regulation	www.fda.gov/medical-devices/postmarket-requirements-devices/quality-system-qs-regulationmedical-device-good-manufacturing-practices
510(k) guidelines and details	www.fda.gov/medical-devices/premarket-notification-510k/content-510k
PMA guidelines and details	www.fda.gov/medical-devices/premarket-approval-pma/pma-application-contents
Information on master files	www.fda.gov/medical-devices/premarket-approval-pma/master-files
In vitro diagnostics (OIVD/CDRH)/EMA	
Starting point	www.fda.gov/medical-devices/products-and-medical-procedures/vitro-diagnostics
List of approved classified IVDs	21 CFR 862, 21 CFR 864, 21 CFR 866 www.accessdata.fda.gov/scripts/cdrh/cfdocs/cfcfr/showCFR.cfm?CFRPart=862 / 864 / 866
Assignment of CLIA categories	www.fda.gov/medical-devices/ivd-regulatory-assistance/clia-categorizations
Regulatory process FAQs	www.fda.gov/medical-devices/ivd-regulatory-assistance/overview-ivd-regulation
Analyte-specific reagents guidance	www.fda.gov/medical-devices/ivd-regulatory-assistance/overview-ivd-regulation#2c
Quality control for IVD devices	www.accessdata.fda.gov/scripts/cdrh/cfdocs/cfStandards/detail.cfm?standard__identification_no=34731
IVDR regulation in EU	https://eur-lex.europa.eu/legal-content/EN/TXT/HTML/?uri=CELEX:32017R0746
Guidance documents for industry on EMA's MDR and IVDR regulations from the Medical Device Coordination Group of the European Commission	https://ec.europa.eu/health/md_sector/new_regulations/guidance_en

Table 6.6 Types of diagnostic tests and their regulation

For manufacturers of in vitro diagnostic tests, there are three potential avenues to access the U.S. market: with an analyte-specific reagent (ASR), a laboratory-developed test (LDT), or an in vitro diagnostic (IVD).

Diagnostic	Regulator	Regulation	Product/service	Typical purchasers (regulated)
Analyte-specific reagent (ASR)	FDA	Medical device regulations	Reagents	CLIA labs, nonclinical labs, other IVD manufacturers,
Laboratory-developed test (LDT)	CMS	CLIA laboratories	Service	Clinics, patients, third-party payers
In vitro diagnostic (IVD)	FDA	Medical device regulations	Diagnostic (device + reagents)	General market, POC, clinics, etc.

FDA: Food and Drug Administration
CMS: Centers for Medicare & Medicaid Services
CLIA: Clinical Laboratory Improvement Amendments
POC: Point of care
Source: www.fda.gov

6.6 Diagnostics: regulatory pathways and NPD considerations

In vitro devices (IVD) are mostly regulated as medical devices and are subject to all the processes for devices outlined above. The specific office that reviews IVDs is the CDRH's Office of In Vitro Diagnostic Device Evaluation and Safety (OIVD). The FDA defines IVD products as:

Those reagents, instruments, and systems intended for use in diagnosis of disease or other conditions, including a determination of the state of health, in order to cure, mitigate, treat or prevent disease or its sequelae. Such products are intended for use in the collection, preparation, and examination of specimens taken from the human body. [21 CFR 809.3]

The IVDR regulation in the European Union also defines IVDs as a reagent, reagent product, calibrator, control material, kit, instrument, apparatus, piece of equipment, software, or system, whether used alone or in combination, to examine specimens in vitro for purposes of providing information on therapy, physiological status, etc.

There are three major groupings of diagnostics through their commercial and regulatory practices as shown in Table 6.6 and summarized in the sections that follow.

A few key differences in the review and approval of IVDs, compared to other medical devices are highlighted in the discussion below.

In addition to CDRH device reviews and controls, IVDs are also subject to the Clinical Laboratory Improvement Amendments (CLIA) of 1988. This law established quality standards for laboratory testing and an accreditation program for clinical laboratories. The requirements that apply vary according to the technical complexity in the testing process and risk of harm in reporting erroneous results. The regulations established three categories of testing on the basis of the complexity of the testing methodology: waived tests, tests of moderate complexity, and tests of high complexity. Laboratories performing moderate- or high-complexity testing or both must be certified under the CLIA and must meet requirements for proficiency testing, patient test management, quality control, quality assurance, and personnel. The OIVD determines the appropriate complexity categories for clinical laboratory devices as they evaluate premarket submissions.

In vitro device premarket submissions (PMA or 510(k)) may also be reviewed by the drug or biologic reviewing divisions (CDER/CBER). Premarket notification 510(k)s, PMAs, and IDEs for medical devices associated with blood collection and processing procedures as well as those associated with cellular therapies and vaccines will be reviewed by CBER. The medical device laws and regulations still apply.

Box 6.8 Intergenetics and a delayed launch

Adapted from a story reported in The Oklahoman, *October 5, 2006, by Jim Stafford*

Intergenetics, a diagnostics company based in Oklahoma had shown that its genetic test could assess a woman's risk of getting breast cancer by analyzing DNA signatures from cells obtained in mouthwash samples. It had developed this test over 13 years and had carried out studies on over 8,000 women to prove that it worked. A few weeks before the product was scheduled to launch at a large number of breast cancer centers nationwide, a new guidance document for such multivariate tests was issued by the FDA. This document essentially raised the bar for multivariate genetic tests, requiring the test to go through PMA approval process.

The summary of the FDA guidance document released September 7, 2006 reads:

> The FDA believes that In Vitro Diagnostic Multivariate Index Assays (IVDMIAs) do not fall within the scope of laboratory-developed tests over which the Agency has generally exercised enforcement discretion. FDA believes that IVDMIAs must meet pre- and post-market device requirements under the Act and FDA regulations, including premarket review requirements in the case of class II and III devices IVDMIAs to assure the public that these tests are safe and effective.

Intergenetics had to postpone the commercial launch, raise more venture capital and put a PMA package together. The company filed an IDE and was able to gain some revenues from launching the test on a limited basis in 7 certified labs nationwide that were able to fulfill the CLIA requirements for the (now deemed to be) higher complexity test. This incident highlights the need to keep in touch with the FDA throughout the development process to make sure that any new regulations or guidances are known to the company in advance.

IVD products also have special labeling requirements and distribution restrictions under 21 CFR 809 as discussed below and in Chapter 7.

Maintaining dialogue with the FDA throughout the product development process is a critical part of the regulatory planning for new IVDs especially ones that are different from already approved IVDs. The example of Intergenetics (Box 6.8) shows the impact that regulatory changes can have on new product development in this area.

6.6.1   IVD – regulatory clearance or approval steps to market:

In summary, IVD products that are exempt from 510(k) or PMA processes, must then go through CLIA categorization and other IVDs that are not exempt must go through the CLIA categorization in addition to the 510K or PMA reviews.

(a)

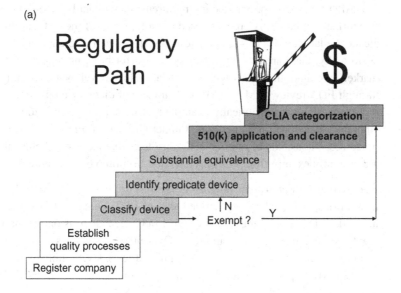

(b)

- FDA regulatory clearance Class I/II diagnostic device (6-9 months)
    - Register company as medical device manufacturer with FDA
    - Establish quality processes - design, packaging, labeling and manufacturing
    - Classify device – Class I exempt, Class I or Class II for some tests. If exempt, apply directly for "CLIA categorization only"
    - Identify predicate device(s) for application
    - Establish substantial equivalence with approved tests
    - Pre-market notification (510(k) submission); CLIA categorization request
    - Post-marketing reporting

Figure 6.10 Regulatory path to market for a medical diagnostic device. (a) Schematic and (b) description of regulatory path steps towards diagnostic approval.

Step 1: Classify the IVD – Class I, II, and III [21 CFR 862/864/866]. The website in Table 6.5 provides a list of approved and classified IVDs and is a good starting point to identify possible classification of the planned product and identifying predicates.

Step 2: Determine path to clearance or approval – 510(k) or PMA or exempt (or analyte-specific reagent, see Section 6.6 for more detail). If exempt (e.g. used only for exploratory purposes or retrospective analysis, not treatment decisions; uses noninvasive sampling), then the IVD does not have to submit an IDE. If there is a significant risk (as defined in 21 CFR 812.3(m)), then an IDE has to be filed and the regulatory pathway is through 510(k) if there is a predicate diagnostic identified or a PMA process if a novel diagnostic. If non-significant risk, then the company needs to provide an explanation as to why there is no significant risk and then submit data as per abbreviated requirements.

Step 3: After approval, special labeling requirements will apply. The IVD test will be categorized in the CLIA process based on complexity of the test and this will dictate the laboratories to which the test can be sold and where it can be performed.
-Historically, there are two regulatory pathways for bringing diagnostic tests to market, one through CLIA regulation of laboratory-developed tests and another through FDA review (PMA or 510K clearance) of diagnostic kits for sale to third-party labs. FDA has historically exempted CLIA approved laboratories and laboratory-developed-tests performed under CLIA from FDA controls or review processes. Thousands of routine laboratory tests that are used in clinical practice are currently exempt from FDA review. See Section 6.6.3 for more details.

Note: Almost all *molecular diagnostic testing* (or genetic tests carried out in laboratories, called laboratory derived tests or LDTs) today is CLIA-regulated and not under FDA. For example, out of the 1,000+ genetic tests on the market only a few have gone through FDA approval. Over 90% of revenues in the cancer diagnostics market come from CLIA-regulated tests. Cystic fibrosis testing, HIV genotyping and phenotyping are examples of commonly performed CLIA laboratory developed tests.

New European regulations (IVDR) now place most diagnostic tests under requirements to be certified by a notified body, a big change from prior practices where most were not required to be reviewed by regulatory bodies. FDA is currently (in 2020) considering (and legislation has been proposed) bringing under its regulation, a certain subset of laboratory developed tests LDTs called IVDMIAs (in vitro diagnostic multivariate index assays), including at-home tests or those sold direct to consumers and a few others. Under the currently proposed legislation, if approved, manufacturers would have a precertification process to determine risk of their tests and only the higher risk tests would have to go through a more rigorous FDA review process.

6.6.2 Preclinical and clinical considerations for IVDs

The FDA review of a 510(k) of an IVD is based on the evaluation of the analytical performance characteristics of the new device compared to the predicate, including:

- The bias or inaccuracy (accuracy) of the new device;
- The imprecision (or precision) of the new device; and
- The analytical specificity and sensitivity

Therefore, preclinical studies must offer valid data to satisfy the reviewer in these areas.

The types of studies required to demonstrate substantial equivalence include the following:

- In the majority of cases, analytical studies using clinical samples (sometimes supplemented by carefully selected artificial samples) will suffice.
- For some IVDs, the link between analytical performance and clinical performance is not well defined. In these circumstances, clinical information may be required.
- FDA rarely requires prospective clinical studies for IVDs (see exception with recent guidance in sidebar example above), but regularly requests clinical samples with sufficient laboratory and/or clinical characterization to allow an assessment of the clinical validity of a new device. This is usually expressed in terms of clinical sensitivity and clinical specificity.

6.6.3   CLIA program

Congress passed the Clinical Laboratory Improvement Amendments (CLIA) in 1988 establishing quality standards for all laboratory testing after inaccurate results on Pap smear led to questions on how labs functioned and what quality control procedures existed. The CLIA program was put in place to ensure the accuracy, reliability and timeliness of patient test results regardless of where the test was performed. 12,000 labs (of some 200,000 registered labs) were in the program in 1988, and over 170,000 labs were regulated as of 2002 under this program. CLIA requires that clinical laboratories obtain a certificate from the Secretary of Health and Human Services before accepting materials derived from the human body for laboratory tests (under 42 U.S.C. § 263a(b)).

The Centers for Medicare & Medicaid Services (CMS) (formerly Health Care Financing Administration) assumes primary responsibility for financial management operations of the CLIA program, which is self-funded by user fees from regulated labs. The CMS thus pays the FDA for CLIA categorization of commercially marketed tests. This task is regulated by the CDRH and CBER divisions at the FDA and the FDA also has an ongoing partnership with the Centers for Disease Control and Prevention (CDC) in this program.

The FDA CLIA program assigns commercially marketed in vitro diagnostic test systems to one of three CLIA regulatory categories based on their potential risk to public health: waived, moderate complexity or high complexity. Laboratories performing moderate- or high-complexity testing or both must meet requirements for proficiency testing, patient test management, quality control, quality assurance, and personnel.

6.6.4   Analyte-specific reagents or "home-brew" tests

In the past, FDA was not actively involved in regulation of in-house (so-called home-brew) tests or in regulation of the building blocks sold and used to create these tests. In 1998, FDA classified the building blocks of in-house tests as analyte-specific reagents (ASRs) and began to regulate them, but still puts very few controls on most ASRs.

From the FDA website: "ASRs are defined as 'antibodies, both polyclonal and monoclonal, specific receptor proteins, ligands, nucleic acid sequences, and similar reagents which, through specific binding or chemical reaction with substances in a specimen, are intended for use in a diagnostic application for identification and quantification of an individual chemical substance or ligand in biological specimens.' 21 CFR 864.4020(a). Most molecular diagnostic tests or nucleic acid tests (NATs) or genetic tests fall under the description of ASRs and are restricted devices. Among the restrictions on ASRs is a requirement that advertising and promotional materials for ASRs may not 'make any statements regarding analytical and clinical performance.' 21 CFR 809.30(d)(4)." See Box 6.9 for an example of this restriction.

In simple terms an analyte-specific reagent (ASR) is the active ingredient of an in-house test. Most ASRs are classified as Class I devices, exempt from the premarket notification process. Thus, ASRs represent a lucrative market opportunity as the product development is relatively short and the product can readily be marketed by meeting the low-burden requirements described below. However, these rapidly developed tests may not have adequate levels of clinical validation or outcome data that would be required to establish adequate levels of reimbursement (see Chapter 8 for detailed discussion).

ASRs are used in conjunction with other general-purpose reagents and general-purpose instruments by a laboratory to set up an in-house ("home-brew") test or laboratory testing service. While specimens can travel to the lab setting up this service, the test itself is not marketed outside of the single lab setting up this service. ASRs must have clear activity as the active ingredients of an in-house test but should be provided without instructions for use or performance characteristics. It is the responsibility of the laboratory using the ASR to develop a recipe for the test at hand and to take responsibility for establishing and maintaining performance.

Both the manufacturers of ASRs as well as the laboratories using them are subject to incremental regulation with controls (based on classification as Class I/II/III ASRs) to:

1. assure the quality of the materials being used to create these tests (Quality System Regulation 21 CFR 820 applies).
2. assure that laboratories preparing these tests were able to establish and maintain performance and understood their responsibility for accomplishing this (test results using Class I ASRs must be labeled as not cleared or approved by FDA).
3. provide appropriate labeling so that healthcare users would understand how these tests were being validated. (As per 21 CFR 809.10(e)(1)(x): "Analyte Specific Reagent. Analytical and performance characteristics are not established.")

The manufacturer of an ASR has to register and list their establishment with the FDA and can only sell the ASR to:

Box 6.9 Example of a warning letter by FDA on incorrect sale of ASRs by a company

Excerpt of letter taken from the FDA website, but identifying names are omitted here.

August 26, 2005

Dear Dr [ ]

The Office of In Vitro Diagnostic Devices (OIVD) has reviewed information on your [ ]™ Brand Internet website.... . Diagnostics that are marketed as analyte specific reagents (ASRs).

Our review indicates that each of these products is a device under section 201(h) of the Food, Drug, and Cosmetic Act (FDCA or Act), 21 U.S.C. 321(h), because it is intended for use in the diagnosis of disease or other conditions, or in the cure, treatment, prevention, or mitigation of disease.

According to your instructions for use/methods for use, each of the gel-based [ ] genetic assays is intended for "in vitro diagnostic use" to detect various human genetic mutations. In addition, a press release issued by [ ] on March 17, 2005 claimed that, "the [ ]™ family of kits for in vitro diagnostic use provide laboratories simple and cost-effective assays for use in genetic screening programs. Industry-leading [ ]™ kits are available for the genetic analysis of human diseases such as cystic fibrosis and cardiovascular disease ..." ...

Based on information on your website, these devices do not adhere to the restrictions on the sale, distribution, and use of ASRs. Your website makes specific analytical and performance claims such as that your devices can detect multiple mutations per device and screen for particular diseases. Statements on your website describing your devices indicate that they are intended for the detection of mutations related to a clinical diagnosis of, for example, ... In addition, the Instructions/Methods for Use supplied for your assays, provide detailed procedures (along with directions and guidelines for the interpretation of results) that are unique for your assays and that constitute analytical and performance claims.... .

A review of our records shows no clearance or approval for your gel-based [ ] genetic assays . ... These devices are therefore adulterated under section 501(f) (1) (B) ...

You should take prompt action to correct these violations. Failure to promptly correct these violations may result in regulatory action being initiated by the FDA without further notice. These actions include, but are not limited to, seizure, injunction, and/or civil money penalties. Also, Federal agencies are informed about the Warning Letters we issue, such as this one, so that they may consider this information when awarding government contracts.... .

Please notify this office in writing within fifteen (15) working days of receipt of this letter, of the specific steps you have taken to correct the noted violations, including an explanation of each step being taken to prevent the recurrence of similar violations.... .

1. In vitro diagnostic manufacturers
2. Clinical laboratories qualified to perform high complexity testing as regulated under the CLIA
3. Organizations that use the reagents to make tests for purposes other than providing diagnostic information to patients and practitioners, e.g., forensic, academic, research, and other nonclinical laboratories

Specific ASRs involved in blood screening are classified as either Class III, or in selected cases Class II devices, and require premarket approval. ASRs used to diagnose life-threatening contagious diseases with high public health impact are also classified as Class III products. Examples of these include tests for HIV and tuberculosis. Manufacturers and healthcare facilities must report deaths and serious injuries that an ASR has or may have caused or contributed to in accordance with 21 CFR Part 803.

Laboratory developed tests (LDTs), which are run by certified clinical test labs, enable ASRs to be used in diagnostic tests without FDA supervision. Many companies choose this route to commercialize their diagnostics as it is easy to get to revenues quickly. However, manufacturers cannot certify the diagnosis, but only state that the result correlates with a likely outcome.

6.7   Regulatory guidelines for co-development of pharmacogenetic diagnostic tests and drugs

Pharmacogenetics and pharmacogenomics are largely interchangeable terms for understanding the influence of variation in genes on the variability of the effects of drugs. In general usage, pharmacogenetics seeks to identify specific genes that affect drug metabolism, and pharmacogenomics covers all genes in the genome that may affect drug responsiveness.

The difference between pharmacogenomic (specific DNA sequence comparison studies) testing and standard genetic testing is described by the FDA:

Pharmacogenetic tests for clinical use . . . aid in selection of certain therapeutics. When sufficient clinical information is available, they may also aid in dosage selection of the therapeutic. Therefore, a pharmacogenetic test target population will typically be composed of candidates for a particular therapeutic. Target populations of genetic tests, on the other hand, will usually be composed of those who are suspected of having, or are at risk of developing, a particular disease or condition.

Additionally, pharmacogenomic (PGx) tests compare inter-individual differences in whole genome patterns (SNPs, mutations, etc.), differentiating this term in practice from the pharmacogenetic tests, which typically test for the presence of specific DNA sequences.

According to the FDA: "The promise of pharmacogenomics lies in its potential to help identify sources of inter-individual variability in drug response (both effectiveness and toxicity); this information will make it possible to individualize therapy with the intent of maximizing effectiveness and minimizing risk."

The area of pharmacogenomic (PGx) testing of patients to help select more responsive groups for clinical trials or as a screening tool for identifying and selecting patients for prescription of a drug is of great interest in the future area of personalized medicine. A PGx test could be used to identify the population most likely to respond to the drug, while another PGx test can help screen out people that are likely to have toxic reactions to the drug; both tests helping to dramatically reduce the number of subjects required for the clinical studies. An example would be a PGx test that identifies people with a rare mutation of a liver enzyme that can slow down the breakdown and hence the clearance of a drug, causing it to remain in the body much longer at higher concentrations, leading to more toxic effects.

There are several business and regulatory issues that make the development of pharmacogenomics-based "personalized" medicines challenging: If the patient population is limited to those identified by the pharmacogenomics tests, will the current pressure on drug pricing make it possible to develop a profitable business in such personalized medicines that only address a small market? How will the use of pharmacogenomic testing (used to help screen study subjects and improve the rates of success) in clinical trials be regulated by the FDA? Although the use of pharmacogenomics is potentially beneficial to developers who could see reduced costs of development for both drug and diagnostic, due to smaller and faster clinical studies, there is uncertainty on how the data will impact the final indication. Will the FDA use this data to limit the final label/indication and make the market so small that it is economically not viable or attractive for the developer? These uncertainties will need to be addressed through adaptive business models and through extensive dialogue with the regulatory and reimbursement agencies. This area of diagnostics is known as "companion diagnostics," tests that are used to select patients for a specific therapeutic product. Joint development of both the companion diagnostic and the therapeutic product means that both products will appear on the label/indication for each product.

The FDA has published several guidance documents in the past few years – starting from 2014 through 2016, in order to clarify the process of developing a pharmacogenomics-based test that is tied to a specific drug therapy, due to the reducing costs of genomic analysis. Mirroring this is the increasing number of guidelines published from the Clinical Pharmacogenetics Implementation Consortium (CPIC), an international group of researchers, clinicians and academicians. As of October 2020, guidelines covering over 20 different genes (biomarkers) related to 66 drugs have been published by CPIC and other groups around the world. These guidelines for clinical practice start with the data obtained during drug development, which is then noted (to various degrees) in the approved drug label. The FDA publishes a biomarkers table at www.fda.gov/drugs/science-and-research-drugs/table-pharmacogenomic-biomarkers-drug-labeling, which shows over 430 pairs of drug-biomarkers where pharmacogenomic information is found in the drug labeling. The labeling for some, but not all, of the products includes specific actions to be taken based on the biomarker information. Pharmacogenomic information can appear in different sections of the labeling depending on the actions.

The regulatory pathway for joint approval of a companion diagnostic with a therapeutic is fairly complex and has taken regulatory bodies some time to catch up

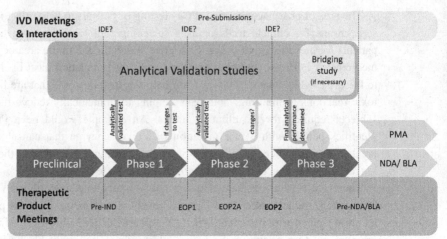

Efficient co-development of a therapeutic product with an IVD companion diagnostic require coordination of the development programs of the two products, including interactions with all relevant FDA review divisions. EOP1 = End of Phase I, EOP2 = End of Phase 2.

Figure 6.11 Regulatory pathway and timings proposed for co-development of personalized medicines, where a diagnostic test is required before a drug prescription is written. The schematic depicts parallel regulatory actions in submissions, meetings and review towards final approval of co-developed drugs and diagnostics. (From the 2016 guidance document *Principles for Co-development of an In Vitro Companion Diagnostic Device with a Therapeutic Product* found at www.fda.gov/regulatory-information/search-fda-guidance-documents/principles-codevelopment-vitro-companion-diagnostic-device-therapeutic-product [last accessed Dec 2020].)

to the practice and science. The FDA released their draft guidance for companion diagnostics in 2016 and EMA released a similar guidance in 2019. Sponsors of both products are jointly required to approach relevant agencies (CDER or CBER) for the therapeutic and (CDRH) for the in vitro diagnostic biomarker. The clinical trials for the therapeutic can incorporate the diagnostic information for positive responder identification prospectively or retrospectively. The timeline for approval could be conjoint, as shown in Figure 6.11.

While an IND is filed for the drug development, an IDE would be filed simultaneously for the diagnostic/device development process. A PMA or 510(k) will be filed concurrently with the NDA/BLA for the drug and the co-approved, co-developed drug and diagnostic will be launched with co-dependent marketing for concurrent use of both. The timings of meetings with the FDA and review periods are shown in this figure for both tracks of development.

The European Medicine Agency (EMA) specifically links the regulatory process for new drugs and companion diagnostics in the Regulation of In Vitro Devices (IVDR) introduced in 2017 (and coming into effect in May 2022). Under the IVDR, companion diagnostics will be classified as Class C devices (the second highest risk level) and the corresponding conformity assessment will necessitate interaction with both a notified body (any one of several private accredited bodies that conduct conformity assessments for medical devices) and the European Medicines Agency (EMA)/National Competent Authorities (NCAs). The notified body has to seek a scientific opinion from the EMA/NCA in this matter requiring interaction between the various regulatory entities for final approval of companion diagnostics.

6.8   Combination products, artificial intelligence and software, genetic materials, and tissues

6.8.1   Combination products – drugs and devices bundled together

A combination product includes (definition from 21 CFR Part 3, Subpart A, Section 3.2 (e)):

* A product comprised of two or more regulated components, i.e., drug/device, biologic/device, drug/biologic, or drug/device/biologic, that are physically, chemically, or otherwise combined or mixed and produced as a single entity;
* Two or more separate products packaged together in a single package or as a unit and comprised of drug and device products, device and biological products, or biological and drug products;
* A drug, device, or biological product packaged separately that according to its investigational plan or proposed labeling is intended for use only with an approved individually specified drug, device, or biological product where both are required to achieve the intended use, indication, or effect and where upon approval of the proposed product the labeling of the approved product would need to be changed, e.g., to reflect a change in intended use, dosage form, strength, route of administration, or significant change in dose; or
* Any investigational drug, device, or biological product packaged separately that according to its proposed labeling is for use only with another individually specified investigational drug, device, or biological product where both are required to achieve the intended use, indication, or effect.

The Office of Combination Products at the FDA helps guide manufacturers who are increasingly designing and developing innovative medical products such as drug-device, drug-biologic, and device-biologic combinations that cross over historical FDA review divisions (CDER/CDRH). While this office has broad regulatory responsibilities for these products, primary regulatory responsibilities for, and oversight of, specific combination products remain in one of three product centers – the CDER, the CBER, or the CDRH – to which they are assigned.

The Office of Combination Products (OCP) (as described on the FDA website, www.fda.gov/about-fda/office-clinical-policy-and-programs/office-combination-products):

• does not review marketing applications for combination products.
• assigns a lead Center (CBER, CDER or CDRH) that will have primary jurisdiction for the premarket review and regulation of a combination or single-entity product
• develops policy for combination product regulation.

- is also responsible for ensuring timely and effective premarket review of combination products by overseeing the timeliness of and coordinating reviews involving more than one agency center.
- ensures consistency and appropriateness of post market regulation of combination products
- is available as a resource to industry and agency reviewers to help facilitate the review process, to help clarify and/or develop appropriate regulatory pathways, or to provide any other assistance as appropriate to OCP's mission.

When a product submission is made to the OCP, a determination is made for assignment to the lead reviewing agency (CBER,CDER or CDRH), based partly on the primary mode of action of the combination product. A guidance document on how the FDA/OCP regulates combination products and makes designation or primary mode of action decisions was posted on the FDA website in September 2006 (www .fda.gov/oc/combination/innovative.html).

Once a lead review agency is appointed, the OCP continues to organize consultative or collaborative reviews. The process for review of the combination product by the OCP is described on the FDA website (www.fda.gov/combination-products):

What is the difference between consultative and collaborative reviews of combination products?
When combination products are assigned to a lead Center, that Center may consult or collaborate with another Center as part of the review process. The consultative review is used to assist the requesting reviewer in making appropriate regulatory/scientific decisions. In contrast, a collaborative review is a review activity in which reviewers in two or more Centers have primary review responsibilities, generally for a defined portion of a submission. Regulatory and scientific decisions will be made by the management of each Center for that portion of the review assigned to it, including the decision to approve or disapprove the product.

Some recent examples of approved combination products are:

- Bioresorbable hemostat that contains collagen and thrombin
- Transdermal patch containing drug for ADHD
- Dental bone grafting material with growth factor
- Paclitaxel-eluting coronary stent system
- Dermagraft Human fibroblast-derived dermal substitute

6.8.2   Software and artificial intelligence in medical products

Software that is used by itself (i.e. not tied to a device) is treated as a medical device (software as a medical device, SaMD). Based on the risk assessment of its intended applications and use, it is classified into a risk class and general and specific controls are applied to assess the software product for approval.

The emergence of artificial intelligence (machine learning or neural networks are other terms, AI/ML) in FDA-approved medical products started approximately in 2016 with the first approval of a software product that mentioned use of artificial intelligence/machine learning (AI/ML). The subsequent FDA approvals of over

29 different software products (mostly through 510k, some through de novo pathway) were on software that had algorithms locked – i.e. the analysis done by the software was fixed at the time of application and thus could be reproduced reliably – if the same input was given, the software would give the same result.

However, the use of AI/ML and adaptive self-learning is accelerating as computing capabilities at the point of service are dramatically increased in computing power and cloud computing services make powerful computing platforms available on the internet on demand. Most of these AI/ML software (SaMD) products are indicated for use in radiology and cardiology for image analysis applications but others for internal medicine or acute care are being utilized to analyze and triage patients or give alerts in real time for more complex diagnoses.

The challenge for regulatory evaluation of the medical use of this AI/ML software is that the adaptive learning nature makes their consistency difficult to gauge, compared to algorithm-locked software programs where the same input will result in the same output. Hence in 2019, the FDA started a process of public dialogue around establishing new regulatory process for AI/ML adaptive software as a device.

A first step by the FDA has been to launch a pre-certification process where FDA will assess the culture of quality and organizational excellence of the sponsor software company and have reasonable assurance of the high quality of their software development, testing, and performance monitoring of their products. This is a life cycle approach for the organization and the software product, where confirming the continuous controls applied to the software design, adaptive training would give patients and healthcare professionals assurance of the safety and quality of the software. Good machine learning principles (GMLP) must be applied during algorithm development and demonstrated as part of the company's quality system.

EMA is also in the process (as of this writing in 2020) of laying the infrastructure for approval of AI software as part of its 2017 MDR regulation. The MDR introduced expanded requirements for post-market surveillance and medical device traceability. Specifically, the definition of "medical device" now specifically includes software for the "prediction" and "prognosis" of disease. The MDR introduces a new classification rule for software such that programs intended to merely "monitor physiological processes" will be classified Class IIa or higher. Most machine learning software–based medical devices will be classified as Class IIa (even if the monitoring information is not intended for diagnosis or therapeutic purposes) or Class IIb if the software monitors vital physiological parameters (e.g. heart rate, blood pressure, and respiration). This implies that most software as a medical device will be subjected to a conformity assessment, including the approval of the technical documentation by a notified body. EMA is not the primary regulator for medical devices, and unlike the centralized FDA in the United States, devices are regulated by the Member States who can designate independent accredited "notified bodies" to conduct the required conformity assessments. In most other countries such China, India, Brazil, or Japan, the regulatory bodies are following the U.S. or EMA framework, which are being harmonized by International Medical Device Regulators Forum (IMDRF) white papers and widespread adoption of ISO standards for medical devices (e.g. ISO 13485). The IMDRF (www.imdrf.org) is a voluntary group of medical device regulators that is continuing the work of the Global Harmonization Task Force on medical devices.

The regulatory review frameworks for approval of AI/ML programs lag behind the actual use of these programs, leading to regulatory uncertainty with risks for manufacturers. The SaMD manufacturers must spend substantial effort and resources to understand how regulations apply in the particular context of these AI/ML based medical devices. This includes working closely with the responsible regulatory authorities to achieve consistency in interpretation.

Among the growing applications of software (including AI and traditional fixed algorithm software) that would need FDA clearance or approval in medical diagnosis and therapy are the following:

- Assisting in disease diagnosis
- Assisting in patient monitoring of specific parameters to note subtle changes in pattern that correlate with disease progression – i.e. during the 2020 pandemic, several companies making wearable devices announced the ability to detect the likelihood of infection a few days before the emergence of any major symptom requiring hospitalization
- Patient triage / risk stratification and augmenting clinical workflow or the "referral chain"
- Software as therapy for psychological or neural disorders
- Intelligent medical devices with software including AI/machine learning with enhanced function or interactivity with caregivers
- Precision medicine – patient selection for clinical trials and drug cocktail selection for patients based on individual history and/or genetic physiometric analysis.

The case described in Box 6.11 is an example of the first FDA-approved video game–based therapeutic software that learns from patient interactions and has demonstrated clinical relevant improvements in the patients.

Numerous other applications for software (including AI/ML) in the healthcare field are continuing to emerge, including medical data analysis for better insurance economics, AI used in drug discovery to decipher hitherto unseen patterns between existing drugs and diseases, or to identify and select new drug candidates from a library based on target and disease profiles.

Box 6.10 Example of regulatory classification of consumer software using artificial intelligence

Is my AI software regulated as a medical device or is it a general wellness software for public use?

A first step for manufacturers or developers intending to market a machine-learning adaptive software product is to determine whether the product is intended as a general wellness product (consumer-grade) or as a medical device (clinical-grade). Software intended "for maintaining or encouraging a healthy lifestyle" that is "unrelated to the diagnosis, cure, mitigation, prevention, or treatment of a disease or condition" does not fall under the medical device definition and is not regulated

Box 6.10 (*cont.*)

by the FDA (*Changes to Existing Medical Software Policies Resulting from Section 3060 of the 21st Century Cures Act* , 2019 FDA Guidance document – at www.fda .gov/media/109622/download). However, Apple's electrocardiogram (ECG) app, intended for use with the Apple's watch to "create, record, store, transfer, and display a single channel electrocardiogram (ECG) similar to a Lead I ECG," and Apple's irregular rhythm notification feature, intended to identify and notify the user of episodes of irregular heart rhythms, are both regulated as medical devices and subject to FDA regulatory controls even though the watch is not a medical device and the apps are "not intended to interpret or take clinical action" and the resulting medical device "is not intended to provide a diagnosis." (From FDA letter response to Apple www.accessdata.fda.gov/cdrh_docs/pdf18/DEN180044.pdf)

Apple used the de novo pathway for regulatory review of its new software as there was no predicate device on the market. Previous ECG devices that had gone through the standard Class II filing or 510(k) process included both software and specific hardware combined into a single product. Apple carried out a large clinical study that compared the ECG app (on the Apple Watch series 4) to the gold standard of a 12-lead ECG obtained at the same time from the subject. Apple used data from the trial to demonstrate low risk and efficacy and that its app met the special controls in the review of specificity and sensitivity for ECG products.

In their letter response (www.accessdata.fda.gov/cdrh_docs/pdf18/DEN180044 .pdf) to the de novo application, the FDA classified the software as a Class II medical device and put forth the following concerns as identified risks:

Identified risks to health	Mitigation measures
Poor quality ECG signal resulting in failure to detect arrhythmia	Clinical performance testing Human factors testing Labeling
Misinterpretation and/or over-reliance on device output, leading to: Failure to seek treatment despite acute symptoms Discontinuing or modifying treatment for chronic heart condition	Human factors testing Labeling
False negative resulting in failure to identify arrhythmia and delay of further evaluation or treatment	Clinical performance testing Software verification, validation, and hazard analysis Nonclinical performance testing Labeling
False positive resulting in additional unnecessary medical procedures	Clinical performance testing Software verification, validation, and hazard analysis Nonclinical performance testing Labeling

Box 6.10 (*cont.*)

The FDA also established several special controls to the ECG software intended for over-the-counter-use such as: demonstration of ECG quality and specificity in a clinical performance trial, and human factors and usability testing to demonstrate user can use device and correctly interpret device output based on device labeling.

Hence it is important to evaluate the intended use of the software even if it is for over the counter use by consumers and discuss this early with the FDA to obtain guidance on classification. Even though the software may not be sold or used by a medical professional, it could still be subject to FDA regulation and oversight.

Box 6.11 Therapeutic software emerging as new class of AI/ML software

In June 2020, the FDA approved, through the de novo pathway, an industry-first, prescription video game that is meant to improve attention function in children with attention deficit hyperactivity disorder. EndeavorRX was the first video game software specifically by the FDA and yet another one of several digital therapeutics that have been commercialized over the past few years.

The video game EndeavorRx, made by Akili Therapeutics Inc., was granted clearance based on data from five clinical studies in more than 600 children diagnosed with ADHD, including a prospective, randomized, controlled study, which showed EndeavorRx improved objective measures of attention in children with ADHD. The game presents specific sensory stimuli and simultaneous motor challenges designed to target and activate the neural systems that play a key role in attention function while using adaptive algorithms to personalize the treatment experience for each individual patient. This de novo authorization creates a new classification for Akili and other companies to bring digital therapy devices for ADHD, defined as software intended to provide therapy for ADHD or any of its individual symptoms as an adjunct to clinician supervised treatment, to market via 510(k) submissions. As a Class II device, the FDA also laid out a series of special controls related to labeling, clinical testing and software verification, validation, and hazard analysis.

A few other examples of software for therapeutic use that have been approved or cleared by the FDA are:

- The company WellDoc has received multiple clearances for their BlueStar software, which guides individuals through the complicated journey of living with diabetes by enabling them to self-manage their conditions and enhancing connections to their healthcare team. The software is reimbursable with its own device code and integrates existing patient devices. The company has demonstrated through multiple studies and peer-reviewed publications that by using the BlueStar software, patients experienced a drop in HbA1c of 1.7 to 2 points on average, improved medication adherence, and better glucose control.

> **Box 6.11** (*cont.*)
>
> - Voluntis has commercialized Insulia®, which is available via mobile app or web portal. It is a prescription-only medical device with digitized clinical algorithms that provide real-time basal insulin dose recommendations for adults with type 2 diabetes based on inputs from the prescribing physician and patient's glucose readings. Voluntis first got 510(k) FDA clearance and CE mark clearance in 2016 for this software as a Type II medical device.
> - Propeller Health provides a combination of sensors that attach to a regular inhaler, and software and services for patients with asthma and COPD. Insights on inhaler/medication use are delivered to the Propeller app on their smartphone, which patients can then share with their clinician to help inform their treatment plan. The app is a 510(k)-approved Class II medical device. The Propeller platform has shown in clinical studies to significantly increase asthma control and medication adherence, reduce asthma-related emergency department visits and hospitalizations by as much as 57 percent, and reduce COPD-related healthcare utilization by as much as 35 percent.
>
> These companies and products represent a new wave of digital health therapeutics that are transforming the healthcare landscape by utilizing software and AI/ML to help patients manage chronic conditions for better clinical health outcomes.

6.8.3   Cellular, tissue, and gene therapies

The Office of Cellular, Tissue, and Gene Therapies was formed in 2002 in CBER, to provide consistent review and develop expertise in with the increase in biological tissue therapeutics. This office works with the Office of Combination Products. The mission of this office is to regulate, review, and develop policy and education on:

- Tissues
- Cellular and TISSUE-based products
- Gene therapies
- Xenotransplantation
- Combination products containing living cells/tissues
- Unique assisted reproduction (ooplasm transfer)

Finally, this office is responsible for assuring the safety, identity, purity, and potency of all products listed above.

Cell, tissue or gene therapy products do not need premarket approval if there is:

- Minimal manipulation
- Homologous use
- Not combined with drug or device
- Exerts NO systemic, or
- Exerts systemic effect, but is

○ Autologous OR
○ Allogeneic in first- or second-degree relative OR
○ For reproductive use

However, Good Tissue Practices (guidance published by Office) must be followed, donor eligibility and screening are required, the establishment must register with the FDA, and compliance with the rules will be determined on inspection.

Regulations that apply (with some guidance) are available at www.fda.gov/cber/rules/gtpq&a.htm (21 CFR Part 1270 and 1271).

For non-tissue products, IND/BLA or IDE/PMA rules apply. The office goes out of its way to initiate a dialogue with the sponsor in order to fully understand the technology and give appropriate guidance on a case by case basis.

Tissue engineered products are treated as combination products and guided to their independent reviews in their respective divisions. The primary center is decided by Office of Combination Products based on the *primary mode of action* of the combination product. If the primary mode of action is dependent on the biologic component, CBER will take the lead, whereas CDRH will take the lead if the mode of action is dependent on the device; and each will consult each other with the Office of Combination Product or Cellular therapies coordinating these consultations and reviews. Finally, it is possible that completely separate reviews will be required for each component.

In 2017, the FDA released a new expedited pathway specifically for investigational regenerative medicine therapies that treat, modify, reverse, or cure serious conditions and have generated promising preliminary clinical evidence of efficacy. These therapies are designated regenerative medicine advanced therapies (RMAT) and get the benefits of the fast-track and breakthrough therapy designation programs with early FDA interactions. Breakthrough designation is for therapies that do not yet have clinical data, whereas RMAT designation requires early clinical proof of efficacy.

In general, combination products usually incorporate novel technologies or use existing technologies in novel ways, which often leads to delays or uncertainty in the FDA review, as the published literature may be sparse or the new area of developing technology is not one that FDA staff scientists are familiar with. This lack of familiarity may result in onerous requirements or guidance statements that require multiple or additional tests. The company scientists, who may be more familiar with the new technology, may not have seen these tests as necessary and may be asked by regulators to repeat several levels of preclinical studies (e.g. prolonged toxicity testing or additional stability testing) or even clinical studies (e.g. increase patient population size, follow patients longer, or make additional measurements) that the company scientists may have thought unnecessary. Box 6.12 illustrates one such story of one of the earliest combination products around the turn of the century that ran into regulatory and development delays. The uncertainty with the FDA reviews and other issues eventually caused the company to fail. The case study in Box 6.12 illustrates that a company with novel products has to work very aggressively not only in product and technology development, but equally diligently on education and passage of regulatory and reimbursement issues. Box 6.13 shows the complexity of the

---

Box 6.12 Novel combination product hit regulatory pitfalls

FDA Issues Felled Advanced Tissue Sciences: How a Biotech with Products on the Market Still Failed

By Andy Stone, Published June 1, 2003 in Genetic Engineering News

(Reproduced here with permission of Genetic Engineering News)

The wound repair industry is littered with defunct, bankrupt companies. Advanced Tissue Sciences (ATS; La Jolla, CA) is but one of the latest victims, a company with good marketed products that nonetheless failed due to a complex array of forces, ranging from the difficulties in getting FDA approvals to the vagaries of the investing public.

The company was based on an innovative technology that enabled the 3-D ex vivo growth of human cells. This technology would eventually give rise to products aimed at the treatment of moderate to severe burns and coverings for diabetic foot ulcers, a market with a potential of about one half million users annually in the U.S., and other products related to the basic 3-D matrix.

But ATS lost $300 million over its nearly 15-year history, which came to an end with bankruptcy and liquidation, beginning in the fall of 2002.

Promising beginnings

The beginnings of ATS, originally named Marrow Tech, seemed promising enough. The company raised its first round of investment through a $5-million IPO in 1988. Marrow Tech was formed on the heels of early biotechnology success stories, such as Amgen, Genentech, and Genzyme, and came to life in an environment in which grants and other forms of early funding were all but nonexistent but where public markets were eager to jump on the biotech bandwagon.

The original business plan of Marrow Tech was for the storage, cryopreservation, and expansion of a patient's own marrow. The benefit was that a small amount of that marrow could be used, instead of the painful normal removal involving 50–100 punch samples from the iliac crest.

It was then shown that administration of growth factors, such as EPO and SCP from Amgen and Genetics Institute in the late 1980s, could release stem cells into the peripheral circulation, where they could be collected through a simple transfusion/separation process.

As a result, Marrow Tech, based on its core tissue-culturing technology, altered its focus to the development of human skin substitutes and changed its name. ATS' first product appeared in October of 1990. Skin2 was a full-thickness, human-derived skin that was used to test consumer products, replacing, for example, rabbit eye and skin tests and allowing companies that produced cosmetics and consumer products to avoid some ethical dilemmas.

Regulatory difficulties

The stage was set, though, for ATS' first run-in with the FDA, in a pattern that would repeat itself over the course of the company's history, with disastrous consequences.

Box 6.12 (*cont.*)

"Skin2 was accepted and well published, but the FDA didn't move quickly to approve the product," says Gail Naughton, former vice chair of ATS and current dean of San Diego State University's business school. As a result, ATS had to abandon Skin2. "The company couldn't afford to work on therapeutic products and move Skin2 through the approval process at the same time," Naughton explains.

Despite this early setback, the company labored on and, by 1996, had entered a joint venture with medical device leader Smith and Nephew (Andover, MA) for the development of wound repair products based on ATS' tissue repair technology.

The first product, Dermagraft, comprised a 3-D matrix of living human cells that was initially intended as a treatment for medium- to full-thickness burns. "It turned out that healing took longer with Dermagraft than with alternatives such as cadaver skin," Naughton says.

The company followed up with a new product, Transcyte, which would be manufactured at Dermaquip, the joint venture GMP-manufacturing facility established by the two companies for the manufacture of tissue engineering products. Transcyte received FDA approval in 1997 as a covering for medium- to full-thickness burns.

The target market for Transcyte was estimated at 40,000 burn victims annually; the actual market potential may have been much lower. "The burn market is not terribly large, with between 5,000 and 15,000 burns per year that require skin grafting," says Michael Lysaght, director of Brown University's Center for Biomedical Engineering.

Despite its effectiveness, Transcyte was not an immediate market success, and a confluence of factors, including the product's high price and slow adoption rate, denied ATS revenues needed to fund further company and product development.

Marketing disappointment

"Transcyte works and is an excellent product that reduces pain and scarring," states Abe Wischnia, former head of investor relations at ATS. "But there were several factors militating against it in the marketplace. The first was cost, and Transcyte was competing with much less expensive gauze bandages, as well as cadaver skin." The price of Transcyte was $1,350 for two 5″-by-71/2″ pieces of product, about three times the cost of cadaver skin and much more expensive than traditionally used gauze with an antimicrobial agent.

Another important factor was the willingness of doctors to try new technologies. "Smith and Nephew had to call on doctors for a year at the Children's National Burn Center in Washington before the doctors would listen to them," Wischnia reports.

In contrast to the cardiovascular market, adds Naughton, "Wound care in general is very conservative, and doctors don't jump easily to new products." The combination of a lack of innovation and few new revolutionary products

Box 6.12 *(cont.)*

severely hampered adoption, and, with sales well below projections, put the company into tough financial straits.

As ATS turned to the development of other products, notably the development of Dermagraft for a much larger diabetic foot ulcer market, the company was forced to return to public markets to fund itself. By the time the company filed for bankruptcy in the fall of 2002, Naughton explains, "We had gone to the public markets at least five times, with over 70 million shares and tremendous dilution."

The fatal blow that befell ATS occurred in 1998, when the company had completed clinical trials for Dermagraft for the treatment of diabetic foot ulcers. In January 1998, an FDA PMA panel approved Dermagraft, 8 to 1, and momentarily it looked as though the product would soon go to market.

Then, in an unexpected and eventually disastrous turn of events, the FDA refused approval and demanded an additional clinical trial, to involve 350 patients. This took an additional three- and one-half years before approval was finally granted in the fall of 2001. In the meantime, the company continued to burn money at a rate of up to $3 million per month, while gaining disappointing revenues from Transcyte and further diluting its ownership.

Retrospective analysis rejected

"The FDA has been consistent in not allowing a retrospective analysis of a subset of data as the basis for approval," Wischnia notes. What ATS did to perform the additional clinical trials of Dermagraft was "analyze its clinical data retrospectively, and what the company realized was that Dermagraft had higher viability after ATS had changed to a more effective cryopreservation process, enhancing the bioactivity of the product with a higher percentage of living cells engaging the wound bed, thus enhancing the product's effectiveness."

ATS decided to look only at the patients treated after the change in the cryopreservation process," Wischnia said, "and these healing results were what was submitted to the FDA. The FDA said that it couldn't approve retrospectively with only a subset." The result was the delay of approval and huge expenses for ATS, as well as lost revenue.

In the meantime, other market forces had come into play that would increase the difficulty ATS faced in gaining money through capital markets. Another tissue engineering company, Organogenesis (Canton, MA), had come to the market with a diabetic foot ulcer product, Apligraf, which had several problems, including a short shelf life and the need for patient immobilization, which slowed acceptance.

Organogenesis, which eventually filed for bankruptcy, failed to meet its financial targets. "Many felt that tissue engineering had become a poor investment prospect," Wischnia says. "The companies had made assumptions that everyone with diabetic foot ulcers would use the product and this showed up in their projections, he notes. "The experience colored the view of the investment community, which ATS similarly had to overcome."

Box 6.12 (*cont.*)

ATS finally got FDA approval for Dermagraft at the end of September 2001, and the product received a reimbursement code from Medicare in March of 2002. But sales didn't grow according to forecast, partially due to the fact that Dermagraft still needed approval from regional care intermediaries for reimbursement.

Smith and Nephew, which handled the marketing for Dermagraft, had only attained coverage for the product in about half of the United States when ATS finally filed a voluntary petition for reorganization under the Chapter 11 bankruptcy code in October of 2002. During the final year of operations, the company's stock plunged from over $4 per share to the $1 range as the public markets took an increasingly sour view on investing in tissue engineering.

On November 13, 2002, ATS decided to undertake an orderly liquidation of assets, a decision based on the difficulty it foresaw in raising additional money to sustain operations, including the $25 million that it would need to match funds from Smith and Nephew for continued operation of the Dermaquip joint venture.

On the eve of the declaration of bankruptcy, the market capitalization of ATS hovered around $14 million, down from approximately half a billion dollars at the peak of expectation in 1996. Since the declaration of bankruptcy, Smith and Nephew purchased ATS' half of the Derma-quip joint venture at a fire sale price of $10 million, while offering employment to about 110 of ATS' 220 employees. . . .

By-product submitted to FDA

As a by-product of the Dermaquip joint venture, ATS developed additional products toward the end of its activities. Derma-quip produced human collagen that was sold to Inamed (Santa Barbara), a marketer of products for aesthetic applications. The product was an injectable collagen used to hide wrinkles and was an alternative to bovine-derived products on the market.

In early 2001, the FDA surprised ATS by requiring skin allergy testing before granting approval for the product. One year and a 400-patient skin trial later, marketing approval was granted, but again, regulatory issues had cost the company badly needed revenues.

Steep learning curve

Overall, it was the steep learning curve in regulatory and reimbursement matters that got the best of ATS. "The company needed to coordinate its regulatory and reimbursement activities" so that both would be optimized and the least amount of time would be lost in getting the products to market, Naughton observes. . . .

Box 6.13 How the first CAR T-cell gene therapy got approval in the United States and Europe

Adapted from Seimetz et al. (2019), and with assistance from Ulrich Granzer, Founder of Granzer Regulatory Consulting & Services, Germany.

Background

In August 2017, KYMRIAH (tisagenlecleucel), developed by Novartis, was the first CAR T-cell gene therapy to receive FDA approval.

KYMRIAH (tisagenlecleucel) is a CD19-directed genetically modified autologous T-cell immunotherapy indicated for the treatment of patients with B–cell precursor acute lymphoblastic leukemia (ALL) that is refractory or in second or later relapse.

Each dose of tisagenlecleucel is customized using an individual patient's own lymphocytes. The patient's T cells are collected and sent to a manufacturing center where they are genetically modified to include a new gene that contains the genetic information for a CAR (an artificial receptor called a "chimeric antigen receptor"). This new genetically built receptor is then expressed to the CAR cell surface that directs the T cells to target and kill leukemia cells that have a specific antigen (CD19), i.e. a "docking station" on the surface. Once the cells are modified, they are infused back into the patient to kill the cancer cells. The treatment was hailed as "a new frontier in medical innovation with the ability to reprogram a patient's own cells to attack a deadly cancer."

Because of the risk of CRS and off-cancer cell toxicities, KYMRIAH is available only through a restricted program under a Risk Evaluation and Mitigation Strategy.

CAR-T cells belong to the regulatory group of advanced therapy medicinal products (ATMPs). Due to the cell-/gene-based complex nature, ATMPs are far more challenging to develop than other, more defined, medicinal products.

Regulatory process and challenges

This ATMP uses the patient's own cells, multiplied and modified external to the body with a foreign gene and reinserted into the patient. The product is not treated as an autologous transplant for regulatory review purposes, but is classified a cell-therapy product since it is functionally modified from the original human cell.

U.S. FDA – steps in the regulatory process

Target indication: Pediatric and young adult ALL

PreIND Meeting April 2013 & March 2014
Special Protocol Assessment (SPA) March 2014
IND submission September 2014
Rare Disease/Orphan Product Designation for ALL January 2014
Breakthrough Therapy Designation February 2016

Box 6.13 (*cont.*)

Pre-BLA (biologics license application) meeting November 2016
BLA submission February 2017
Breakthrough Therapy Designation April 2017

This product was a breakthrough in this new class of gene/cell therapy and the regulatory interactions were a learning process for both sides (sponsor and FDA), especially since there were no clear published guidances from the FDA or any other regulatory body.

In fact, in March 2017, the U.S. FDA introduced the new Regenerative Medicine Advanced Therapy (RMAT) designation, thus recognizing the enormous potential of these medicines and the need for efficient regulatory tools to accelerate their development and their commercial availability. The development of regenerative medicines is very challenging because of their complex and unique nature, especially to the rather inexperienced small- and medium-sized developing enterprises. With the new RMAT designation, the FDA aims at providing intensive support to companies developing cell- and tissue-based therapies, tissue-engineering products, and combination treatments. Interestingly, for RMAT designated applications, FDA often asks the companies for meetings. In one case this added up to a total of 34 meetings between the FDA and the biotech company prior to submission of a BLA.

Issues found by the FDA during program review that had to be addressed:

1. Replication competent lentivirus (RCL) – where the virus carrying the antigen gene into the cell starts replicating itself, causing a viral infection – but to date, no RCL has been detected in any clinical trials using a lentiviral vector-transduced cell product
2. Insertional mutagenesis, where the new gene carried into the cell is inserted into an incorrect location in the genome – this issue has been addressed through vector design and a limited copy number per cell. However, there was still one case where a cancer gene was triggered via the experimental gene therapy during the early days of development leading to a rare type of leukemia in patients affected.
3. There were several major concerns during pre-license inspection of the manufacturing facility which were duly addressed

Key non-clinical studies conducted for KYMRIAH to address concerns included:

- Evaluation of the specificity of the CD19-binding domain using a human plasma membrane protein array,
- Assessment of in vivo antitumor activity of KYMRIAH in mouse xenograft tumor models,
- Evaluation of selected toxicology parameters, cell distribution, and persistence of KYMRIAH in tumor-bearing mice, and
- Genomic insertion site analysis of lentiviral integration into the human.

Box 6.13 (*cont.*)

Clinical development of Kymriah

In summary, Novartis conducted four Phase II trials and further planned one Phase I trial, including long-term follow-up and managed access program, as well as two Phase III trials, including event-free survival.

The pivotal trial submitted for registration was conducted under a Special Protocol Assessment. The SPA review allows FDA to provide input into the design of certain studies critical to marketing approval prior to initiation. By meeting prior to the start of a study, FDA and sponsors can streamline the approvals process because the scientific and regulatory requirements have already been agreed upon. A SPA, in this respect is a de facto contract between FDA and the applicant. If the applicant develops the drug as agreed with FDA and comes to a positive result, usually FDA will accept such a trial as pivotal for later approval.

Regulatory tools

KYMRIAH was granted "orphan designation" in the United States and the EU. In the United States it also received "breakthrough therapy designation," and KYMRIAH additionally received "rare pediatric disease" designation to qualify for a Rare Pediatric Disease Priority Review Voucher

European Union review at EMA

KYMRIAH applied under EMA's PRIority MEdicines (PRIME) scheme and received positive opinions from the Committee for Medicinal Products for Human Use (CHMP). The voluntary scheme provides early and enhanced scientific and regulatory support to medicines that have the potential to address, to a significant extent, patients' unmet medical needs.

An early nomination of a CHMP/CAT rapporteur at EMA often leads to a more conventional development being followed, which may or may not lead to an acceleration of development. The EMA-HTA Parallel Scientific Advice procedure (now called "Parallel Consultation with Regulators and Health Technology Assessment Bodies") facilitates the initiation of early dialog between medicines developers, regulators, and health technology assessment bodies to discuss and agree on a development plan that generates data that both parties may use to determine a medicine's benefit–risk balance and value.

---

regulatory approval pathway of a novel cell therapy that modifies the patient's own cells to enable them to effectively recognize and kill cancer cells, but this time resulting in a successful regulatory and commercial outcome for the patients and the company.

6.9   Summary

Devices, drugs, and diagnostics have two general regulatory paths: Innovative, novel products (first time in human) have a longer regulatory path requiring at least two pivotal clinical trials and more time and money than the process for products (generic drugs, devices with predicates) that are similar to other marketed products. These "similar" products have a shorter, less expensive regulatory path, where they have to show some form of equivalence to the products to which they are most similar. Low-risk or well-established products are exempt or waived from FDA reviews, but still have to adhere to some regulations. These products are OTC drugs, exempt devices and most ASR diagnostics products. The three main divisions in the FDA (CBER, CDER, and CDRH) review market applications and interact with manufacturers in a process that is driven by scientific principles.

The goal of the regulatory bodies worldwide is to ensure the safety of the health of their countries' public by getting safe and effective products to markets.

Exercises

Assuming a novel product is being developed:

6.1. Select the indication for your product and state the desired product label claims.

6.2. Identify and describe the clinical endpoint(s) for the product clinical trials.

6.3. If drug, then describe points of contact with FDA/IRB/IACUC and specific topics for emphasis and attention in these interactions/submissions.

6.4. If device, then discuss rationale for classification and pathway for marketing approval.

6.5. Feedback regulatory pathway information and rethink product development.

6.6. Generate timeline showing potential interactions (meetings and submissions) with the FDA mapped onto the product development stages.

6.7. What data will you submit to the FDA at those interaction points? Describe key data that will verify the mode of action and support the indication for the product, for each submission. Identify the timeline for these submissions and review processes.

References and additional readings

Amato, SF, and Ezzell Jr, RM. (editors). (2014). *Regulatory Affairs for Biomaterials and Medical Devices.* Elsevier Science and Technology. ISBN: 9780081015339

Beck, JM, and Vale, A. (2004). *Drug and Medical Device Product Liability Deskbook.* New York: Law Journal Press. ISBN: 1-58852-121-4

Benjamens, S, Dhunnoo, P, and Meskó, B. (2020). The state of artificial intelligence-based FDA-approved medical devices and algorithms: an online database. *npj Digital Medicine* 3, 118.

Cauchon, NS, Oghamian, S, Hassanpour, S, and Abernathy, M. (2019). Innovation in chemistry, manufacturing, and controls – a regulatory perspective from industry. *Journal of Pharmaceutical Sciences – Review* 108(7), p2207–2237.

Daniel, A., and Kimmelman, E. (Compilers) (2008). *The FDA and Worldwide Quality System Requirements Guidebook for Medical Devices.* ASQ Quality Press. ISBN: 9780873897402

Flannery, EJ, and Danzis, SD. (2010). *In Vitro Diagnostics: The Complete Regulatory Guide.* Food and Drug Law Institute. ISBN: 9781935065227

Harnack, G. *Mastering and Managing the FDA Maze: Medical Device Overview: A Training and Management Desk Reference for Manufacturers Regulated by the Food and Drug Administration*, 2nd ed. New York: ASQ Quality Press. ISBN: 9780873898874

Pines, WL, and Kanovsky, SM. (editors). (2020). *A Practical Guide to FDA's Food and Drug Law and Regulation*, 7th ed. Published by Food and Drug Law Institute. ISBN: 9781935065876

Pisano, D., David Mantus. D. (editors). (2004). *FDA Regulatory Affairs: A Guide for Prescription Drugs, Medical Devices and Biologics.* New York: CRC Press. ISBN: 1-58716-007-2

Royzman, I, and Shah, K. (2020). 10 years of biosimilars: lessons and trends. *Nature Reviews Drug Discovery* 19, 375.

Salminen, WF, Wiles, ME, and Stevens, RE. (2019). Streamlining nonclinical drug development using the FDA 505(b)(2) new drug application regulatory pathway. *Drug Discovery Today* 24(1), 46–56.

Seimetz, D, Heller, K, and Richter, J. (2019). Approval of first CAR-Ts: have we solved all hurdles for ATMPs? *Cell Medicine* 11.

Tobin, JT, Walsh, G. (2021). *Medical Product Regulatory Affairs: Pharmaceuticals, Diagnostics, Medical Devices.* Wiley-VCH Verlag GmbH. ISBN: 9783527333264

Van Norman, GA. (2016). Drugs and devices: comparison of European and U.S. approval processes. *JACC: Basic to Translational Science* 1(5), 399–412.

Websites of interest

510(k) regulatory pathway – find predicates, guidance documents, application guidance at www.fda.gov/medical-devices/premarket-notification-510k/content-510k

FDA guidance document on early collaboration meetings for medical device development: Early Collaboration Meetings Under the FDA Modernization Act (FDAMA) Docket Number: FDA-2020-D-0957. Feb 2001, discusses how to prepare and what to expect from these meetings with the FDA. Final guidance document is available at www.fda.gov/regulatory-information/search-fda-guidance-documents/early-collaboration-meetings-under-fda-modernization-act-fdama-final-guidance-industry-and-cdrh

Principles for Co-development of an In Vitro Companion Diagnostic Device with a Therapeutic Product found at www.fda.gov/regulatory-information/search-fda-guidance-documents/principles-codevelopment-vitro-companion-diagnostic-device-therapeutic-product.

Companion Diagnostics List at www.fda.gov/medical-devices/vitro-diagnostics/ companion-diagnostics. Latest guidance document at www.fda.gov/regulatory-information/search-fda-guidance-documents/principles-codevelopment-vitro-companion-diagnostic-device-therapeutic-product

Visit https://shreefalmehta.com/csbtbook for additional enriching readings around the topics covered in the book, topical updates on the content and for industry viewpoints and news.

Foreign regulatory bodies websites

European Union: EMEA – The European Agency for the Evaluation of Medicinal Products: www.ema.europa.eu/en

Canada, Health Canada: www.hc-sc.gc.ca/index_e.html

Japan, Ministry of Health and Welfare: www.mhlw.go.jp/english/policy/health-medical/pharmaceuticals/index.html

China, National Medical Products Agency: www.nmpa.gov.cn

India, Central Drugs Standard Control Organization (CDSCO): https://cdsco.gov.in/opencms/opencms/en/Home/

UK – MHRA (Medicines and Healthcare products Regulatory Agency, UK): www .gov.uk/government/organisations/medicines-and-healthcare-products-regulatory-agency

Australian Therapeutics Goods Administration: www.tga.gov.au/

https://cdsco.gov.in/opencms/opencms/en/Home/

7 Manufacturing

Plan	Position	Pitch	Patent	Product	Pass	**Production**	Profits
Industry context	Market research	Start a business venture	Intellectual property rights	New product development (NPD)	Regulatory plan	Manufacture	Reimbursement

---

Learning points
- When do you begin planning for commercial scale manufacturing?
- What does it mean to be at commercial scale of manufacturing?
- How is cost of goods (cost of final product) estimated?
- What controls and systems are needed to comply with regulatory oversight of manufacturing processes?
- What standards apply to manufacturing processes?
- What are the timing and quantity requirements for scale up of production for drugs devices and diagnostics?
- What key issues typically arise in the production or scale-up processes for biological drugs?
- How does the design for manufacturability and assembly prepare for device manufacturing?
- What are challenges to watch out for when transferring new products from the R&D stage into commercial manufacturing operations?

The transition from the new product development environment of early product versioning and animal testing, to manufacturing environment "operations" is a significant one for a project team. It means that the results to date have shown significant reduction in technology risk and the product has demonstrated feasibility, safety, and efficacy in living systems.

7.1 Technology transfer to manufacturing operations (drugs, devices, and diagnostics)

Technology transfer is the systematic means of conveying knowledge (product, process, equipment, and method), documentation, and skills between two parties. In manufacturing companies, the term "technology transfer" covers the process of converting R&D small scale processes or early design and laboratory prototypes to larger scale bulk manufacturing processes and specifically involves the transfer of the product and production know-how from the R&D group to the commercial operations group ("Operations"). It is important to note that in the context of intellectual property transactions and in the general business environment of new ventures, "technology transfer" refers to the process of licensing of intellectual property and bringing patented technology in from the university labs to a commercial product development environment.

In scaling up from R&D synthesis processes or from laboratory prototypes, the information that was developed in R&D may not have the same relevance in this new, larger-scale environment that is concerned about reproducibility and reduction of defects per thousand/million products made, while adhering to the highest degree of regulatory compliance and accountability in highly documented processes. Operations organizations value compliance, reliability, predictability, and efficiency. In contrast, R&D organizations value scientific innovation, creativity, flexibility, and development speed. There is not only a potential culture-clash in communication between R&D and Operations but also some willful ignorance about the parameters and boundaries within which manufacturing operations organizations must operate. Box 7.1 contains illustrative notes from the biopharmaceutical industry regarding technology transfer between process development in R&D and manufacturing engineers. These differences have historically caused great trouble in the technology transfer process.

Why do "technology transfers" frequently fail?

- "Over the fence" transfer (sharp transfers)
- Lack of respect
- Lack of engagement of either party
- Lack of ownership following transfer
- "Not invented here" syndrome
- Lack of common goal
- Too many transfers
- Incomplete, inadequate documentation

Three main models of technology transfer are (1) push (let R&D drive the transfer); (2) pull (let Commercial Operations participate and share "ownership" early in the product development process); and (3) personnel transfer. Personnel transfers from R&D to Operations are usually the most successful. However, the R&D employees transferred to operations may suffer as their values may be better suited to the R&D organization. Some organizations have adopted a pull model where R&D and Operations co-develop the manufacturing process and product in the final stages. This integrated model embeds Operations technical functions into the product development (NPD) process to learn about the manufacturing technology and the product. This pull model also enables R&D to take the product through approval and Operations to successfully validate and sustain production following launch.

Box 7.1 Technology transfer in a biopharmaceutical manufacturing company

From an interview with Dan VanPlew, currently the Executive Vice President and General Manager, Industrial Operations and Product Supply at Regeneron Pharmaceuticals, Inc., with past experience managing manufacturing at Chiron Biopharmaceuticals and Crucell Holland BV. The comments in this text are not intended to be understood as specific references to any particular company or organization, including without limitation Crucell, Regeneron, or any of their collaborators.

Experiences in technology transfer from R&D to manufacturing

Here is a roadmap of key issues and some learnings from R&D to commercial activity with biological drug molecules. Some of the less obvious things smaller companies may miss:

- My first suggestion in considering transition to commercialization activities is that companies should *conduct a freedom to operate search on their patents before scaling up* for commercial production and use this understanding of FTO risk to ensure they are making informed decisions that take into account their own intellectual property rights as well as those of third parties.

   As one example, when you are starting the commercialization stagere with your biological drug, a significant "forgotten" technical hurdle could be a patent infringement claim from a third party. Companies with an interest in the same indication or with similar technology may aggressively act against any potential unrelated party by filing infringement lawsuits in various countries, whether valid or not. Almost every drug launch I have been involved with has been affected by FTO analysis and actions, sometimes obviously frivolous claims filed just to delay the launch. If the global manufacturing site for the new drug is located in a country impacted by legal wrangling, the entire commercialization engine and even R&D can be impacted, including in countries where the company may be free to sell. There are countries (e.g., Singapore and Ireland) where it seems easier to preserve your freedom to operate and this plays a role in why many companies go there for manufacturing scale up. Hence, I recommend that one of the first planning steps after early clinical trials, should be to complete a Freedom to Operate study of the patents that may relate to the molecule.

- *Respect assay transfer from R&D to manufacturing.* Assays are as important as the actual production process. Biotechnology drug production frequently relies on qualitative assays to derive a quantitative number, which results in some interesting control strategies for assays. As only one example, I recall a situation some time ago where we had a higher particulate level than we expected. We had dozens of people working on this to find the solution and made many trial changes to the process. But after a lot of digging, it turned out the assay was not being operated correctly. Specifically, it turned out the samples weren't being handled correctly and were experiencing more freeze-thaws than expected, and the samples were too near the elements of the freezer motor which exacerbated the situation. Another time, I had a product showing greater degradation than expected and it caused a significant delay to the program's launch. We

Box 7.1 (*cont.*)

discovered the product samples were being exposed to a special type of bulb used in the freezers. We only discovered the root cause after we learned the assay was not detecting what we expected. In yet another example, I have seen staff at manufacturing sites run an electrophoresis gel, keep it for staining and then leave the room to browse the internet (losing track of time) and then return to process the gel. Results often show a highly marked or amplified noise signal, beyond the allowed level, potentially failing the qualification test or requiring the test result to be questioned or repeated, delaying the lot processing. Qualitative assays that are commonly used in biological production assessments must have careful documentation and process transfer protocols as to the conditions, color development, time, or intensity of staining, etc., before being applied in the manufacturing environment.

- *BLAs (biologics license applications; sometimes called "marketing authorizations").* As a generic term, BLAs have different lead and cycle times by country and companies can have as many as 40 BLAs in progress over a few years. These staggered global *BLA approvals, separated by months or years, lead to subsequent issues with SKU (stock keeping units) proliferation and complex manufacturing lot controls.* These filings have different timelines, expectations, the establishment of process and specification requirements. All of these result in country-specific inspections, and 40 different sets of expectations for language and data analysis.

  Various countries' regulatory bodies may have differing requirements – e.g. for lot qualification thresholds and variance levels permitted. Staggered approvals can result in some unintended and frequently illogical consequences – e.g., some manufactured lots may not pass one country's regulatory requirements but may be just fine for others. For example, Japan is seen as having very stringent testing requirements whereas others perform some different tests with wider leeway on variances. Practically, manufacturing quality controls should argue for the widest quality control variances possible that can achieve human clinical benefits without increasing any risk. BLA molecules are complex to analyze and tests are varied and sometimes qualitative. There is continuous learning over time with many batches of production on how the various inputs and process controls affect test variances and clinical outcomes. In the meantime, as BLAs are approved in various countries, the process shifts that occur over this learning need to be carefully considered in the context of regulators' allowable parameters. And all of the countries regulators' exchange information amongst themselves, so if you report an issue to ANVISA (Brazil) – for example – then EMA (Europe) will have their antenna raised.

- *Actively assay your incoming materials.* One time we had media going from a 70% acceptance rate on our incoming testing to a 30% acceptance, clearly a sign that something was not right at the supplier's production line. However, many of

Box 7.1 (*cont.*)

the specific types of testing of incoming materials are only added progressively with learning. If you don't know what to look for you don't know how to test. For example, I know of one organization that learned about a drop in yield due to a Vesivirus 2117 contamination, but they had no reason or expectation to be testing for that virus before they found the problem. And as they learned about the problem and how to remediate it, the regulatory guidance in the industry changed with that learning and forced all other companies to change their approaches, causing companies to invest more in their viral control strategies. Regulatory inspectors can make requests that can take couple of years to put in place, So, it is very important to *keep track of ongoing trends in industry*. In fact, I have people whose main job is to scour regulator inspection reports (form 483), warning letters, consent decrees, fines, industry or trade publications, conference abstracts etc.

- Make efforts to *broaden your proven* and *acceptable ranges for your materials and products*. For example, *carry out hold time lots to get broader specifications* on your product and production controls. The FDA permits manufacturers to test product quality by holding qualification production lots for specific periods of time in process. The availability of this particular step is not known by many small companies as they are initially purely science-led. The hold-time lot process step is a good way to get broader specifications on your clinical lots for regulatory approval, and help maintain surety (and quality) of supply.

This issue of manufacturing controls and allowable variances on assay results has significant impact on the supply chain controls and costs of goods on the product. As a first-time producer, where the goal is to get a product approved and on the market in the fastest time possible, the production lots are made in a narrowly defined range of specification values in order to make sure the lots meet regulatory approval. However, approval with a narrow range of production values has an impact up stream, as each of your incoming materials have to fit even narrower specifications which may lead to rejection of lots at later times, causing untenable pressure on producers or cost increases as producers struggle to consistently meet stringent and narrow variances. On the other side, broadening the specifications of the product lots must be balanced and gated by clinical outcomes, ensuring that the drug product not cause undesirable side effects or allergic reactions in the patient.

I recall a situation some time ago where we only had to make two lots in order to get the drug through all of the clinical trials. So, the entire BLA – which relies on clinical experience – was dependent on just those two lots to set specifications, etc. You will make a drug hundreds of times over its life and you will live with the narrowness of those two initial lots for years. You will not fully know the magnitude of your mistakes until the supply chain has run for a while – the memorialization of your decisions in the BLA will remain with you for a long time. It took that organization over a decade to fix the gaps associated with that

Box 7.1 (*cont.*)

terribly limited dataset in the initial filing – throughout that entire time, they had to manage incredibly tight specifications that did not represent the true robustness of their process. This resulted in many lost lots, many sleepless nights, and weaker negotiating positions due to an underlying fear when dealing with vendors.

- Manage your supply chain with humility and a "firm handshake." In the biotech industry, many raw materials (like soy) are purchased in tiny volumes compared to typical production volumes in the massive industrial food processing facilities. Hence, insisting on lot specifications and GMP [good manufacturing practices] quality and cleanliness in these large-scale facilities, while aiming at lowest costs is highly unlikely. Instead, you can focus on controlled lots that can be delivered at a reasonable cost. Many first-time manufacturing engineers are surprised to learn that the entire supply chain for their materials is not as pristine as their own ultra-clean sanitized facilities for GMP production of the biological drug. For example, in one telling story, I remember going to a vegetable production facility in the Midwest to review the processing plant. During the unloading of the vegetables from a large dump truck into containers fitted with grates (to sift out large rocks), a worker went around banging the side of the dump truck with a hammer. On asking him why he was doing that, turns out he was trying to scare the rats to get out of the truck as the vegetables were lowered in.

- You have to remember *that manufacturing biological drugs in cells is like taking care of a living breathing human body*. The sparging of the bioreactor with oxygenated air has to have bubbles of the right size, mixed at the right speed and agitated to ensure good distribution without damaging the cells, right amount of food to the cells has to be given and the CO_2 removal at the top has to be done with correct flow rates. Cells are fragile and react in unexpected ways to subtle changes. I remember a situation where a change had been made to a smaller bioreactor (second step in a multi-step scale up Seed Train process) due to a repair. A seemingly innocuous alteration to an impeller in that smaller reactors and subtle damage to the cells in that second step reactor resulted in anomalies many steps away from the exposure. The cells have a real memory, and if you "kick" them they will remember.

- *Technology transfer is truly a contact sport*. In one company I have seen the process development (R&D group that designed the original production processes) and the manufacturing engineers (in charge of scaling up) in shouting matches and almost coming to blows with the frustrations in communication and the feeling that the other side did not respect them or the context of their concerns. An analogy that tries to explains the differences in thinking and activity in the process of transfer from R&D to manufacturing is that it is similar to the differences in going from a DMV training driver to driving for Fedex. The other example to explain the process scaling context is that of a muffin being

Box 7.1 (*cont.*)

made one at a time using a quick bake mix and a light bulb in a plastic oven to then scaling to a kitchen where you have to make dozen muffins using raw ingredients and a real oven. In this step, the way that the single muffin was made is indicative but not informative on how to do it at the scale of a dozen with different input materials and tools. And then consider going to a catering event with hundred muffins being needed and the mixture changes with different tools, again requiring learning to get the results close to the original single muffin and to do it consistently and reliably.

A story that I remember from one of my previous companies, shows the reason I choose the above examples. When we visited the process development team to see why our methods was not producing the same results and yields (titers) as their developed process, we found that they had selected the "golden" medium (cell culture food) lot over multiple lots of medium they received from the manufacturer and proceeded to use only that one lot of medium for all process development. When we explained the real-life challenges and reality that incoming medium has some variability that needs to be taken into account else we will be at a 10% acceptance rate if we only insisted on each lot of the medium fitting exactly within the specifications of their "best" lot, their initial resistance was replaced by collaboration.

I consider a few specific steps as a model that I've seen yield 100% success in transferring R&D products to manufacturing scale. Comingling is a very important activity between R&D process scientists and manufacturing engineers. Ideally, the manufacturing engineers will work closely with R&D experts, and attend data meetings in person with the process scientists; and, the R&D process scientists should participate in the manufacturing team's technical reviews, watch the process scale up trials and generally work side by side to trouble shoot. Over time, such site visits and personal interactions can create an atmosphere of trust and improved appreciation of the challenges each side faces. As these interactions improve, I have seen the process scientists start to make calls to counterparts in manufacturing to review problem-solving choices as the process scientists proceed in their development experiments. It is also important for the physical plant – e.g., equipment – receiving the technology to match as closely as possible to the innovator's equipment, laboratories, and clean spaces.

Today, thankfully, with integrated, multidisciplinary teams formed early in product development, there is better planning and design for manufacturability than previous historic siloed attitudes of "throw it over the wall to the engineers and let them figure out how to make thousands/millions of product – we're done with our final design and testing and it worked when it left our hands...". In fact, it is increasingly clear that design for manufacturing and manufacturing engineers should be engaged much earlier in the design process as advisory roles, so long as they can provide input about potential manufacturability issues rather than leading the design iteration process.

7.2   Regulatory compliance in manufacturing

Regulatory oversight and adherence to industry standards are key considerations in the final product development stage of final manufacturing. Most manufacturing sites are visited by the FDA before final approval of a new product (pre-approval inspection) and these visits continue post-approval. Most manufacturing companies also seek to bring best practices and efficient, high-quality performance into their organizations through gaining certification under certain standards like the International Standards Organization (ISO), which require site inspection. These various site inspectors will want to see accurate and detailed record-keeping, careful consideration to the maintenance of hygienic or sterile conditions (as appropriate) throughout the process, and most importantly, the adherence to ensuring reproducibility and reliability of the manufacturing process. Compliance with manufacturing regulations and certain standards is thus a critical part of getting a safe and effective product through approval to market. The FDA regulations and controls are together referred to as current good manufacturing processes (cGMP).

7.2.1   Current good manufacturing practices

Current good manufacturing practices (cGMP) are defined as "a set of current, scientifically sound methods, practices or principles that are implemented and documented during product development and production to ensure consistent manufacture of safe, pure and potent products."

The cGMP regulations are published in the Federal Register in 21 CFR:

Parts 210 & 211	Drugs
Part 820	Medical Devices
Part 606	Blood Products
Part 110	Food

The cGMP regulations:

–   Apply to both manufacturing process and facilities
–   Are expected to be in place throughout clinical studies
–   Controls increase as the process progresses to final commercial stage
–   Include specifically, sterility assurance & validation
–   Include implementation of a quality assurance (QA)/quality control (QC) program
–   Require everyone to document, document, document . . .

The cGMP controls thus cover the following activities (not a comprehensive list):

–   Record-keeping, documentation
–   Personnel qualifications, training
–   Raw materials selection, purchasing
–   Equipment verification
–   Process validation

- Specifications and quality control
- Sanitation, cleanliness
- Complaint handling

In particular, there has been a strong emphasis on establishing quality and design-based systems in pharmaceutical and medical device manufacturing. These quality systems have a bearing on all processes, including management processes. A quality system involves quality control (QC) and quality assurance (QA).

Quality control (QC) usually involves (1) assessing the suitability of incoming components, containers, closures, labeling, in-process materials, and the finished products; (2) evaluating the performance of the manufacturing process to ensure adherence to proper specifications and limits; and (3) determining the acceptability of each batch for release. *Quality assurance* (QA) primarily involves (1) review and approval of all procedures related to production and maintenance, (2) review of associated records, and (3) auditing and performing/evaluating trend analyses.

The cGMPs are constantly being revised to accommodate advances in technology (the discovery of protein A resin for purification of monoclonal antibodies, for example) or to address specific issues that were not know before (e.g., presence of hantaviruses in some raw materials used in the manufacturing process). Changes to cGMPs occur via new initiatives from regulatory authorities or from the combined knowledge of the field inspectors who visit each production facility. Their assimilated inspection histories are the basis upon which all facilities are inspected and subsequent changes recommended.

The FDA (CDER and CBER) issued a guidance document, *Guidance for the Industry, Q7 Good Manufacturing Practice Guidance for Active Pharmaceutical Ingredients*, in 2018. This guidance document was developed over the years by an expert working group (Q7) of the International Conference on Harmonization of Technical Requirements for the Registration of Pharmaceuticals for Human Use" (ICH). The regulations and guidelines in this document are adopted by the regulatory bodies in the United States, EU and Japan. This guidance attempts to standardize GMPs across the world.

7.2.2 Validation

Validation is defined as "the act of proving that any process, procedure, equipment, material, activity or system leads to expected results." Process validation is a part of current good manufacturing practice and is required in the United States and EU for a manufacturing license. A validated manufacturing process has a high level of scientific assurance that it will reliably and consistently produce acceptable product. The emphasis on this process (or sets of processes, as it can be a multistage activity) is a significant part of the activity of synthesizing the product. As an example, the process of cleaning and validating equipment from one batch to the other in biological drug manufacture can sometimes take a few days, and production of one batch can take a similar length of time. The validation process is even more extensive (weeks) if a different drug is to be made in the same production equipment. Validation in devices includes checking the product characteristics against initial customer and design inputs, and can be completely different types of processes in the context of different types of biomedical technologies.

7.2.3   Drug manufacture regulations – control systems reviewed for compliance

In August 2002, the FDA launched an overhaul of the pharmaceutical GMP regulations with a new focus on integrating quality systems and risk management approaches due to a high number of prescription drug recalls (354 in 2002, up from 176 in 1998), largely related to manufacturing quality issues. This initiative was called the "Pharmaceutical cGMPs for the 21st Century" and had a goal to encourage industry to adopt modern and innovative manufacturing technologies.

Compliance with manufacturing regulations and standards requires that six distinct control processes and systems must be in place at any organization:

1. Materials system
2. Production system
3. Packaging and labeling system
4. Laboratory systems
5. Facilities and equipment system
6. Quality system

The specific points evaluated by the FDA inspectors for each system are reviewed in excerpts from the FDA Inspectors Manual in Appendix 7.1 and makes for informative reading. The six recording and process control systems which are required to demonstrate regulatory compliance are inter-dependent, with the quality system providing the foundation for other systems within it, as seen in the schematic in Figure 7.1 and as discussed in further detail in Appendix 7.1.

7.2.4   Device and diagnostic manufacture regulations – control systems reviewed for compliance

A quality systems approach to medical device design and manufacture (also includes diagnostics) was implemented (in 1996) in the Quality System Regulation (21 CFR Part 820) revision to the current good manufacturing practice (cGMP) requirements. The QSR rules for device development (see Section 6.5.7) and the cGMP rules for pharmaceuticals both have a common goal – to ensure that the manufacturer has put a set of reproducible processes in place to safely and reliably make commercial devices on which people's lives depend.

An important component of the revision was the addition of design controls which have been discussed in Chapter 5. Design controls are a system of checks and balances that make systematic assessment of the design an integral part of product development.

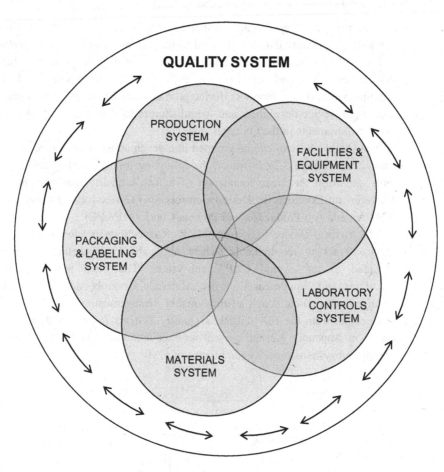

**MANUFACTURING PROCESSES AND CONTROLS
CGMP REGULATORY COMPLIANCE**

Figure 7.1 Quality system in pharmaceutical GMP controls and systems is the foundation for all other controls. Figure adapted from the FDA document mentioned in Appendix 7.1.

As described on the final FDA ruling published in the Federal Register (www .govinfo.gov/content/pkg/FR-1996-10-07/html/96-25720.htm):

"A Government Accounting Office (GAO) review of medical device recalls during a six-year period in the 1980s found that about 44 percent of quality problems were attributable to errors or deficiencies in the device design and could have been prevented with proper design controls. Design-related defects have been found in such critical products as heart valves, catheters, defibrillators, pacemakers, ventilators, patient chair lifts and laboratory tests. . . .

"The [Quality System] regulation requires that various specifications and controls be established for devices; that devices be designed under a quality system to meet these specifications; that devices be manufactured under a quality system; that finished devices meet these specifications; that devices be correctly installed, checked and serviced; that quality data be analyzed to identify and correct quality problems; and that complaints be processed. Thus, the QS regulation helps assure that medical devices are safe and effective for their intended use. The Food and Drug Administration (FDA) monitors device problem data and inspects the operations and records of device developers and manufacturers to determine compliance with the GMP requirements in the QS regulation."

The Quality System can be grouped into seven subsystems and related connected subsystems (Figure 7.2); however, the following four are considered major subsystems and form the basic foundation of a firm's quality management system: (i) Management Controls, (ii) Design Controls, (iii) Corrective and Preventive Actions (CAPA), and (iv) Production and Process Controls (P&PC).

The Medical Device Reporting (MDR), Corrections and Removals, and Medical Device Tracking requirements (where applicable) are satellite systems that are included in the overall CAPA subsystem. The three remaining subsystems (Facilities and Equipment Controls, Materials Controls, and Document/Records/ Change Controls) cut across a firm's quality management system.

Excerpts from the FDA/CDRH's *Quality Systems Inspection Manual* are reproduced in Appendix 7.2 and present an excellent summary of the requirements of various subsystems towards compliance of GMP.

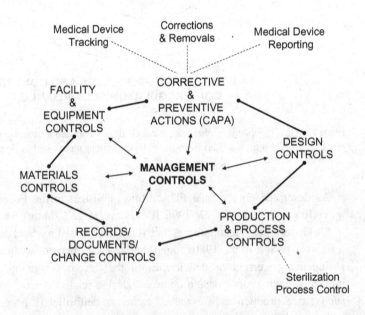

Figure 7.2 Medical device manufacturers have to put several subsystems and a few satellite systems in place as process controls and documentation controls in order to make their facilities and processes GMP compliant. The management controls focus on the quality system which runs through all the main subsystems shown here.

Box 7.2 Importance of written procedures in GMP compliance
Contributed by Lawrence Roth.

Written procedures are essential to compliance with GMP regulations

One central tenet to compliance with GMP regulations is to develop and maintain written procedures. The concept of "document what you do then do what is documented" is central to the guiding principles of the GMP regulations. Written documentation of the manufacturing process is an integral component of any GMP system. This documentation provides the trained operator with clear instructions on how to perform a process; i.e., equipment required, set-up procedures, step-by-step processing, components/raw materials to be used, and in-process inspections. This documentation is created during the development of a product and is finalized as part the Device Master Record (DMR) following design validation prior to transfer to full manufacturing. The objective is to ensure that the product manufactured in volume is equivalent to the product that supported the information submitted to FDA for approval and that the manufacturer has adequate control over the production to ensure unit-to-unit consistency. Lack of adequately documented written procedures is considered a violation of the GMP regulations.

A search of the CDRH database provides numerous examples of Warning Letters issued to companies for failure to develop adequate procedures for documenting its manufacturing procedures. A Warning Letter is issued only for the most serious offenses identified by FDA during a routine inspection of a manufacturer. For example, in September of 2004, Daavlin Distribution Company was issued Warning Letter (Warning Letter CIN-04-22469) for numerous GMP infractions. Daavlin is a leading manufacturer of phototherapy products sold throughout the United States and in over thirty other countries around the world. In addition to offenses related to sale of an unapproved device, the FDA cited the following infractions related to documentation of the manufacturing process:

> Failure to maintain device mater records (DMRs), as required by 21 CFR 820.181. For example, there are no established and implemented DMRs for any of the medical devices manufactured, which includes the 3 Series Full Body Phototherapy Device, the Spectra Series of Phototherapy devices and DermaPal.
>
> Failure to establish and maintain procedures to ensure that the device history records (DHRs) for each batch, lot, or unit are maintained to demonstrate the device is manufactured in accordance with the DMR and the requirements of this part, as required by 21 CFR 820.184. For example, there are no established and implemented DHRs for any of the medical devices manufactured, which includes the 3 Series Full Body Phototherapy Device, the Spectra Series of Phototherapy devices and DermaPal.
>
> Failure to develop, conduct, control, and monitor production processes to ensure that a device conforms to its specifications, as required by 21 CFR 820.70(a). For example, process control procedures for the 3 Series Full Body

> **Box 7.2** (*cont.*)
>
> Phototherapy Device, the Spectra Series of Phototherapy devices and DermaPal have not been implemented. Also, the procedures are not signed and approved by management and QA.
>
> Establishing and maintaining written procedures is an important component of systems required for manufacturing a medical device. Failure to comply with the GMP regulations related to manufacturing can have serious effects on the business including loss of goodwill with FDA and customers, loss of sales, and potentially, criminal proceedings.

7.3   Manufacturing standards

7.3.1   What are standards and what is their purpose?

Standards are "a prescribed set of rules, conditions, or requirements concerning definitions of terms; classification of components; specification of materials, performance, or operations; delineation of procedures; or measurement of quantity and quality in describing materials, products, systems, services, or practices" [National Standards Policy Advisory Committee, Dec. 1978]. Standards, if correctly established and tested without bias, are useful tools to communicate between groups and also can be used to set thresholds of acceptance. Standards are important to demonstrate compliance to regulations that reference them and to give a certain comfort to the users who see a certification of having met a recognized standard, on the product.

For our purposes, standards can be classified into:

Process standards

These are standards established for validating specific manufacturing processes.

Testing standards

These are methodologies that must be followed to test products or materials – e.g. some standard methods for testing the viscosity of fluids or compression test for a hip implant are already established by professional societies and regulatory authorities who have accepted those standards will question the tests if they do not follow the established standard methodologies

Product or service standards

The product must perform to an established standard. E.g. purity of a drug compound or sterility of a device or the reproducibility of a diagnostic test are also expected performance standards.

7.3.2   Who sets standards?

Various professional bodies in different fields set the standards. In the United States alone, approximately 30,000 current voluntary standards have been developed by more than 400 organizations. The International Organization for Standardization (ISO) probably produces the largest number of internationally recognized Standards. ISO is a non-governmental organization, and is composed of a network of the national standards institutes of 157 countries, on the basis of one member per country, with a Central Secretariat in Geneva, Switzerland, that coordinates the system. More than 20,000 experts from all over the world participate annually in the development of ISO standards. Two of the more well-known ISO standard families are the ISO 9000 and ISO 14000 families, which are "generic management system standards." From the ISO website: "ISO 9000 is concerned with 'quality management' and covers the organization's activities for enhancing customer satisfaction by meeting customer and applicable regulatory requirements and continually to improve its performance in this regard. ISO 14000 is primarily concerned with 'environmental management.' This means what the organization does to minimize harmful effects on the environment caused by its activities, and continually to improve its environmental performance." More information is available at www.iso.org.

Professional societies have committees that meet regularly to set and update standards. For example, the American Society for Testing of Materials (ASTM) has committees of professional engineers who meet to set and update standard methods for testing of materials. The CDRH, in particular, references multiple standards from different sources in its guidance for specific classes of products and many of those standards are from the ASTM, defining certain limits or bounds of performance that must be met by the device or by materials used in the device. See www.fda.gov/medical-devices/device-advice-comprehensive-regulatory-assistance/standards-and-conformity-assessment-program for further details on how the CDRH evaluates and applies various standards from the organizations listed below to assess quality and compliance from manufacturers.

Some examples of organizations that publish standards (not a comprehensive list) include the following

AAMI	Assoc. for Advancement of Medical Instrumentation
ACR	American College of Radiology
ADA	American Dental Association
ANSI	American National Standards Institute
APA	American Psychiatric Association
ASA	Acoustical Society of America
ASME	American Society of Mechanical Engineers
ASQC	American Society of Quality Control
ASTM	American Society for Testing and Materials
CENELEC	European Committee for Electrotechnical Standardization
CGSB	Canadian General Standards Board

HPS	Health Physics Society
IEC	International Electrotechnical Commission
IES	Illuminating Engineering Society of North America
IEEE®	Institute of Electrical and Electronics Engineers
INMM	Institute of Nuclear Materials Management
ISO	International Organization for Standards
LIA	Laser Institute of America
NCCLS	National Committee for Clinical Laboratory Standards
NEMA	National Electrical Manufacturers Assoc.
RESNA	Rehab. Engineering and Assistive Tech. Soc. of N. A.
SAE	Society of Automotive Engineers
UL	Underwriters Laboratories, Inc.®

7.3.3   Which of the thousands of standards apply to my product?

The best practical ways to find the standards that should apply to the product being developed:

- Check FDA special controls and other guidance documents.
- Check industry association for any offered training courses or seminars.
- Examine existing marketed products and associated literature.
- Talk to people in the business – by asking fellow professionals or attending meetings.
- Hire experienced employees or consultants who have employed. standards in product development or organization planning.
- Speak to insurance companies as to their requirements to insure your facility or your products.

7.3.4   What are "clean room" standards?

An important segment of standards applies to the manufacturing environment. Most manufacturing facilities have sterile/aseptic processing areas which contain "clean rooms" that are used in specific critical processing steps. Clean rooms are frequently used in final manufacture of biomedical products. A clean room is an environmentally controlled area that has a low level of environmental pollutants such as dust, microbes, aerosol particles and vapors. Clean rooms have special air handling systems where outside air is filtered and the air inside is recirculated through special filters (high-efficiency particulate arrestance; HEPA) to remove internally generated contaminants. The clean room environment can be a small room or can be an entire manufacturing plant.

In the United States, clean rooms are defined by the U.S. Federal Standard 209. The class of a clean room is determined by measuring the number of particles greater than 0.5 microns in 1 cubic foot of air. The clean room classifications (under the U.S. standards) range from Class 1 (1 particle/cu-ft of air) to Class 100,000 (100,000 particles/cu-ft of air).

The European Union (EU) established a standard in 1997 and it has been widely adopted in Europe as a part of EU GMP. The clean rooms are classified as A through D, with Class A being the "cleanest."

An ISO (International Standards Organization) standard (ISO 14644) also exists for classification of clean rooms. The classifications range from ISO Class 1 (0 particles >0.5 microns/cu-m of air) to ISO Class 9 (35,200,000 particles).

Most pharmaceutical or device manufacturing processes use clean room technologies that are ISO Class 5, Class 7 or Class 8. For example, tableting and oral liquid preparation facilities should meet at least ISO Class 8 standards, while aseptic filling and manufacture/packaging of implantable medical devices would be done in an ISO Class 5 standard facility.

Check FDA or other regulatory body guidance documents for clean room requirements for aseptic manufacturing facilities.

7.3.5   ISO 13585 standard for medical device manufacturers

ISO 13485 is the ISO standard for Quality Systems that medical device manufacturers have to have in place, in order to comply with regulations internationally. This standard helps medical manufacturers design and implement a management system and an effective process for maintaining quality of their products, giving customers confidence in their ability to bring safe and effective products to the market. This internationally accepted standard is also used by certifying bodies and by the manufacturer to bring its supply chain partners to conform.

Most manufacturers of all types of products (non-medical) certify their quality systems under the ISO 9001 standard, which is focused on delivering customer satisfaction. The ISO 13485 based quality system, required for medical device manufacturers internationally, includes most of the requirements of ISO 9001 but adds a specific focus on documentation and safety requirements in the quality system and related management controls.

7.4   Manufacturing in drug development

The biologically active drug molecule is known as the Active Pharmaceutical Ingredient (API) (small molecule or large molecule biologic drug candidate).

The final dosage form includes the API and various other excipients that make up the liquid, powder, pill, or capsule that is to be delivered by oral or parenteral route. Many manufacturers make the API in one location and then ship the API to another facility that specializes in final blending, granulation, tableting, packaging or filling (in bottles, gels, syringes, etc.) and testing the various dosage forms (e.g. 5 mg, 10 mg, 25 mg pills; injectable fluid formulation). This provides the manufacturer with flexibility to make multiple types of products (children's medicine, adult dosage form) from one starting point of the active drug moiety. The FDA will evaluate validation data from the final dosage form in addition to validation of the API production.

Pharmaceutical manufacturing for small molecule (synthetic chemistry) drugs entails different technology and process platforms compared to manufacturing large molecule (biologic) drugs. Small molecule drugs need chemical reactors where controlled chemical reactions are used to synthesize the API in a step-wise fashion, whereas biologics need large bioreactor tanks where cells (which produce the biologic API) are grown under carefully monitored conditions. However, the business risks involved in the investment decision to manufacture in large scale, and the timing of this scale up and size of investments are very similar for both types of drugs. As discussed in the following sections, a robust market exists for contract manufacturers globally who serve multiple clients and make APIs and finished packaged product.

Small molecule drugs are made using large-scale (bulk) chemical reactors, filters, and dryers in chemical factories. Elements to monitor in the final output are: by-products of the chemical reactions (usually well known), removal of all solvents, and other properties of the API molecule, such as appropriate chirality or appropriate mix of enantiomers (which can have an impact on biological activity of the drug), crystalline forms of compound, purity of final product, and other parameters.

Biological drugs are typically made using bacterial or mammalian cells, with a selected cell line grown in a tailored and controlled environment. A whole bioreactor is filled with these cellular mini-factories, each cell secreting multiple quantities of the desired protein. A key process step and quality control issue in biomanufacturing is the separation process, to get the API out of the medium that the cells were kept in, while eliminating all other proteins that the cell produces. Other issues also include monitoring the cell phenotype and genotype to make sure the same protein is reproducibly being made by the cells with the same post-translational modification and initial sterility of the bioreactor. The API protein's function is defined by its composition (linear sequence of amino acids), and its three-dimensional structure. This 3D structure and hence the function of the protein can be altered by a single point mutation (change in just one amino acid among hundreds that comprise the protein), by extra crosslinks in the protein (post-translational modifications) or by other subtle changes in the large molecule drug (a protein drug usually contains hundreds if not 1000s of amino acids). These subtle changes make it necessary to closely watch many parameters of the production process as the biological behaviour of the final protein can be strongly influenced by these subtle changes in the manufacturing and final production processes. *Biologicals are often said to be defined by their process.*

Some of the advanced disciplines that are useful in the complex environment of a bulk drug production facility are given in Table 7.1.

The complex manufacturing process and difficulty in assessing these subtle changes in the final product have been brought up as the main reason that generic biological drugs must go through similar development and testing processes as the original innovator product, a much rigorous and onerous requirement compared to small molecule generic drugs. The often-quoted example that demonstrates the significance of paying careful attention to the manufacturing process is that of J&J and its biological drug Eprex (see Box 7.3 for more details). A subtle change in the final

Table 7.1 Disciplines used in biopharmaceutical drug manufacturing

Microbiology	Finance and accounting
Cell biology	Chemical engineering
Food science	Biochemical engineering
Analytical chemistry	Pharmaceutical chemistry/science
Mechanical engineering	Business management
Biochemistry	

Box 7.3 Case study: small manufacturing changes have a big impact

In 1998, the EU regulatory body directed J&J to stop using human serum albumin (HSA) as a stabilizer in Eprex (human erythropoietin) due to concerns of "Mad Cow" disease. It was subsequently determined over the next few years, that an increased number of patients receiving Eprex developed a severe loss of red blood cells (aplasia) which required transfusions. The exposure-adjusted incidence of this aplasia per 100,000 patient years was estimated to be 18 in Eprex without HSA, 6 in Eprex with HSA and 0.2 in Epogen (product of Amgen, from whom the Eprex manufacturing process was transferred to J&J). Some intense detective work by J&J (the drug was not taken off the market as the benefits to patients outweighed the risks) showed that the replacement stabilizer (polysorbate 80 replaced HSA) reacted with the long-used, uncoated rubber stoppers of the single injection tube in which Eprex was stored, causing plasticizers to leach into the solution (McCormick [2004]). These plasticizers acted as adjuvants, stimulating a very small percentage of patients to mount a strong immune response to the recombinant injected erythropoietin. These antibodies then attacked the patient's own erythropoietin, leading the body to shut down all red blood cell production and leading to severe anemia. In 2002, J&J switched to PTFE-coated rubber stoppers which halted the leaching problem and reduced the immune response. This example and the continued difference in incidence of these immune responses is often cited when arguments are made against allowing biosimilars (biological generics) to come on the market without substantial clinical testing to demonstrate biological safety and efficacy, essentially going down the same path of tests as the original drug.

The complexity and subtlety of changes and the difficulty in characterizing biological drugs is apparent in a continuing article on this matter, published in 2006 (Schellekens and Jiskoot [2006]). This article argues that product aggregate formation may in fact be the cause of the differences in behaviour between the two products rather than adjuvant-induced immune reaction.

---

manufacturing process of Eprex (erythropoietin; increases red blood cells) led to an increase in serious cases of aplasia (lack of red blood cells). On the other hand, Amgen changed its manufacturing process for its version of erythropoietin after launching Phase III studies and successfully demonstrated equivalence and efficacy for the new process and product in a series of clinical trials.

Box 7.4 Importance of culture in building a world-class biopharmaceutical manufacturing organization

From an interview with Dan VanPlew, currently the Executive Vice President and General Manager, Industrial Operations and Product Supply at Regeneron, with past experience managing manufacturing, quality, and process sciences at Chiron Biopharmaceuticals and Crucell Holland BV.

Culture in a manufacturing facility. These are a few points that I continue to reinforce with personal involvement at the incident level if needed.

(i) I don't want uncontrolled creativity from personnel while they are working on the floor. If they have ideas, I ask them to write them up and submit them, but to not ever deviate from established process steps. This is a big difference from R&D – where innovation is king – which is why it is important to respect the fact the cultures and focus of personnel in those two parts of the company as the goals and methods require and reward different behaviors and personalities. However as discussed in the technology transfer discussion earlier, they need to learn to respect each other's strengths and skills without deriding the differences. By the way, each year we have almost 100% of our manufacturing site employees enter improvement ideas, the vast majority of which are implemented. This type of engagement has led to our facility having competed for and won the Shingo Prize (recognition for organizational excellence) in 2019.

(ii) In operations you do things to manage and condition everyone's psyche – including your own. It is difficult for people to turn behaviors on and off. To that end, as soon as you step foot onto one of my sites you are immediately expected to adhere to strict standards of behavior and attitudes. As only one, simple example, with a goal to achieve a low microbial footprint which in biotech is core to maintaining your right to operate, I have been extremely strict with all areas of our buildings regardless of whether or not the area is in bioreactor clean room spaces. Everyone entering our campuses have to adhere to highly controlled standards even if they are in non cGMP common spaces as soon as they step onto the campus, including departments that will never even be in a clean space. I have gone to extremes to enforce these types of policies. Some people view my behaviour as draconian (e.g., having people who forget to wear their booties mop entry ways on rainy days) but sticking to my laid-down policy with no exceptions has paid off for us, with one of the lowest microbial counts among any manufacturing facility globally. Similarly, we have a sterling inspection history with global regulatory agencies, which is a core symptom of a well-controlled environment.

(iii) Give everyone a chance to learn from their well-intended actions resulting in mistakes. You want a degree of tension, but you don't want it to be paralyzing. I use the example of driving down the road. Most people don't drive the strict speed limit of 65MPH on the expressway, but you stay close

Box 7.4 (*cont.*)

enough to the legal speed limit because you fear a police officer may be ahead of you and tracking your speed – i.e., tension. But the tension isn't so great that you are afraid to drive at all. Mistakes can be very expensive but I have never fired someone for making even a multimillion-dollar mistake as long as their behaviors and intentions were good. Instead, I have charged them with putting in place processes that will prevent that mistake from happening again. Paradoxically, I have fired people for seemingly trivial issues when that person shows bad intentions and behaviors. For example, I kept someone on staff that made a multimillion-dollar mistake while terminating two people for splattering spitballs on someone's desk and personal belongings.

In this environment, people learn quickly to be professional, compliant, and respectful without feeling paralyzed by overbearing management. Many of a pharmaceutical company's controls rely heavily on document handoffs and operator recorded and interpreted data. For GMPs to truly work, the employees must have firm handshakes throughout the process. And people need to genuinely trust and respect their peers for GMP compliance and product quality and reliability to truly shine. As a simple illustration of these handshakes, the trainer must trust the student. The student must then be trusted to follow the process. The reviewer of the student's work must trust the student and the reviewer. And so on. One weak handshake and you could harm the very people you are trying to help

(iv) As a manager you must take a long view, and walk the floors while making your door feel truly open. You want your employees to feel safe and supported, and their belief in you – catalyzed by your attitude and actions – to drive their trust, confidence, and engagement. Not just physically, but emotionally. I want my people to be with me for years. These people will spend more time with each other and me than they will their own family. I believe it cannot be a one-way street where the company takes and takes. Over a long enough horizon, life events happen– divorce. marriage, death, birth. I expect my people to support me as much as I support them, and I know it is necessary to provide deeper support to some folks at different times (e.g., I had an employee who lost his wife shortly after adopting two children. I asked him how we could help – truly help, not just lip service – and he told me cooking was anxiety producing because he didn't know how. We set him up with delivered meals 4 nights a week and a cooking instructor to work with him 2 nights a week. And he was "on his own" for the remaining night. We did this for a few months and weaned him to self-sufficiency). Philosophically this guides me daily. If someone has a request, I try to fill it. And I can assure you – now with over 5,000 employees – the cost/benefit analysis of this is quite heavily skewed to the benefit side of the equation. Over the course of the year, the requests are things like more lights in the dark parking lots, emphasis on starting "green" programs, etc. which aren't expensive but have a huge

> **Box 7.4** (*cont.*)
>
> impact (e.g., my team has played a big role in the company being listed on the Dow Jones Sustainability World Index). People will naturally do their best while working on their passions, and this makes it even more economical. For example, I have a handful of environmentally conscious employees that helped convert us to compostable-centric food containers and utensils. This is fun for the employees and drives strong commitments. All of this leads to "firm handshakes" due to pronounced trust, transparency, and teamwork.

7.4.1   Process validation before approval

Pharmaceutical products are approved only after submission of data that validates the manufacturing process (often called equipment or process qualification studies). At the time of the pre-approval inspection (PAI), the companies must have validation protocols written and available for review. Validation data is often submitted as part of the first annual report after approval. The FDA could require production of one or more (usually three) conforming batches of material and submission of those records and analyses to the FDA prior to approval.

The FDA has defined process validation as "establishing documented evidence which provides a high degree of assurance that a specific process will consistently produce a product meeting its predetermined specifications and quality attributes" (from www.fda.gov/files/drugs/published/Process-Validation–General-Principles-and-Practices.pdf). In addition to process validation, firms must also perform analytical method validation, facility and equipment validation, software validation and cleaning validation. The final product quality is assured when all these elements are combined with other elements of GMP established by the FDA. All supporting data and analyses are part of the CMC (chemistry manufacturing and controls) section of the NDA/BLA package submitted to the FDA for approval.

However, advances in process analytical technology (PAT) in the past few decades have resulted in improved validation of manufacturing processes with real-time feedback and ability to measure multiple parameters online (see Box 7.5 for more details on PAT). The FDA responded to these technological and engineering advances by issuing new GMP guidelines for batch validation of manufacturing processes.

Some extracts from the FDA reports such as *Pharmaceutical cGMPs for the 21st Century* (www.fda.gov/media/77391/download) and others found at the FDA website (www.fda.gov/about-fda/center-drug-evaluation-and-research-cder/pharmaceutical-quality-21st-century-risk-based-approach-progress-report) are reproduced here:

[The FDA has] ... begun updating ... [its] current thinking on validation under a Cross-Agency Process Validation workgroup led by CDER's Office of Compliance Coordinating Committee with participation from CDER, CBER, [and others]. ... [The] FDA began this process [by] issuing a compliance policy guide (CPG) entitled *Process Validation*

Box 7.5 Process analytical technologies

Process analytical technologies (PAT), as defined by the FDA, is a system for designing, analyzing, and controlling manufacturing processes through timely measurements of critical quality and performance attributes of raw materials and processes with the goal of ensuring final product quality. The primary use of PAT is to generate product quality information in real time. This type of real-time quality analysis is most useful in a continuous product manufacturing environment like a chemical manufacturing process – i.e. pharmaceuticals.

Most pharmaceutical facilities are adopting PAT in order to reduce operating costs and improve manufacturing efficiencies. It is being increasingly utilized in non-complex, though critical, manufacturing steps – like the purification of water and generation of water to be used to constitute products that will be injected into humans. The introduction of automated TOC analyzers has reduced the cost of water generation tremendously.

The term "analytical" could include chemical, physical, microbiological, mathematical and risk analysis, all of which must be conducted in an integrated manner, requiring a thorough understanding of the manufacturing process for the best implementation and use of PAT.

Conventional process monitoring often involves taking samples from the production line into the testing (QA/QC) laboratory, whereas PAT relies on in-line testing. Examples of online/inline analytical instrumentation are:

Spectroscopy (UV, NIR, turbidity, refractivity)
Chromatography (HPLC)
Electrochemical (pH, DO, conductivity)
Chemical
Physical (temperature, pressure)

Through the use of strategically placed probes in the process, critical endpoints at specific stages of the process can be pinpointed to a high degree of certainty. Sampling errors (and more importantly, failed batches) can be minimized.

PAT is not a regulatory requirement. The 2004 FDA guidance document *Pharmaceutical cGMPs for the 21st Century*, (www.fda.gov/media/77391/download) encourages the use and adoption of risk-based approaches to the development of automated process control systems in the pharmaceutical industry.

Requirements for Drug Products and Active Pharmaceutical Ingredients Subject to Pre-Market Approval (CPG 7132c.08, Sec 490.100).

[The CPG states that] agency drug product pre-market review units may approve applications for marketing before a firm has manufactured one or more conformance batches at commercial scale (also sometimes referred to as "validation" batches). The revised CPG again recognizes certain conditions under which a firm may market batches of drugs while gathering data to confirm the validity of the manufacturing process. . . . The document clearly signals that a focus on three full-scale production batches would fail to recognize the complete story on validation.

Advanced pharmaceutical science and engineering principles and manufacturing control technologies can provide a high level of process understanding and control capability. Use of these advanced principles and control technologies can provide a high assurance of quality by continuously monitoring, evaluating, and adjusting every batch using validated in-process measurements, tests, controls, and process endpoints. For manufacturing processes developed and controlled in such a manner, it may not be necessary for a firm to manufacture multiple conformance batches prior to initial distribution.

Thus, products manufactured in more innovatively designed, efficient and updated plants with better process control and analysis technology will benefit from lowered costs of compliance to cGMP.

7.4.2   Bulk drug scale-up and production stages

Figure 7.3 highlights the various scale-up requirements of material used for various tests and stage of product development. The scale-up requirements are similar for biologics and small molecules. The following text puts Figure 7.3 into context.

After the molecule moves into advanced preclinical development, the earliest scaled-up material is made for animal toxicological studies to determine safety. Up to this point, a few milligrams of material for in vitro studies or a few grams for animal efficacy studies is typically produced using lab-bench production techniques in the laboratory. However, in advanced preclinical development, tens of grams to a kilo material will typically be required (depending on the length of time and range of dosages of the particular drug to be tested). At this stage, the manufacturing process used should yield product close to the purity (but not necessarily as pure as) ultimately desired. This is the pilot-manufacturing or first-stage scale-up process. While optimizing the scaling-up process, it is critical that the final manufacturing process not introduce additional contaminants or product variants that would cause a repeat of the toxicology studies. Therefore, the ultimate manufacturing process should yield material that is at least as pure as the material used for the toxicology studies.

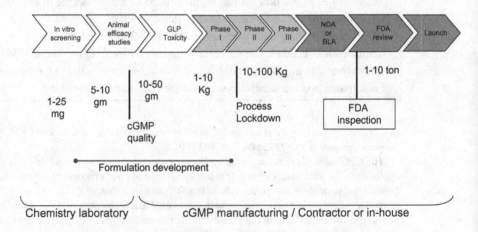

Figure 7.3 Pharmaceutical manufacturing needs and scale-up timing.

At this stage of first scale-up and pilot manufacturing, it is a good idea to test the cost of production and project it out to the final commercial scale of production. This calculation (along with input from reimbursement specialists – see Chapter 8) will allow for (1) a projection of whether the product can be made cost effectively on a commercial scale or whether alternative manufacturing procedures should be investigated and (2) the development team to optimize those parts of the manufacturing process that contribute substantially to the overall manufacturing cost.

Another important step at this point of first-stage or pilot production is to establish the characteristics of the ideal product as they apply to manufacturing issues; e.g., how it is packaged, what its shelf-life is, what the formulation is. Most small companies typically outsource this part of the manufacturing process.

As the program moves ahead to commercial level manufacture in kilogram or ton quantities, it is essential the process be scalable and produce material that is comparable to early toxicology testing material in its purity profile, stability, bioactivity, or other characteristics of the molecule. Thus, the process of technology transfer between R&D, pilot manufacturing and large-scale manufacturing must be carefully and actively managed.

7.4.3 Commercial manufacturing planning

During the initial selection of a product for development, a rough estimate of the cost of goods is made. Revenue estimates are also created based on the potential market size and early market surveys. In the absence of defined supply chains and commercial manufacturing processes, rough estimates of the cost of goods are made based on experience with similar products.

As the product progresses through proof on concept (Phase I/II), estimates of market size and product dosing significantly improve. With this information, commercial manufacturing plans are developed. A typical commercial manufacturing plan includes:

- Product development/approval timeline
- Commercial Forecast (number patients, dosage/s, regional requirements)
- Commercial product definition (marketed package or SKU)
- Target commercial manufacturing concept (processes, scale, batch size)
- Supply chain (including make vs. buy, single/multi-source, strategic relations)
- Facility requirements (scope and capital estimate)
- Organizational requirements
- Cost of goods
- Commercialization estimate
- Risk management plan (including mitigation plans for highest risks)

The commercial manufacturing plan is used as a guide for the development of manufacturing processes for Phase III and commercial sales and for making capital investment decisions. As mentioned above, it is critically important that the Phase III manufacturing process and product are representative of the ultimate commercial process and product. If changes are made to the process or product after Phase III, it is likely that additional clinical trials will be required with an associated delay of the product development timeline (and at significant added expense). This is especially true for biologics.

The decision on implementing specific scale of manufacturing is also affected by the cost of goods (COGs or bill of materials) requirements of the product. COGs include the cost of materials, labor for manufacturing, testing, and release, facilities (depreciation on capital), utilities, and overhead. The typical COGs for drugs and biologics ranges from 5% to 25% of the sales price.

A major issue arises towards the beginning of Phase III studies, especially when developing biologicals – continue to contract out production or build a production plant? The risk has been reduced with a positive Phase II proof of concept result, and the company has to decide on the best use of its resources on the path forward – an expensive Phase III trial has to be balanced against the investment in building a dedicated manufacturing facility. The key issue here is the large investment needed ($15–$50 million) and the still-lingering uncertainty (the product still has an industry average of 50% chance of success at this point). The build or buy decision is discussed in greater detail in Section 7.7. It is also important at this stage to have strong market research (see Chapter 2) to predict the scale of production needed and either prepare contractors or build out to compensate for that volume, else, as Immunex learned to its chagrin (see Box 7.6), a success can quickly become a bust. These issues of scale up in biological drug manufacturing and in buy (CMO) versus build (in house) manufacturing are discussed by an industry veteran in Box 7.7.

Box 7.6 Case study: Immunex and manufacturing planning

This case summary is based on a BusinessWeek *article and several other online commentaries on Immunex at the time.* ("The Corporation: Biotech and the Spoils of Success," 2001)

Immunex' rheumatoid arthritis (RA) drug, Enbrel, was approved in November 1998 as the first biologic to treat the disease. Immunex initial sales estimates focused on the 25% of RA patients who failed traditional therapies and so when the FDA approved the drug for children in 1999 and then as a first-line defense for RA patients in 2000, the company was unprepared for the demand. Only 75,000 of the 1 million patients who might benefit from the drug could get Enbrel, resulting in lost sales of over $200 million (that's $500,000 or more lost every day). Immunex had contracted Boehringer Ingelheim Pharma of Germany to manufacture Enbrel – but that wasn't enough. Immunex and its partner American Home Products (now Wyeth) started a new plant in 1999 at the cost of $450M, but the plant would not be ready until the end of 2002. This left the door open to competitors, like J&J's Remicade, approved a year after Enbrel. The CEO of Immunex said in early 2001 that if he could recreate history, he would have prepared more aggressively.

Box 7.7 Scale–up manufacturing of biologic drugs and the buy vs build decision

Notes from an interview with Dan VanPlew, currently the Executive Vice President and General Manager, Industrial Operations and Product Supply at Regeneron, with past experience managing manufacturing, quality, and process sciences at Chiron Biopharmaceuticals and Crucell Holland BV.

Discovery and development stage

In the early stages of in vitro testing, the biological protein of interest is typically supplied from up to the 500 liter bioreactor scale. This quantity provides enough material for GLP studies that are researching efficacy, determining dosing, evaluating formulation, developing production processes, etc. To speed products to the clinic, in the past, the product was usually made using a stable cell line with a low productivity of material. In parallel, work would go on to maximize the output of the cell line as the product progressed in its commercialization development. However, it is becoming increasingly common for those initial cell lines to produce at a commercially viable scale without slowing down the regulatory filing. The amount of material typically used for this pre-GMP phase is in the tens of grams range.

Preclinical IND enabling and Phase I studies:

At this point, before moving to IND enabling toxicity tests, the product has to be made under GMP, and production is transferred to the commercial manufacturing site. Typically, at least two lots of 2,000 liters, or greater, are run to yield a total of about 5–15 kg of protein which will be used for IND-enabling studies and first clinical study. When you look at the benchmark costs of production internally for most companies versus CMO costs, a 40–50% premium is typically paid to the CMO. The initial 5–15 kg of material is used for long-term toxicity and for Phase I clinical trials, but a large amount of material is also used to run initial cleaning and validation studies, assay development and stability testing. It is also important to note the much larger cost to the company comes from extended timelines and lagging responsiveness – i.e. internal capacity usually affords more flexibility which greatly influences speed to clinic and market.

Clinical studies: Phase I, Phase II, and Phase III

The manufacturing site works to establish reliable and reproducible production, along with higher titers and maximized yields and will phase in this cell line for making drug product for Phase III. At this point, 2,000 L or larger bioreactors are typically used for making tens of kilos of drug as per size of trial.

Some thoughts on using a CMO or building manufacturing in house for biologics

In my experience, for smaller companies (50–250 staff) the split of companies doing bulk production at a CMO versus making it in house or at a collaborator is roughly fifty–fifty. Production of the early-stage bulk product is at, up to 500

Box 7.7 (*cont.*)

L batch sizes for discovery and preclinical work. For later stages of growth, if it is a first biological product, a larger percentage of companies will look to scale with CMOs or collaborators who already have the infrastructure.

Here are a few economic arguments and examples for building (in-house production) versus buying (CMO), specifically for biologic drugs:

(i) As a rule of thumb, one can expect a 50% reduction in cost of goods (COGs) with the learning that takes place in the first 3 to 5 years of production. If the company is using a CMO, the chances are that the CMO keeps the majority of that savings and continues to offer only year-over-year price changes negotiated by the customer for subsequent batches. Typically, smaller companies do not have in-house commercial manufacturing, quality, analytical, and regulatory experience, and may not have a good enough insight into the process parameters and controls.

(ii) Experience and learnings gained from processing multiple lots and subsequently proactively managing the lot production parameters can improve the profitability and affordability of the drug as it faces more competition in the market. However, with a CMO, they are often not well incentivized to pass those learnings and thus savings to the customer, but to maximize their profit from each lot. As an example, in one drug production process, the yield started to drop and after some intense investigation over 2 years, it was found that a previously unmeasured amino acid in the incoming material was the factor, so all incoming material was then screened and lots with the right amount of the key amino acid were selected. If this had happened at a CMO, chances are high that the CMO would keep this information to themselves and add a significant charge for the (now simple step) of screening the incoming material.

Another important part of planning manufacturing activity is to understand the context of the manufacture for drugs – for example, barbiturates must be manufactured in a highly controlled and regulated environment, with only a few plants in the world currently configured to make those kinds of drugs. Similarly, for cytotoxic drugs used for chemotherapy, the product must be made in a facility with extremely high levels of protection for personnel, with the physical plant configured to allow for completely closed environments for production, formulation, and final packaging of the toxic product. These special manufacturing considerations will significantly increase the manufacturing cost of the product and could have an unforeseen impact on the profit margins of the business.

The steps of packaging and filling usually require specialized plants and apparatus for these steps, and it is common to outsource these last steps of making the final oral (capsule, gel, tablet) or parenteral preparation to specialized filling facilities.

Finally, the FDA will inspect the manufacturing site before approval of the drug for marketing. Typically, this PAI (pre-approval inspection) is announced, giving a company enough time to prepare. The inspectors go over minute details in the operations, evaluating the books of standard operating procedures (SOPs) and training SOPs that are vital for GMP process implementation and subsequent FDA certification. Outcomes of an inspection can range from including comments in an inspection report, specific recommendations to change or improve processes, or if the noncompliant process or system is found to potentially affect product quality, the inspector can even issue an injunction or restraining order to close down the plant or to close down the production of that specific product until the situation is satisfactorily rectified and the plant is in compliance again.

7.5   Manufacturing in devices and diagnostics

Owing to the highly varied nature of products in the device sector, the comments here are fairly general but relevant to most device manufacturing processes as prototypes are scaled up to production level.

After the first few prototypes have been made in machine shops based on early sketches (Figure 7.4) and used for proof of concept testing and design optimization, the production needs turn to pilot manufacture or full-scale manufacturing for in vivo application and testing.

Pre-manufacture planning steps include identifying and selecting suppliers, developing a pilot run plan, a manufacturing strategy (build or buy see Section 7.7), identifying costs and timelines and other plans to prepare for transfer of the technology from R&D to production. As shown in Figure 7.4, pilot production needs could be met without full GMP processes, but a GMP-compliant production facility will need to make validation batches of products before the FDA will approve marketing the device.

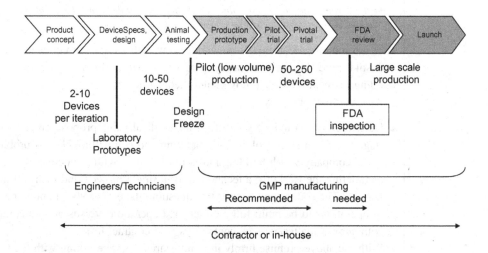

Figure 7.4 Device manufacturing needs and scale up.

At this pre-manufacture stage, the following processes and considerations are useful for management and the development team to review as part of the design process before investment in pilot production or full-scale production:

- Verify and validate (V&V) customer needs
- Verify and Validate (V&V) needs translated to design requirements and performance
- Material selection
- Packaging & sterilization
- Preclinical testing
- Use of engineering analyses for production, assembly, and testing planning

Good engineering analysis at this stage can point the way to a manufacturing prototype and investment in analytical planning and virtual (computer-aided) design helps:

- Accelerate development
- Minimize prototype iterations
- Avoid design mistakes

but is not a replacement for full prototyping and testing.

The case study in Box 7.8 shows the decision making and scale-up planning implemented by a startup company as they went from design to approved integrated development environment (IDE) for human clinical trials for a Class III invasive cardiovascular medical device. This case example also highlights the production scale of prototype and production devices required at each stage of development.

Box 7.8 Case study: Cardiovascular device startup – lessons learned in the early planning and scale up of manufacturing

Written from interviews contributed by the co-founder and CEO of the NewCo cardiovascular startup, currently operating in "stealth" mode.

Here are some key learnings from an interview with a serial entrepreneur who built and sold a $1 billion respiratory pharmaceuticals company, and is currently building his next one in cardiovascular medical devices.

Key learning takeaways:

- Determine and convince yourself of the medical value proposition at concept stage – are you really solving an important and compelling clinical problem?
- Is the company worth building into a full business with a pipeline, or just completing the R&D for a technology or a single project, and then sell or license? The sooner you answer this question, the earlier you'd know the scope of operations to be built: full products and operations versus build a company with subject matter experts, contractors, and contract firms.
- With the above premise firmly in mind, start doing everything with the view to building a robust product, including documenting the progression of the product and plan expenses prudently, keeping patient safety primarily in mind.

Box 7.8 (*cont.*)

- Focus on finding the right subject -matter experts and overcoming scientific, technical, and manufacturing challenges. Don't invest in operations until you have reached some milestone of manufacturability and technical feasibility
- Once you have settled the clinical efficacy feasibility and scope of the technology to reach commercial full value, it would then make sense to invest in building a company infrastructure versus continuing in outsourcing mode.
- Contract firms should be treated like an extension of your company in operational thinking and working, so it is important to spend time up front in selection process. This will also help the startup be prepared to bring the work in house, should the contract firm not scale-up or perform over time.

Background and startup activities

The founder had a learning experience early on in his career in pharmaceuticals that has influenced his thinking today for successful scale up from prototype to scale up in GMP production. He had been hired to fix a problem where a device–drug combination had reached Phase III clinical trials, but newly issued FDA regulations on device–drug combination meant the product would not be approved even if the clinical trial was successful, based on its current test and performance data. The founder and his team had to redo many aspects of design and testing in order to collect data to satisfy the new regulations. The manufacturing controls and device design trials were not adequate, and the drug-delivery system had the potential of underperforming in the field. This led to an extensive redesign and rework project that took millions of dollars and many months for the team to complete. The drug commercialization was paused while some clinical evaluations had to be repeated to collect data with the updated product. Paying attention to quality and reliability details at the early design stage, which may seem overly expensive in time and money to a young company, were the learnings that the founder applied in his new medical device startup, Newco.

Early decision to focus on product reliability

When the founder co-founded Newco, he and the other founder and investors had already decided that their product had transformative potential. If it worked through early stages of development, they were committed to building this company out to full scope. So, the first milestone was clear – show robustness of the technology and demonstrate in vivo safety and efficacy (animal and human) trials. The investments were deliberately streamlined to achieve this goal. However, even in that early decision making, the founder made key contract hires keeping both, the long-term goal of building a robust product for a full commercial launch, and the short-term goal of getting a product made with sufficient quality and reliability that they could run trials in animals and humans with confidence. These contract-hires included structural testing, design, and manufacturing personnel, with total experience of about 100 years. Their task was to find failure-points in reliability

Box 7.8 (*cont.*)

tests and feed this knowledge back into the early design process. With these first hires, the founder also established a quality system early on in the company to ensure adequate documentation for each stage of development this device, to ensure and record that the device would be considered safe for use in patients.

Rather than build out operations for fabrication of prototypes during the early design prototyping and iteration phase, Newco looked for a contract-firm that would work through R&D and scale up with them. They interviewed several contractors and found one that could work closely with the Newco's small engineering group and start the design prototyping and iterations. The goal was to find a contract group where Newco staff and the contactors technicians and engineers could work as though the contractor was a seamless extension of Newco. A contract signed with the chosen entity delineated clearly in the work-for-hire agreement that all IP created was owned by the sponsor (Newco). The contractor was also chosen for the commercial production experience that its founders had, including having a clean room for next stage GLP and GMP manufacturing.

Transitions from design prototyping to human trials and from GLP to GMP manufacturing

It took 4 years to patiently go through the various stages of prototyping various components of the device design and early animal studies. While the team had confidence in the value proposition of the proposed product concept and the severe medical need it was addressing, proper technical evaluation approaches had to be developed, and safety and feasibility of the chosen designs confirmed with actual data. Using the early capital with variable cost-structure of a contractor with the right capabilities was the correct decision, instead of having an expensive, fixed operational cost during this period. Their chosen contractor had an experienced design engineer and a commercial manufacturing process engineer as two of their founders, and Newco solicited their early inputs, ensuring that their prototype designs could transition to manufacturing process readiness.

Newco had also put a quality system in place early on. When the required quality tests for collecting process control and quality data became repetitive and reached high volumes, it was too expensive to conduct this at the contractor. The quality engineer hired by the company set up testing at the company's own facilities while other R&D work continued at the contractor. The contractor's existing clean room facility was used to manufacture devices under GLP (good laboratory processes) for animal studies. A major focus in the animal trials was confirming device robustness and risk-free interaction with vasculature so as to identify any risk of failing or catch any other safety issues before starting human testing. Extensive safety, usage, histopathology data was collected from each animal in the study and with this data set in hand, the company conducted initial human trials in a country which had a far more rapid clinical study approval process than the United States and Europe. The company made this important step to start clinical stage tests as it had confidence in the extensive safety and reliability

Box 7.8 (*cont.*)

testing in live animals and in the high quality of their device design. In addition, this proof-of-concept data in humans would drive investor confidence in the project (and be useful to further convince the FDA for IDE permission for pivotal clinical trials). When the first data from humans showed the successful deployment of the full scope of the technology in various indications planned, the company went ahead with investment in operations and built a clean room and GMP production and testing facility. Methods for device component fabrication and sterilization were established and maintained at certain contractors.

Scaling up volume, controls, and quality issues

As a metric to understand what it takes to bring a device through design and human trials, Newco elected to make hundreds of devices during early GLP-stage preclinical testing, and several hundred more under GMP to complete human clinical studies. This count includes product used internally for reliability testing, production lot sampling and units kept aside for stability testing. This served multiple purpose: gain confidence in design to achieve the desired safety and device-performance in animal and human testing, understand manufacturing process parameters, improve process consistency such that potential of design or manufacturing process change after clinical work could be minimized.

Issues with contactor as scale-up nears

GLP work was done in their first contractor's clean room facility, but that contractor was acquired by a larger firm, whose management was not as committed to R&D stage contract work. The manufacturing was moved to a second contract firm, whose founder was very committed to R&D work. After helping the Newco accomplish additional design, manufacturing, and clinical and animal testing work, this contract firm landed a much larger manufacturing contract with a major company, and deprioritized the Newco's work.

After these two unsatisfactory experiences with contract firms, having lost time in transitioning processes to their clean rooms and having already proven the value of early in-house design, quality, and testing in animal and clinical evaluation, Newco decided to build its own clean room facility. Two early decisions helped make this a rapid, cost-efficient activity – (i) Newco had bought much of the manufacturing equipment needed by its second contract firm, and this equipment could be moved very quickly to the company's own clean room; (ii) the investments in reliability and quality made early on that had seemed too expensive for those stages – e.g. reliability engineering being a major part of the device design concept iterations, involving structural and manufacturing engineers in the prototype concept design stages, and investing in hiring experienced quality engineering staff early on – all paid off as the manufacturing was already quite robust with high yields.

However, transitioning from R&D line to production is never without hiccups and despite attention to detail, issues will crop up. After moving the equipment

---

Box 7.8 (*cont.*)

from the contractor to their own facility, the team found that one piece of equipment had a slight drift in its precision over time, and new process controls for that station had to be put into place to ensure manufacturing quality. In fact, this issue came up at a critical time that highlights one more learning on dealing with regulators. The company had just completed all its testing and the data package for an IDE application was ready to submit when this production machine drift was discovered. Rather than delaying the IDE to fix the drift, the company chose to submit the IDE, in which it disclosed the previous robustness data, described the equipment issue along with a plan to fix it. Toward the end of the review period for the IDE, a regulator from the FDA inquired about the matter and the company was able to submit a package of data showing they had fixed the machine production process issue in the time since submission of the IDE, and had tested enough devices to demonstrate equipment consistency. This not only satisfied the FDA reviewer, but also likely created confidence in the company's commitment to quality and safety. The IDE received full approval in first review, saving the company several weeks in precious time.

Material selection is another important iterative pre-manufacturing step that must be considered from multiple perspectives, beginning with the end use and stepping through all the steps and processes used in larger scale manufacturing. This phase of "technology transfer" between R&D prototype product and production prototype is a critical stage where many development failures occur. For example, starting with "medical grade" pure materials does not put an end to the technology transfer issues or the assessment of materials quality and characteristics – it is important to consider the overall processing conditions at each step of production that the entire device will go through, including sterilization and packaging, and also including specific forms of usage possible at the site of delivery. This analysis is partly included in the validation stage mandated in quality systems for cGMP compliance (see Chapter 6 and Appendix 7.2). Certain standards in ISO 10993 also lay out specific paths for materials testing in development stages, including the toxicity testing discussed in Chapter 5 and additional testing (leachables, extractables) for each component used in the manufacturing process that interacts with the product.

In this stage of planning final manufacture and final design reviews, the key issues of *packaging and sterilization* are interconnected as:

- Packaging depends on the product design
- Sterilization methods depend on product design – e.g. a biological product cannot be sterilized with gas or steam methods
- Packaging depends on sterilization method – e.g. packaging must allow sterilization to penetrate (gas sterilization) or must not degrade (irradiation sterilization)

7.5.1   Design for manufacturability

Important issues that can derail a product development process in production stages due to increased cost or time need to be considered in the design phase. Specifically, keeping manufacturing issues in mind, assures that a design can be repeatedly manufactured while satisfying the requirements for quality, reliability, performance, availability, and price. The design should be reviewed with a "manufacturability" lens early in the iterative design process to make it less costly by the following considerations, wherever possible:

- Simplify design to use fewer parts; eliminate nonfunctional parts and reduce functional parts.
- Use simple production processes with low defect rates in process.
- Design in higher quality and reliability.
- Make it easier to service.

7.5.2   Design for assembly

The final steps of assembly, packaging and sterilization can be made easier by specifically keeping requirements of assembly of parts in mind by the following points:

- Checking overall design concepts keeping assembly processes in mind
- Component mounting connections
- Test points accessible at various stages of assembly
- Stress levels and tolerances of connections points well tested
- Printed circuit boards standardized or component connections clearly identified

After delivery of the first production batch to the consumer, continue to monitor field performance and correct or improve manufacturing process or device design. In particular, the case with recall of Sulzer hip implants (Box 7.8) points out the need to monitor field use of the final products for manufacturing or production problems that may have escaped notice in final testing.

7.6   Manufacturing in diagnostics

The following section is adapted from the FDA guidance document: Guideline for the Manufacture of In Vitro Diagnostic Products.

In vitro device product characteristics are defined during the pre-production process. Apart from parameters such as *physical characteristics, chemical composition and microbiological characteristics*, critical performance characteristics such as accuracy, precision, specificity, purity, identity, and sensitivity are also specified. These established specifications will determine the appropriate production and process controls such as mixing and filling processes, sterilization, or lyophilization needed to

---

> **Box 7.9** Sulzer hip implant recall – subtle changes in manufacturing causing large problems
>
> *Adapted from Lefevre (2011), and website www.Sulzer.com*
>
> In 2000, Sulzer Orthopedics recalled about 17,500 hip replacement acetabular shells. These shells fit into the hip socket and interfaced with the stem that fit into the femur. These products were part of a batch of 25,000 that had been manufactured between 1997 and 1999. In fall of 2000, orthopedic doctors began reporting problems with the hip implant. Sulzer voluntarily recalled the implant after researching patient records, surgical techniques, and the product itself. On reviewing the production process, it was found that some mineral oil-based lubricant had leaked into the machine coolant and left an oily residue on the acetabular shell. The shell went through the established cleaning process but that did not remove the oily residue which went undetected. The acetabular implant was made to snap into the hip socket so that bone would grow into it. The oily residue interfered with the bonding of the bone and allowed the implant to loosen and fail, frequently requiring another surgical procedure to fix the problem.
>
> In February 2002, Sulzer agreed to settle patients' lawsuits for $1 billion.

manufacture the IVD. The specifications established for the IVD will also determine the appropriate environmental controls needed, in conjunction with the manufacturing process, to ensure that product specifications are consistently met. The consistency of these product attributes is not "tested into" the finished product, but is achieved through the establishment of adequate product specifications; and by ensuring that these specifications are met through product and process design, process validation, process controls, manufacturing controls, and finished product testing.

The specifications for the product, the manufacturing process, and the environment are maintained as part of the device master record (DMR), as required by CFR 820.181.

7.6.1   Labeling requirements for IVDs

Product developers should read the specific guidelines and controls for labeling of the final product, described in 21 CFR 809.10(b), and summarized here. The label includes the following details:

- Proprietary and established names
- Intended use(s)
- Summary and explanation of test
- Principle of procedures
- Information on reagents
- Information on instruments

Box 7.10 A diagnostics startup pivots and plans out its manufacturing and development

Written from interviews kindly contributed by Jonathan Romanowsky, co-founder and Chief Business Officer of Inflammatix, and member of StartX.

Manufacturing in diagnostics development

Inflammatix is building novel molecular diagnostics based on their abilities to interpret, using advanced bioinformatics techniques, how the immune system responds to diseases. The company's first test (the InSep™), an acute infection and sepsis test, informs on the presence, type, and severity of acute infections for patients seen in the emergency department.

The product consists of (i) an instrument (called Myrna™; note: the name is a play on "My RNA" since the technology focuses on human immune response gene expression analysis), (ii) specific assay cassettes (microfluidics, all analytes and reagents contained in the cassette), and (iii) the software in the instrument used to analyze the assay readings. Specifically, the software applies machine learning–based algorithms on the expression levels of multiple genes and generates results that are clinically actionable. Based on conversations with FDA to date, their InSep test and Myrna instruments are likely to be Class II and will require FDA clearance through the 510(k) route.

The company's initial commercialization plan was to make their tests run on existing molecular diagnostic instrumentation which was already in place in many laboratories or hospitals. They would then analyze the readouts with their proprietary algorithms software to combine the expression levels of their host response biomarkers into clinically actionable results. However, the company found two main impediments to this plan as they went further – the assay process on existing equipment did not meet all of the three key criteria that Inflammatix knew were critical for success in their field of diagnostic indications: (1) rapid turnaround time, (2) quantitative gene-expression readouts, and (3) high multiplex (the ability to measure the expression level of dozens of genes at once). In addition, the business terms (royalty rates, licensing fees, time to launch) proposed by the potential partner platform companies were not attractive to provide an acceptable return for Inflammatix venture investors. Given these learnings, Inflammatix decided to make their own instrumentation platform ("build instead of buy").

They also had to design and make their own cassettes with microfluidics and reagents to process patient blood samples for their multigene expression–based assays in a self-contained assay assembly. It was clear that the core innovation of the company's tests lay in their proprietary algorithms for recognizing patterns among a set of biomarkers of the host-response readouts. Hence the instrument design and development team were directed to use as many off-the-shelf materials as possible and to limit additional technical risk, keep costs low and move speedily to commercial readiness.

Box 7.10 (*cont.*)

Inflammatix interviewed several medical device R&D and manufacturing contractors for their instrument development partner before choosing one that met their criteria for having an ability to go from prototyping to commercial production scale up and having the interest and capability to do small-lot manufacturing. They used a similar process to identify a contractor to build their cartridges. The company had already done much of the algorithm development for their software in the early days of the startup but they also initiated adaptive learning features from their analyses of clinical outcomes datasets for continuous improvement of their software diagnostic capability. Ultimately, the algorithm software that would be loaded on the instrumentation would be locked (not adaptive) for all clinical use. New versions of the algorithm would need to undergo new FDA clearance for them to be put into clinical use.

The company assigned a key person internally to lead the two hardware projects (instrumentation and assay [cassette] development) and several system engineers and assay development teams to form the rest of the team. This company team worked actively and closely with the contractors' engineers, technicians and assay development teams with multiple weekly calls and exchanged test data and documents.

In the planning and prototyping phase, individual components and sub-systems of the instrument and the cassette were designed, fabricated, and tested in an iterative process. Throughout the prototype design phase, the company built a handful (<10) of full instruments and tens (<100) of assay cassettes. Even though they tried to use proven materials and methods, several issues required more intense trouble shooting to resolve – one example was the finding that the blister pack material on the cassettes influenced the assay performance, so they had to look for new neutral materials and develop novel blister adhesion methods.

They also had to empirically determine the right pressures and valve systems for optimal flow of materials in the cassette channels. Toward the end of the prototyping phase the company made about a dozen instruments and several hundred assay cassettes in order to prove out the manufacturing process, reduce the variability and prove out the viability to be able to scale.

At the time of the interview, the InSep test and Myrna instrument are in the "Development" stage after recently completing the "Plan and Prototype" stage. Inflammatix addressed the following key questions at the "stage gate" in order to advance:

- Are further inventions required?
- Will it be possible to make the product at the price point we need?
- Did we build a working prototype and prove technical feasibility?
- Have we determined a budget and schedule for the entire project?

At this point, the manufacturing processes will need to move to a GMP process in order to get the three GMP production lots that the FDA would need to see for

Box 7.10 (*cont.*)

approval and the instruments and cassettes used for the clinical trials had to be made in the same manner as the final planned commercial product in order to meet regulations. The reagents and analytes are not currently being manufactured under GMP. Some of them are FDA cleared. Thus, most of the analytes and reagents are being validated by Inflammatix prior to utilization of that lot /batch received. Inflammatix plans to have some vendors validate themselves and if the data is acceptable, Inflammatix can proceed with their validation steps.

In this phase at the time of this interview, the company anticipates it will need to make between 50 and 100 or so instruments and several thousands of cassettes that they plan to use for the following: (1) clinical trial sites, (2) internal analytical validation testing, and (3) some for early users (pre-commercial launch) in order to allow some hospital sites to continue to conduct their own internal validation studies.

Note on manufacturing budgeting: Mr. Jonathan Romanowsky, a co-founder of the company, in his interview for this case study, recommended that founders would be well advised to budget thoughtfully for the number of manufactured units planned in the three phases of prototyping, manufacturing scale up with beta units, and first GMP production runs with the rule of thumb that early manufactured versions (small batches, handmade fabrication and assembly) of the instruments/ devices cost three to five times the eventual target cost at moderate volume production. This is usually an underbudgeted item for most entrepreneurs who have not gone through the launch of a new assay instrument.

Inflammatix had to plan for low-volume, first-production runs due to additional considerations unknown supply chain process or delays through distributors to buyers and 6-month expiry dates of the cassettes, potential for slow adoption rate by sites for assay utilization, and slow sales ramps without known reimbursement metrics.

Having met the challenges of taking a new diagnostic assay and related instrumentation from prototyping to scale up, Inflammatix is now in its final pre-commercialization sprint and has to navigate the planning of production volumes, pricing, and reimbursement discussions while carrying out its pivotal clinical trials.

- Information on specimen collection and preparation
- Procedures
- Results
- Limitations of the procedure
- Expected values
- Specific performance characteristics
- Bibliography
- Name and place of business
- Date of the package insert

Box 7.11 Seriousness of noncompliance to FDA manufacturing and quality regulations

Abbott, a large diagnostics manufacturer (which also makes and sells drugs and devices), had its diagnostics manufacturing plant shut down in 1999 by the FDA when inspections found that they were not in compliance with cGMP or QSR. The letter from the FDA to the public explaining the reason for this serious step is reproduced here in excerpts (from www.fda.gov/cdrh/ocd/abbottletter.html):

Dear Colleague:

On November 2, 1999, the Department of Justice, on behalf of the Food and Drug Administration (FDA), entered into a consent decree of permanent injunction with Abbott Laboratories and responsible officials. This action involves the company's diagnostic devices division. We are writing to explain the significance of this action, and how it may affect the healthcare community.

Reason for seeking an injunction

We took this action because of the firm's long-standing failure to comply with FDA's Good Manufacturing Practices (GMP) regulation, now called the Quality System regulation, and its failure to fulfill commitments to correct deficiencies in its manufacturing operations.

These failures go back to 1993, when our inspection of the facilities where Abbott Diagnostics Division manufactures diagnostic products showed non-compliance with the GMP requirements. Areas of non-compliance included process validation, corrective and preventive action and production and process controls. The company's failure to comply with these requirements increases the likelihood that the diagnostic products produced at these facilities may not perform as intended.

Subsequent inspections, including one as recent as July 1999, showed little improvement in the company's compliance, despite three warning letters from FDA. Since 1995, FDA repeatedly encouraged the company to achieve compliance voluntarily. Despite assurances by the company that it would correct the manufacturing problems, the firm failed to bring its manufacturing operations into compliance. Ultimately, in order to protect the public health, FDA sought action by the Department of Justice.

Public health significance of non-compliance

It is important to understand the public health significance of FDA's Quality System regulation. When a manufacturer complies with the regulation, there is a level of assurance that the product has been properly designed and manufactured in a consistent way and will perform as intended. Conversely, a manufacturer who fails to comply is less likely to produce a product that performs as intended. We are especially concerned about Abbott Diagnostic Division's long-standing non-compliance with these accepted manufacturing principles because they represent the minimum requirements for manufacturing quality.

> **Box 7.11** (*cont.*)
>
> What does the firm's failure to comply with the Quality System regulation mean for the users of its diagnostic products? It does mean that users have less assurance of successful performance than they would have had if these products had been manufactured properly.

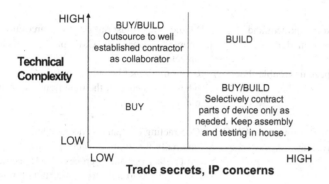

Figure 7.5 Buy or build decision – one set of strategic considerations.

7.7   Buy or build

In drug or device manufacture, the decision to buy (contract out) or build a manufacturing facility for a new product can get clouded with debates over political, financial, emotional, and strategic issues. However, there are some clear strategic and financial discussion points that should be primary considerations. Figure 7.5 exemplifies one such set of parameters that should be evaluated in making this decision.

In manufacturing biomedical products, building one's own facility for a new product is a typically large and high-risk investment. Many companies start by using a contract manufacturing organization (CMO) (buying the capacity as needed) and once approval is obtained, start building their facility and transfer the process from the CMO. Some of the other reasons are discussed briefly below, with one set of strategic issues laid out in Figure 7.5. This type of matrix can be used to compare and contrast other parameters that are listed in Table 7.2.

In general, if capacity and capability are available, the discussion focuses on company strategy. For example, if a product that addresses a very large market is being developed right behind a product designed for a smaller market, it is possible the company will decide to reserve capacity for the product still in development and outsource the current product. If the manufacturing process is very sensitive to minor

Table 7.2 The buy or build decision – pros and cons of contract manufacturing

Pros	Cons
Cost	
Using a CMO avoids the high up-front investment and conserves cash	Must pay a mark-up
The CMO spreads the overheads among many projects and can achieve economies of scale that would not be possible for a small company with only one product	Cancellation of contract carries penalty payment
Time	
Faster to use a CMO – avoid recruitment and building time, improve speed to market	Management must put aside resources/time and people to manage the CMO process, timeline, and objectives
Experience at CMO might speed up trouble shooting in process	CMO might not be able to provide capacity whenever needed in the time frame wanted by the company
Capability	
Able to access experienced staff through the CMO for specific needs and time periods	Contracting company is not building the expertise internally for future projects/products
	Proprietary optimized process (trade secret) will be disseminated to industry competitors through CMO potentially losing competitive edge
Capacity	
Volume required for production can be configured from existing large volume facilities at CMO	Smaller capacity needs may get lower priority in the waiting list with CMOs.
Uncertainty in product development success can be restrictive in building large capacity in house, allowing developers to plan for production capacity with CMO	

changes or is very complex, the decision to keep manufacturing in house may be the right one. Another reason to not use a CMO is if you perceive a risk that your project may get less attention if a much larger project comes up to the CMO. This is always a risk in working with a large CMO that has the capacity flexibility small companies are looking for, but the CMO is also looking for the larger projects that take up more guaranteed capacity and thus will later assign lesser priority to these smaller jobs. There are also other more general transaction costs, such as increased time from management usually required with the smaller companies. Some of these considerations and others are discussed in Table 7.2. An example of two device companies that made selective and strategic arrangements with CMOs for their final production illustrates (Table 7.3) that the arrangements can be very flexible to suit the company's strategic needs and capabilities.

Table 7.3 Examples of virtual and semi-integrated CMO arrangements

Virtual outsourced example	
Percardia Inc.	*Vendors*
Defines product requirements	Component Manufacturing
Works with clinicians to refine design	Device Assembly
Performs testing to verify performance	Packaging
	Sterilization
Semi-integrated outsourced example	
Focal, Inc.	*Vendors*
Defines product requirements	Component Manufacturing
Works with clinicians to refine design	Formulation & Vial Fill
Synthesizes polymers	Sterilization
Assembles devices	
Packages system	
Performs testing to verify performance	

7.8   Summary

The manufacturing process requires a rather different mindset in its organization and people than was prevalent through the product development process to date. Compliance to regulations, reproducibility, operational precision, and efficiency all become the center of focus of operations.

Manufacturing organizations must establish standard operating procedures (SOPs) that adhere to international standards and must also have well-documented operational processes that show compliance to FDA regulations. Successful scale-up from laboratory prototypes or small-scale production batches requires intense oversight and good management practices.

Exercises

7.1. Project and identify the timing of scale-up of various volumes of product into the overall product development plan.

7.2. Project the total volume needed during peak sales of the product, highlighting key assumptions. Review the assumptions frequently during further product development activities.

7.3. Develop a manufacturing plan for the product, including a summary discussion and rationale of the decision to build or buy. Identify the level of GMP compliance and the biosafety levels needed for manufacturing and packaging, the projected product life cycle and the need for product diversification from one core manufacturing template (e.g. different dosage forms from one API or diagnostics developed for multiple assay platforms).

7.4. Assess the product for any steps in synthesis, production or assembly that might be unusually complex or have a high cost of materials. Example: special medical-grade rare-metal alloys, or specific injection molding techniques, or expensive starting chemicals.

(*cont.*)

7.5. Decide on final packaging and delivery needs – is it important to keep the item away from light at all times, is it necessary to keep the product at a specific temperature during storage or shipping?

7.6. Get cost and timeline estimates for pilot scale-up and for large-scale/bulk production from suppliers in the industry – this is product dependent and most suppliers/CMOs (contract manufacturing organizations) will sign confidentiality agreements in order to receive product details and give back quotes/proposals.

7.7. Refer to the industry association that your company belongs to in order for training courses in GMP and other regulatory issues

References and additional readings

Avis, KE, Wagner, CM, and Wu, VL. (2019). *Biotechnology: Quality Assurance and Validation (Drug Manufacturing Technology Series)*. CRC Press. ISBN: 978-0367400255

Bhatia, S. (2012). *Microfabrication in Tissue Engineering and Bioartificial Organs (Microsystems)*. Springer Press. ISBN: 978-1461373865

Business Week Online. (August 13, 2001). The corporation: biotech and the spoils of success. (*author unknown*).

Bunn, G. (2019). *Good Manufacturing Practices for Pharmaceuticals: A Plan for Total Quality Control*, 7th ed. CRC Press. ISBN: 978-1498732062

Davim, JP. (2012). *The Design and Manufacture of Medical Devices*. Woodhead Publishing. ISBN: 978-1907568725

Gee, A. (editor). (2009). *Cell Therapy: cGMP Facilities and Manufacturing*. Springer Press. ISBN: 978-0387895833

Lefevre, G. (January 17, 2001). Hip replacement patients may face more surgery. *CNN*.

Levin, M. (editor). (2011). *Pharmaceutical Process Scale-Up*. CRC Press. ISBN: 978-1616310011.

McCormick, D. (2004, November). Small changes, big effects in biological manufacturing. *Pharmaceutical Science and Technology News*, 16 (accessed online August 2006).

National Standards Policy Advisory Committee. (December 1978). *National Policy on Standards for the United States and a Recommended Implementation Plan*. Washington, DC, p. 6.

Schellekens, H, and Jiskoot, W. (2006). Eprex-associated pure red cell aplasia and leachates. *Nature Biotechnology*, 24(6), 613–614.

Shuler, ML, Kargi, F, and DeLisa, M. (2017). *Bioprocess Engineering: Basic Concepts*, 3rd ed. Pearson Press. ISBN: 978-0137062706

Signore, A, and Jacobs, T. (editors). (2016). *Good Design Practices for GMP Pharmaceutical Facilities*, 2nd ed. CRC Press. ISBN: 978-1482258905

Singh, S, Prakash, C, and Singh, R. (2020). *3D Printing in Biomedical Engineering*. Springer Press. ISBN: 978-9811554254

Vogel, HC, Todaro, CC, and Press, WA. (2014). *Fermentation and Biochemical Engineering Handbook: Principles, Process Design, and Equipment*, 3rd ed. Noyes Publications. ISBN: 978-1455725533

Visit https://shreefalmehta.com/csbtbook for additional enriching readings around the topics covered in the book, topical updates on the content and for industry viewpoints and news.

Appendix 7.1 Pharmaceutical GMP

This manual was developed by the FDA/CDER for inspectors of GMP pharmaceutical facilities. It is available in its whole form at www.fda.gov/inspections-compliance-enforcement-and-criminal-investigations/inspection-references/investigations-operations-manual (last visited Nov 25, 2020).

COMPLIANCE PROGRAM GUIDANCE MANUAL PROGRAM 7356.002F CHAPTER 56 – DRUG QUALITY ASSURANCE

Appendix E:

Quality system

Assessment of the Quality System has two phases. The first phase is to evaluate whether the Quality Unit has fulfilled the responsibility to review and approve all procedures related to production, quality control, and quality assurance and assure the procedures are adequate for their intended use. This also includes the associated recordkeeping systems. The second phase is to assess the data collected to identify quality problems and may link to other major systems for inspectional coverage. ...

. . .

For each of the following, the firm should have written and approved procedures and documentation resulting therefrom. The firm's adherence to written procedures should be verified through observation whenever possible. When this system is selected for coverage in addition to the Quality System, all areas listed below should be covered; however, the actual depth of coverage may vary from the planned inspection strategy depending upon inspectional findings.

Facilities and equipment system

1. Facilities
   - Cleaning and maintenance.
   - Facility layout, flow of materials and personnel for prevention of cross-contamination, including from processing of non-drug materials.
   - Dedicated areas or containment controls for highly sensitizing materials (e.g., penicillin, beta-lactams, steroids, hormones, and cytotoxics).
   - Utilities such as steam, gas, compressed air, heating, ventilation, and air conditioning should be qualified and appropriately monitored (note: this system includes only those utilities whose output is not intended to be incorporated into the API, such as water used in cooling/heating jacketed vessels).
   - Lighting, sewage and refuse disposal, washing and toilet facilities.
   - Control system for implementing changes in the building.

- Sanitation of the building including use of rodenticides, fungicides, insecticides, cleaning and sanitizing agents.
- Training and qualification of personnel.

2. Process Equipment
   - Equipment installation, operational, performance qualification where appropriate.
   - Appropriate design, adequate size and suitably located for its intended use.
   - Equipment surfaces should not be reactive, additive, or absorptive of materials under process so as to alter their quality.
   - Equipment (e.g., reactors, storage containers) and permanently installed processing lines should be appropriately identified.
   - Substances associated with the operation of equipment (e.g., lubricants, heating fluids or coolants) should not come into contact with starting materials, intermediates, final APIs, and containers.
   - Cleaning procedures and cleaning validation and sanitization studies should be reviewed to verify that residues, microbial, and, when appropriate, endotoxin contamination are removed to below scientifically appropriate levels.
   - Calibrations using standards traceable to certified standards, preferably NIST, USP, or counterpart recognized national government standard-setting authority.
   - Equipment qualification, calibration and maintenance, including computer qualification/validation and security.
   - Control system for implementing changes in the equipment.
   - Documentation of any discrepancy (a critical discrepancy investigation is covered under the Quality System).
   - Training and qualification of personnel.

Materials system

- Training/qualification of personnel.
- Identification of starting materials, containers.
- Storage conditions.
- Holding of all material and APIs, including reprocessed material, under quarantine until tested or examined and released.
- Representative samples are collected, tested or examined using appropriate means and against appropriate specifications.
- A system for evaluating the suppliers of critical materials.
- Rejection of any starting material, intermediate, or container not meeting acceptance requirement.
- Appropriate retesting/reexamination of starting materials, intermediates, or containers.
- First-in / first-out use of materials and containers.
- Quarantine and timely disposition of rejected materials.

- Suitability of process water used in the manufacture of API, including as appropriate the water system design, maintenance, validation and operation.
- Suitability of process gas used in the manufacture of API (e.g., gas use to sparge a reactor), including as appropriate the gas system design, maintenance, validation and operation.
- Containers and closures should not be additive, reactive, or absorptive.
- Control system for implementing changes.
- Qualification/validation and security of computerized or automated process.
- Finished API distribution records by batch.
- Documentation of any discrepancy (a critical discrepancy investigation is covered under the Quality System).

Production system

☐ Training/qualification of personnel.
  - Establishment, adherence, and documented performance of approved manufacturing procedures.
  - Control system for implementing changes to process.
  - Controls over critical activities and operations.
  - Documentation and investigation of critical deviations.
  - Actual yields compared with expected yields at designated steps.
  - Where appropriate established time limits for completion of phases of production.
  - Appropriate identification of major equipment used in production of intermediates and API.
  - Justification and consistency of intermediate specifications and API specification.
  - Implementation and documentation of process controls, testing, and examinations (e.g., pH, temperature, purity, actual yields, clarity).
  - In-process sampling should be conducted using procedures designed to prevent contamination of the sampled material.
  - Recovery (e.g., from mother liquor or filtrates) of reactants; approved procedures and recovered materials meet specifications suitable for their intended use.
☐ Solvents can be recovered and reused in the same processes or in different processes provided that solvents meet appropriate standards before reuse or commingling.
  - API micronization on multi-use equipment and the precautions taken by the firm to prevent or minimize the potential for cross-contamination.
  - Process validation, including validation and security of computerized or automated process
  - Master batch production and control records.
  - Batch production and control records.

- Documentation of any discrepancy (a critical discrepancy investigation is covered under the Quality System).

Packaging and labeling system

☐ Training/qualification of personnel.
  - Acceptance operations for packaging and labeling materials.
  - Control system for implementing changes in packaging and labeling operations
  - Adequate storage for labels and labeling, both approved and returned after issued.
  - Control of labels which are similar in size, shape, and color for different APIs.
  - Adequate packaging records that will include specimens of all labels used.
  - Control of issuance of labeling, examination of issued labels and reconciliation of used labels.
  - Examination of the labeled finished APIs.
  - Adequate inspection (proofing) of incoming labeling.
  - Use of lot numbers, destruction of excess labeling bearing lot/control numbers.
  - Adequate separation and controls when labeling more than one batch at a time.
  - Adequate expiration or retest dates on the label.
  - Validation of packaging and labeling operations including validation and security of computerized process.
☐ Documentation of any discrepancy (a critical discrepancy investigation is covered under the Quality System).

Laboratory control system

☐ Training/qualification of personnel.
  - Adequacy of staffing for laboratory operations.
  - Adequacy of equipment and facility for intended use.
  - Calibration and maintenance programs for analytical instruments and equipment.
  - Validation and security of computerized or automated processes.
  - Reference standards; source, purity and assay, and tests to establish equivalency to current official reference standards as appropriate.
  - System suitability checks on chromatographic systems.
  - Specifications, standards, and representative sampling plans.
  - Validation/verification of analytical methods.
  - Required testing is performed on the correct samples and by the approved or filed methods or equivalent methods.
  - Documentation of any discrepancy (a critical discrepancy investigation is covered under the Quality System).
  - Complete analytical records from all tests and summaries of results.

- Quality and retention of raw data (e.g., chromatograms and spectra).
- Correlation of result summaries to raw data; presence and disposition of unused data.
- Adherence to an adequate Out of Specification (OOS) procedure which includes timely completion of the investigation.
- Test methods for establishing a complete impurity profile for each API process (note: impurity profiles are often process-related).
- Adequate reserve samples; documentation of reserve samples examination.
- Stability testing program, including demonstration of stability indicating capability of the test methods.

Appendix 7.2 Device and diagnostic GMP

This manual was developed by the FDA/CDRH for inspectors of device and diagnostic facilities. It is available in its whole form at www.fda.gov/inspections-compliance-enforcement-and-criminal-investigations/inspection-guides/quality-systems   (last visited Dec 2020).

CDRH: QUALITY SYSTEM INSPECTION TECHNIQUE

This Guide to Inspections of Quality Systems (QS) provides instructions for conducting medical device quality system/ GMP inspections. It is to be used in conjunction with the compliance program entitled Inspections of Medical Device Manufacturers (7382.845).

Most device firms are inspected more than once. By probing different subsystems, different devices or different processes each time, FDA will eventually have covered most of the firm's quality system. ... As a general rule of thumb, one day should be sufficient to cover each subsystem when using the "top-down" approach described within this document.

Management controls

The purpose of the management control subsystem is to provide adequate resources for device design, manufacturing, quality assurance, distribution, installation, and servicing activities; assure the quality system is functioning properly; monitor the quality system; and make necessary adjustments. A quality system that has been implemented effectively and is monitored to identify and address problems is more likely to produce devices that function as intended.

A primary purpose of the inspection is to determine whether management with executive responsibility ensures that an adequate and effective quality system has been established (defined, documented and implemented) at the firm. Because of this, each inspection should begin and end with an evaluation of this subsystem.

Quality policy

The firm must have a written quality policy. It means the overall intentions and directions of an organization with respect to quality. The firm is responsible for establishing a clear quality policy with achievable objectives then translating the objectives into actual methods and procedures. Management with executive responsibility (i.e. has the authority to establish and make changes to the company quality policy) must assure the policy and objectives are understood and implemented at all levels of their organization.

Management review and quality audit procedures

[T]he manufacturer has written procedures for conducting management reviews and quality audits and there are defined intervals for when they should occur. The firm's quality audits should examine the quality system activities to demonstrate that the procedures are appropriate to achieve quality system objectives, and the procedures have been implemented.

Quality plan

The firm must have a written quality plan that defines the quality practices, resources and activities relevant to the devices that are being designed and manufactured at that facility. The manufacturer needs to have written procedures that describe how they intend to meet their quality requirements. Quality plans may be specific to one device or be generic to all devices manufactured at the firm. Quality plans can also be specific to processes or overall systems.

Quality system procedures and instructions

The term "quality system" ... encompasses all activities ... necessary to assure the finished device meets its predetermined design specifications. This includes assuring manufacturing processes are controlled and adequate for their intended use, documentation is controlled and maintained, equipment is calibrated, inspected, tested, etc.

Management controls inspection steps

1. Verify that a quality policy, management review and quality audit procedures, quality plan, and quality system procedures and instructions have been defined and documented.
2. Verify that a quality policy and objectives have been implemented.
3. Review the firm's established organizational structure to confirm that it includes provisions for responsibilities, authorities and necessary resources.
4. Confirm that a management representative has been appointed. Evaluate the purview of the management representative.
5. Verify that management reviews, including a review of the suitability and effectiveness of the quality system, are being conducted.

6. Verify that quality audits, including re-audits of deficient matters, of the quality system are being conducted.
7. Evaluate whether management with executive responsibility ensures that an adequate and effective quality system has been established and maintained.

Design controls

The purpose of the design control subsystem is to control the design process to assure that devices meet user needs, intended uses, and specified requirements. Attention to design and development planning, identifying design inputs, developing design outputs, verifying that design outputs meet design inputs, validating the design, controlling design changes, reviewing design results, transferring the design to production, and compiling a design history file help assure that resulting designs will meet user needs, intended uses and requirements.

The design control requirements of Section 820.30 of the regulation apply to the design of Class II and III medical devices, and a select group of Class I devices. The regulation is very flexible in the area of design controls.

Inspection of design controls

If design control requirements are applicable to the operations of the firm, select a design project ... that provides the best challenge to the firm's design control system. This project will be used to evaluate the process, the methods, and the procedures that the firm has established to implement the requirements for design controls.

Utilize the firm's design plan as a road map for the selected design project. Plans include major design tasks, project milestones, or key decision points and must define responsibility for implementation of the design and development activities and identify and describe interfaces with different groups or activities. Verification and validation activities should be predictive rather than empiric. Acceptance criteria must be stated up front. Design verification activities are performed to provide objective evidence that design output meets the design input requirements. Verification activities include tests, inspections, analyses, measurements, or demonstrations.

Review how the design was transferred into production specifications. Review the device master record. Sample the significant elements of the device master record using the Sampling Tables and compare these with the approved design outputs. These elements may be chosen based on the firm's previously identified essential requirements and risk analysis.

Corrective and preventive actions (CAPA)

The purpose of the corrective and preventive action subsystem is to collect information, analyze information, identify and investigate product and quality problems, and take appropriate and effective corrective and/or preventive action to prevent their recurrence. Verifying or validating corrective and preventive actions, communicating corrective and preventive action activities to responsible people, providing relevant information for management review, and documenting these activities are essential in dealing effectively with product and quality problems, preventing their recurrence, and preventing or minimizing device failures. One of the most important quality system elements is the corrective and preventive action subsystem.

Inspection objectives for CAPA

1. Verify that CAPA system procedure(s) that address the requirements of the quality system regulation have been defined and documented.
2. Determine if appropriate sources of product and quality problems have been identified. Confirm that data from these sources are analyzed to identify existing product and quality problems that may require corrective action.
3. Determine if sources of product and quality information that may show unfavorable trends have been identified. Confirm that data from these sources are analyzed to identify potential product and quality problems that may require preventive action.
4. Challenge the quality data information system. Verify that the data received by the CAPA system are complete, accurate and timely.
5. Verify that appropriate statistical methods are employed (where necessary) to detect recurring quality problems. Determine if results of analyses are compared across different data sources to identify and develop the extent of product and quality problems.
6. Determine if failure investigation procedures are followed. Determine if the degree to which a quality problem or nonconforming product is investigated is commensurate with the significance and risk of the nonconformity. Determine if failure investigations are conducted to determine root cause (where possible). Verify that there is control for preventing distribution of nonconforming product.
7. Determine if appropriate actions have been taken for significant product and quality problems identified from data sources.
8. Determine if corrective and preventive actions were effective and verified or validated prior to implementation. Confirm that corrective and preventive actions do not adversely affect the finished device.
9. Verify that corrective and preventive actions for product and quality problems were implemented and documented.
10. Determine if information regarding nonconforming product and quality problems and corrective and preventive actions has been properly disseminated, including dissemination for management review.

Medical Device Reporting (MDR) – CAPA satellite system

The Medical Device Reporting (MDR) Regulation requires medical device manufacturers, device user facilities and importers to establish a system that ensures the prompt identification, timely investigation, reporting, documentation, and filing of device-related death, serious injury, and malfunction information. The events described in Medical Device Reports (MDRs) may require the FDA to initiate corrective actions to protect the public health. Therefore, compliance with Medical Device Reporting must be verified to ensure that CDRH's Surveillance Program receives both timely and accurate information.

Reports of Corrections and Removals – CAPA satellite system

The Corrections and Removals (CAR) Regulation requires medical device manufacturers and importers to promptly notify FDA of any correction or removal initiated to reduce a risk to health. This early notification improves FDA's ability to quickly evaluate risks and, when appropriate, initiate corrective actions to protect the public health.

Medical Device Tracking – CAPA satellite system

The purpose of the Medical Device Tracking Regulation is to ensure that manufacturers and importers of certain medical devices can expeditiously locate and remove these devices from the market and/or notify patients of significant device problems.

Production and process controls

The purpose of the production and process control subsystem is to manufacture products that meet specifications. Developing processes that are adequate to produce devices that meet specifications, validating (or fully verifying the results of) those processes, and monitoring and controlling the processes are all steps that help assure the result will be devices that meet specifications.

Inspection objectives for production and process controls

1. Select a process for review based on:
   a. CAPA indicators of process problems;
   b. Use of the process for manufacturing higher risk devices;
   c. Degree of risk of the process to cause device failures;
   d. The firm's lack of familiarity and experience with the process;

  e. Use of the process in manufacturing multiple devices;
  f. Variety in process technologies and Profile classes;
  g. Processes not covered during previous inspections;
  h. Any other appropriate criterion as dictated by the assignment
2. Review the specific procedure(s) for the manufacturing process selected and the methods for controlling and monitoring the process. Verify that the process is controlled and monitored.
3. If review of the Device History Records (including process control and monitoring records, etc.) reveals that the process is outside the firm's tolerance for operating parameters and/or rejects or that product nonconformances exist:
  a. Determine whether any nonconformances were handled appropriately;
  b. Review the equipment adjustment, calibration and maintenance; and
  c. Evaluate the validation study in full to determine whether the process has been adequately validated.
4. If the results of the process reviewed cannot be fully verified, confirm that the process was validated by reviewing the validation study.
5. If the process is software controlled, confirm that the software was validated.
6. Verify that personnel have been appropriately qualified to implement validated processes or appropriately trained to implement processes which yield results that can be fully verified.

Sterilization process controls – process control satellite system

The purpose of the production and process control subsystem (including sterilization process controls) is to manufacture products that meet specifications. Developing processes that are adequate to produce devices that meet specifications, validating (or fully verifying the results of) those processes, and monitoring and controlling the processes are all steps that help assure the result will be devices that meet specifications. For sterilization processes, the primary device specification is the desired Sterility Assurance Level (SAL). Other specifications may include sterilant residues and endotoxin levels.

8 Reimbursement, marketing, sales, and product liability

Plan	Position	Pitch	Patent	Product	Pass	Production	**Profits**
Industry context	Market research	Start a business venture	Intellectual property rights	New product development (NPD)	Regulatory plan	Manufacture	Reimbursement

---

Learning points

- Who are the purchasers and who are the payers in the health system in the United States?
- Flow of products and payments in the U.S. healthcare system for devices and drugs
- What are technology assessments and who uses them?
- What steps can a biomedical product company (drugs, devices, diagnostics) take to maximize revenues in the U.S. healthcare system?
- What specific payer perspectives should be taken into account during product development / how can product development be planned for maximum reimbursement benefit?
- How do new products gain recognition and reimbursement in the healthcare payment system?

The healthcare system of payments and reimbursements in the United States is a complex system and causes much confusion for those who are not familiar with the processes and the various players. Most other developed countries have a simpler single-payer system, where the government is the dominant provider and the payer. A careful reading of this chapter along with active reference to other texts and the Centers for Medicare & Medicaid Services (CMS) website will help the reader gain the fundamental understanding required before launching a product in the United States. Reimbursement framework in other countries is similar to that in the United States, with some differences that are mentioned as examples throughout this chapter.

Reimbursement planning typically begins during early clinical development of the biomedical product. This planning is used (1) to determine whether the product can deliver appropriate returns, (2) to develop specific clinical data on performance and benefit, for differentiation of the product for payers and hospital and clinical customers, and (3) to identify possible strategic business relationships for product launch.

8.1   Healthcare system in the United States

8.1.1   Economic impact of the healthcare system

The healthcare payments to providers of services and goods in the healthcare system have been increasing faster than the gross domestic product (GDP, an aggregate figure that is used to assess the overall economy of a country) and are projected to reach 19.7% of GDP by 2028 (data from CMS.gov) (Table 8.1 and Figure 8.1).

Table 8.1 National healthcare costs – part of GDP and per capita

	1965	1980	2004	2018
National healthcare spend ($billions)	$41	$246	$1 878	$3 600
Percentage of GDP	5.7%	9.1%	16.0%	17.7%
Per capita amount	$205	$2 821	$6 280	$6 280
Source of funds	*Percentage of total*			
Private	75.1%	59.6%	54.9%	52%
Public	24.9%	40.4%	45.1%	48%

Source: CMS, Office of the Actuary, U.S. Dept of Commerce, Bureau of Economic Analysis, U.S. Bureau of the Census

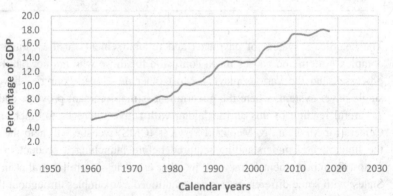

Figure 8.1 National healthcare spending is increasing faster than the GDP. Data and chart from CMS, Office of the Actuary, National Health Statistics Group.

8.1.2   Insurance coverage of the U.S. population

In the United States, 92% of people had insurance, wherein most people (55.4% of the population) were covered by a health insurance plan provided by their employer for some or all of 2019 and of the rest, 12.5% purchased private insurance directly and the rest were covered by public health insurance (Medicare, Medicaid, or other programs) in 2019 (Figure 8.2 and Table 8.2). A large percent (8%) of the population was uninsured for at least a portion of the year in 2019, as reported by the U.S. Census Bureau (census.gov) (Figure 8.2). The various public programs are described in more detail in Section 8.1.3.

Table 8.2 Public and private coverage in United States by age and other economic groupings

Age or other grouping	Dominant form of insurance coverage
65 and older	Predominantly public insurance with private self-purchased or employer paid
Needy and indigent and other select groups who would not be able to buy private care (end-stage renal disease patients, children)	Public insurance cover
Below 65 years of age	Predominantly private insurance through employers or direct purchase + some public coverage (if in above groups)

Note: A significant percentage of the population is uninsured in the United States.

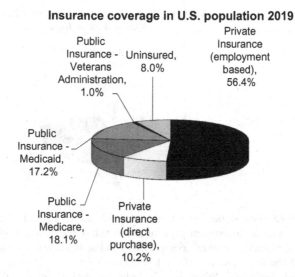

Insurance coverage in U.S. population 2019

Figure 8.2 Health insurance coverage of the U.S. population in 2019 shows that a majority of the population is covered by employer based private insurance. (Data from U.S. Census.)

8.1.3   Who pays for the national healthcare expenditures?

Eventually, we each pay for healthcare, either directly or indirectly. However, in the U.S. healthcare system, the payments to providers of services and goods are made by third parties – public and private payers. Examples of select European and Asian country's systems are given later in this chapter.

In the United States, public payers include the U.S. government's Department of Health and Human Services (DHHS), and the main agency, Centers for Medicare & Medicaid Services (CMS), whose programs (Medicare and Medicaid, described below) account for about a third of all payments in the healthcare system. If the Veterans Administration and other programs are added in, the state and federal governments pay almost half of the healthcare bills (Figure 8.3).

Medicare is a federal insurance program that covers most individuals over the age of 65, the disabled (who satisfy certain statutory requirements) and end stage renal disease sufferers (totaling over 37 million people in 2019). In 2020, 62.6 million people were enrolled in one or both of Parts A and B of the Medicare program, and 25.0 million of them chose to participate in a Medicare Advantage plan (details from CMS). Total expenditures for Medicare in 2019 were $796.2 billion.

Medicare has three main parts (programs) that have evolved over time. Each part has a different annual deductible and co-insurance structure. Parts A and B reimburse the provider based on a prospective payment system, having evolved from an original fee-for-service payment structure (see Section 8.4.3 for details on payment system structures). Medicare benefits in parts A and B are administered by locally contracted entities. Legislation passed in 2003 reduced the number of Medicare carriers and fiscal intermediaries from over 50 to approximately 12 Medicare Administrative Centers

Nation's Health Dollar Calendar Year 2018:
Where it came from

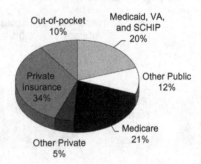

Figure 8.3 The distribution of payments across various payers in the U.S. healthcare system shows that the government pays for almost half of the healthcare bill. Other Public includes programs such as worker's compensation, public health activity, Department of defense, Department of Veterans Affairs, Indian Health Service, State and local hospital subsidies and school health; Other Private includes industrial in-plant, privately funded construction and non-patient revenues, including philanthropy. (Data and figure source: CMS, Office of Actuary, National Health Statistics Group.)

(MACs). Durable medical equipment (DME) claims are processed by 4 specialty "DME MACs," and home health and hospice claims, are processed by 4 specialty MACs that also process part A and B claims. This process has significant coverage implications on manufacturers of biomedical technology and their product development choices (see Section 8.4.1).

Part A – Hospital Insurance: This covers inpatient hospital care; including critical access hospitals, and skilled nursing facilities (not custodial or long-term care). It also helps cover hospice care and some home healthcare. Beneficiaries must meet certain conditions to get these benefits and pay certain deductibles and co-insurance for each separate site of care. Medicare Part A pays the hospital. Most people don't pay a premium for Part A because they or a spouse already paid for it through their payroll taxes while working.

Part B – Supplementary Medical Insurance: Beneficiaries, if they choose to enroll in Part B, may be required to pay a monthly premium for Part B. Medicare Part B covers doctors' services (in all settings), durable medical equipment, clinical laboratory services and outpatient care, and IV and injectable drugs. Medicare Part B pays the doctor.

Part C – Medicare Advantage Program (optional): This provides beneficiaries enrolled in both part A and B, to have the choice of receiving their Medicare benefits through local approved private health plans. These private health plans must provide the same coverage as in traditional Medicare (Parts A & B) and can also offer additional benefits such as prescription drug coverage. Beneficiaries pay an extra premium to cover additional benefits. Most plans also offer Part D drug coverage.

Part D – also part of the Supplementary Medical Insurance program (optional): This provides subsidized outpatient prescription drug coverage. Individuals must pay a monthly premium, co-insurance and annual deductible for this coverage and is available to everyone with Medicare. Beneficiaries can choose either a stand-alone drug plan from a private company or a comprehensive benefit program consisting of Parts A, B, and D from a private company.

Medicaid is a state-administered assistance program that covers people below 65 years of age with income below 138% of the federal poverty level, in specific eligibility groups that vary from state to state, and a large managed care population (total covered were about 73.7 million in 2019). The Medicaid program is administered by state health services and each state sets its own guidelines regarding eligibility and services. The federal government shares in the cost (from 50% for higher-income states to over 70% for lower-income states). In FY 2019, net outlays for the Medicaid program (Federal and State) were an estimated $629.4 billion, and are projected to reach $886.4 billion by FY 2025 (data from CMS).

Private insurers, who are paid premiums by employers and individuals, include companies like Blue Cross Blue Shield of Massachusetts, Anthem, Aetna, Cigna, and Wellpoint, and health maintenance organizations like Kaiser Permanente. These private insurers usually dominate regionally and cover all contracted healthcare services. Insurance policies have gone beyond their original intention and use which was to safeguard against high-cost hospitalization events and now cover routine medical care as well. Coverage is provided using a range of plan types that offer varying degrees of provider choice. Preferred provider organizations (PPOs) and health maintenance organizations (HMOs) are the two most common plan types.

This mix of private and public insurance coverage in the United States, with majority payments being made by private insurance and combination of Medicaid or private insurance and Medicare (for those above 65) is a unique mixed system among industrialized countries as discussed further in Section 8.9. Figure 8.3 shows the source (payers) of the nation's healthcare spending dollars.

Medicare is often the most important payer for medical procedures in the patient population above 65 years of age. This trend will only strengthen as the U.S. population demographic shifts towards a more aged population. Additionally, Medicare reimbursement policies greatly influence the private payers.

8.2   Flow of payments and distribution models for products and services

In the U.S. third-party payer system, the flow of funds from multiparty payers is schematically described in Figure 8.4. Wholesalers who buy and stock products get reimbursed for these products, not from the patient, but predominantly from the providers – the hospitals, clinics and other providers. The government and private insurance payers reimburse the providers, and the patients only pay their assigned co-pay, unless they are uninsured. Thus, the payment system for healthcare products and services is a third-party payment system. Medicare is the predominant payer for patients above 65. Private insurance payers are the predominant payers for patients under 65 unless the patient falls into one of a number of special category groups (e.g., low income; veteran, end-stage kidney disease).

An important conclusion from this schematic (Figure 8.4) describing flow of payments is that the purchasing decision is influenced by a variety of stakeholders, and so *most biomedical product companies will have to market their product and its context-specific benefits to different groups at the same time.* For example, an insurance company needs to be shown how the innovative product can reduce the overall costs of healthcare for a given health problem; a hospital, pharmacy or physicians' clinic needs to know how the innovative product can increases efficiencies or operating margins compared to total reimbursement they are given by the payers; and the patients must understand why the new product is better for them, especially if they can influence the usage decision. Physicians are the most important influencers of the purchasing decision for drugs, devices, and diagnostics, but in some instances, patients also have significant influence, as seen by drug companies' efforts to drive drug product purchasing or usage decisions by direct-to-consumer (DTC) marketing.

The immediate market for manufacturers typically comprises of the wholesalers and group purchasing organizations, while many device and diagnostic companies sell their product through distributors or directly to the hospitals and physicians. While

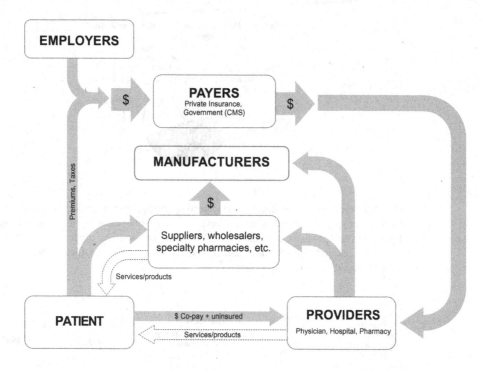

Figure 8.4 Flow of payments in the third-party payer U.S. healthcare system.

drugs are typically purchased from the manufacturer by wholesalers, devices are usually purchased from the manufacturers by hospitals or physicians directly or through group purchasing organizations (which negotiate discounts). The end-providers – the hospitals, the clinics, the hospital pharmacies, the retail pharmacy organizations –get reimbursed by the public or private insurers. *Given this flow of payments, most manufacturers' reimbursement and sales efforts are driven toward making sure the hospitals and providers or direct purchasers get adequately reimbursed for purchasing the drugs or devices.*

Figure 8.5 shows the various groups in the healthcare systems to which the U.S. healthcare dollar gets paid out – the providers of services and products and the intermediary distributors and wholesalers.

8.3   Distribution and payment flow for biomedical product types

8.3.1   Drugs /biologics product payment and distribution model

Drugs are distributed to the patient by two means: (1) self-administered drugs delivered through pharmacies, and (2) provider-administered or infused drugs (most biologics) delivered through hospital inpatient or outpatient settings or physician's office (Figure 8.6). Pharmaceutical manufacturers typically sell drugs in bulk to either

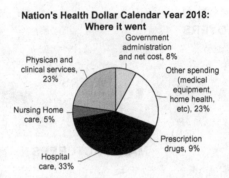

Figure 8.5 Percentage of healthcare payments to specific groups of service and good providers in 2018. Other spending includes dentist services, other professional services, home health, durable medical products, over-the-counter medicine and sundries, public health, other personal healthcare, research and structures and equipment. (Source: CMS, Office of the Actuary, National Health Statistics Group.)

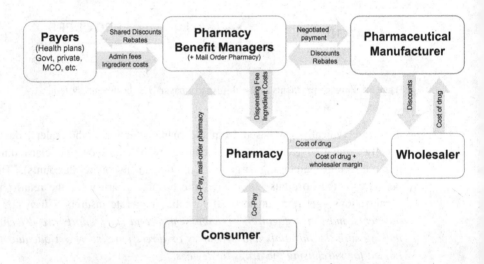

Figure 8.6 Flow of payments back to drug manufacturer (schematic adapted from CMS Office of Research Development and Information). See text for more details.

wholesalers or large pharmacy groups. In the case of infused drugs, the purchasers are specialty pharmacies or pharmacies at care-provider institutions or the physician's offices (Figure 8.7).

Thus, the payments flow as shown in Figure 8.6, from the payers through the providers or retail outlets (pharmacies, hospitals) and through the wholesalers to the manufacturers. When a prescription is filled at a pharmacy, the pharmacy collects any co-pay and then bills the insurance company/payer to get reimbursed for the drug.

Manufacturers offer discounts and rebates to the pharmacy benefit manager (PBM) companies that manage formularies and drug reimbursement programs for many

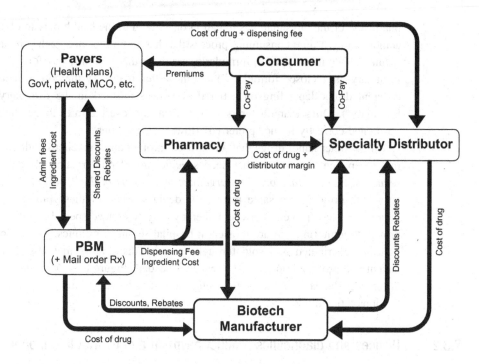

Figure 8.7 Payment flow for infused drugs (biologics and other drugs that need to be delivered by caregiver) through a specialty distributor. The PBM does not play a major role in this process.

different payers. These PBMs emerged in the period from 1995 to 2006 as discount formulary negotiators and now influence the providers and payers (hospitals and private insurers) to select specific drugs into a formulary. The various discounts and rebates (discounts for early payments, volume discounts, pre-negotiated rebates, etc.) offered to the supply chain parties (wholesalers, pharmacies, PBMs) make drug pricing a complex field. Large institutional pharmacies or chains of pharmacies and specialty pharmacies that have their own warehousing logistics can order direct from the manufacturer and thus pay the manufacturer directly. This specific flow of payments from pharmacy direct to manufacturer is also shown in Figure 8.6.

A *formulary* is a select menu of drugs (and devices/diagnostics) that a managed care organization or insurer (Medicare, private insurers) will cover under its plan. The drugs are selected by a Pharmacy and Therapeutics (P&T) Committee in hospitals or at insurance plans. The P&T committee is composed of physicians, pharmacists and administrators. The formulary not only lists drugs but can also contain additional prescribing guidelines. The P&T committee reviews drugs based on parameters such as clinical safety and efficacy (compared to other products in that class), indications for use, dosages, method and route of administration, net cost, physician demand, and impact on total care costs and quality of life (pharmacoeconomic data).

Formularies are effectively used today as a formal system that assures the selection of cost-effective, affordable medications in quality patient care. If a physician wants to prescribe a drug outside the formulary, either the patient has to pay full price at the pharmacy, or the physician has to call in or write in to the health plan to obtain prior authorization, a time-consuming process that has to be repeated each time a refill is required. Two main types of formularies are those that will only pay for drugs on the formulary list ("closed formulary") and those that may vary the co-pay required for different drugs depending on a tiered system established in the formulary ('open formulary'). Drug manufacturers collect data and exert efforts to get their drugs positioned correctly in the "preferred" (lowest co-pay) tier.

On the other hand, drugs that are infused through an intravenous catheter directly into the patient's blood or that need specific provider interaction for delivery *are reimbursed to the care-provider organization, as part of the procedure* of administering the drug, in the same manner as devices. Most of these infused drugs are biological drugs (proteins) or chemotherapy drugs that need special handling or care-provider monitoring and are infused in hospital settings or clinics. The biologics are typically distributed and sold through specialty distributors that also educate the patient and provide home-based services if needed (Figure 8.7). More details on the differences between infused and self-administered drugs (pills) are discussed in Sections 8.6 and 8.7.

8.3.2   Devices and diagnostics product payment and distribution model

The flow of payments and distribution of product is similar for devices and diagnostics (and for provider-administered infused drugs). For devices, infused drugs and diagnostics, the provider (hospital if in the hospital, or physician's office if outside) purchases the product and submits a claim to the payer charging for the services and the product. The service costs are derived as a combination of effort and facilities costs, making the site of delivery a major variable in the reimbursement payment. Thus, *device reimbursement is usually bundled with the overall procedure reimbursement and payment level is dependent on site of delivery,* creating different dynamics in sales and product pricing for this product type, compared to self-administered drugs.

Medical devices are typically sold through specialized distributors or through a direct sales force to hospitals or clinics or to their purchasing consortia. The device and related procedures are delivered in the hospital or clinic and the payments from insurers or the patient are collected by both, the hospital and the physician. The hospitals and clinics pay for the devices at purchase and typically hold inventory, getting discounts that have been pre-negotiated through group purchasing organizations or directly with the manufacturer. The insurers and government payers typically reimburse the care provider (who purchased the device; usually the hospital) or the specialized distributor, or else the payers reimburse the customer directly (not common) if the patient themselves paid entirely for the product or treatment (e.g. for home medical products or out of network costs). Care providers can try to get expensive, or novel, devices reimbursed separately from the all-inclusive cost of the specific procedure, but have to meet specific criteria (as per Medicare guidelines) for their product to be considered for "add-on" payment. Most payers will want to bundle the cost of the device in the cost of the procedure or other billing group. In this situation, not getting reimbursed adequately for an expensive device will result in pressure from the finance division of the hospital to discontinue or reduce use of the new device that was chosen by the physicians. This will obviously impact sales significantly. Many organizations also have a managed device formulary, similar to the drug formulary described earlier. For cases where there is no reimbursement or if the device is not on the payer's or organization's approved formulary, the hospital or clinic bills the patient directly.

Clinical diagnostic tests that are cleared through the 510(k)/CLIA process, or approved via the PMA process, are usually sold to large, centralized laboratories or to one of thousands of local clinical or hospital-based diagnostic labs. These products (reagents, diagnostic kits, etc.) are purchased by the laboratory or service provider organization within which the laboratory is housed (e.g. hospital) and reimbursement is claimed directly from the insurers/payers per use as part of a procedure claim for the specific test or panel of tests based on the patient condition diagnosis. Medicare determines payment rates for clinical laboratory tests reimbursed for under the Clinical Laboratory Fee Schedule (CLFS).

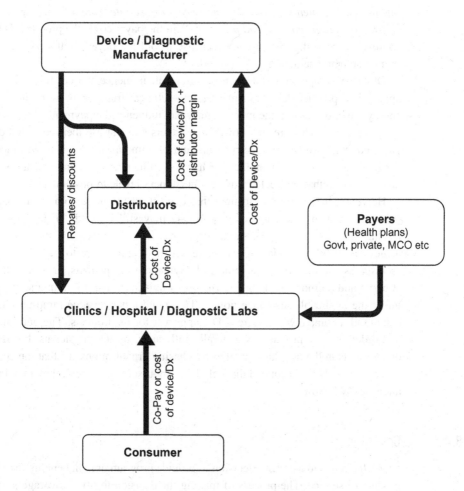

Figure 8.8 Payment flow for devices and diagnostics sold through distributors or directly through service providers to consumer.

8.4   Components of the reimbursement process

The previous sections gave a brief overview of the flow of payments and the distribution structure for the industry. *Reimbursement* is the key to healthy sales revenues in the biomedical marketplace as most purchasers will buy more of your product if there is an adequate third-party (insurance) reimbursement available to those purchasers (hospitals, individuals, care providers) for those products. A few key questions that drive and control the flow of payments will be discussed in the rest of this chapter:

- Is the control on prices influenced more by the manufacturer, the third party payer, or the purchasers?
- How does one get agreement for reimbursement for a newly approved drug or device from the various payers? In other words, how does a newly approved biomedical product get adopted and accepted into the reimbursement system?

The *three components that must be in place for adequate reimbursement* are discussed below – *coverage, coding, and payment*. Will the payer cover the product? Is there an appropriate code that the provider can use to bill the payer? Is the payment rate in reimbursement adequate for the provider?

Of these components, the manufacturer can influence the coverage decision by appropriate product development planning (clinical trial design and data collected) and by payer education; the manufacturer can influence the payment rates, as they set the price they wish to receive for their products and can use their collected data and payer education to achieve the right amount of reimbursement which will incentivize the purchasers; but the manufacturer has less influence on coding and the setting of relative values that will determine reimbursement levels to physicians.

However, it is important to note that even if all the three components described below are not in place, approved products may still be sold. In the case that the product does not fit into the standard process for reimbursement, the sales will not achieve their full potential, as each caregiver will either have to pay for the product themselves and then risk collecting full payments from patients, or they will have to struggle and do more work than normal to convince a payer to reimburse the product under the applicable insurance policy. These and other revenue ramp-up challenges can derail a company even after FDA approval for its products. The decision to use and order specific products is generally influenced by the physicians, but if finance departments in the hospitals or other purchasing organizations see that the cost of the product is not being recouped through the reimbursement process, they may intervene and sales will slow.

8.4.1   Coverage

Coverage refers to a payer's decision to provide program or plan benefits for a specific product or service. The process of making the decision to offer coverage is typically one of clinical, technical, and (depending on the payer) economic evaluation of an FDA-approved product or technology. Most payers will not accept a product into full review before it is approved by the FDA (Box 8.1 discusses CMS coverage decision before FDA approval or clearance for devices).

Since CMS/Medicare is the largest single healthcare payer for most patients over 65 (the most medically needy population) and because their review process is public, the coverage decision by the CMS can have a significant impact on the coverage decision by other payers in the industry.

The FDA is a *regulator* and principally puts weight on safety and efficacy (see Chapter 6). Historically, FDA and CMS have not communicated with each other, and

---

Box 8.1 Can a new device be covered for reimbursement before FDA clearance or approval?

A new device can achieve reimbursement before it gets FDA approval, but only after it has passed an IDE (see Chapter 6 for details of an IDE). The device also must be categorized as Non-experimental/investigational (Category B) device by the FDA:

> A "Category B" device refers to a device believed to be in Class I or Class II, or a device believed to be in Class III for which the incremental risk is the primary risk in question (that is, underlying questions of safety and effectiveness of that device type have been resolved), or it is known that the device type can be safe and effective because, for example, other manufacturers have obtained FDA approval for that device type. [As in 42 CFR 405.201].

Once the device is categorized as a Category B non-experimental/non-investigational device by the FDA, this categorization is reported to CMS and care-providers who are using this device in a clinical trial can appeal for coverage and reimbursement to the local Medicare insurance provider. Coverage evaluation will be made using standard criteria of "reasonable and necessary." If a positive coverage decision is made, the manufacturer could get paid for the use of the device in the clinical trial as the provider will get reimbursed, albeit the reimbursement will be at the rate established for previously approved similar devices and related procedures.

In September 2020, Medicare proposed a new rule that would establish a new national coverage pathway to innovative medical devices that are designated as "breakthrough" by the FDA, starting directly from the date of FDA market authorization and continuing for four years. This new Medicare Coverage of Innovative Technology (MCIT) coverage pathway is now (Jan 2021) an established rule that will accelerate Medicare beneficiaries' access to innovative new medical devices in a more uniform manner.

For further details, see FDA guidance document at www.fda.gov/cdrh/d952.html *and related article from Device Link Online at* www.devicelink.com/mddi/archive/03/05/018.htm.

while that has changed recently for specific products designated as breakthrough by the FDA, obtaining coverage is still a sequential process that follows FDA approval. Drugs (self-administered and provider-administered) are usually covered and reimbursed for their FDA-approved indication, mainly due to the high quality of data generated during the clinical trials with placebo and randomized groups. However, if the clinical study did not cover specific patient groups that are a large component of a prescription drug plan, some insurers could drop the drug from their formulary. Over 85% of devices are usually cleared through the 510(k) process (see Chapter 6 for details) but the data required for clearance may not be rigorous enough to satisfy payers, whereas data generated for a device approved through PMA process (Chapter 6) has more rigorous clinical trial data.

CMS is a *purchaser* that pays for services and products and heavily weighs the outcomes of clinical usage when making coverage decisions. Specifically, the Social Security Act [Section 1862(a)(1)(A)] mandates that the government agency (CMS) should only make payment if the treatment is *"reasonable and necessary* for the diagnosis or treatment of illness or injury or to improve the functioning of a malformed body member."* CMS applies the criteria of 'reasonable and necessary' and evaluates whether there is adequate evidence to conclude that the item or service is (1) safe and effective, (2) not experimental or investigational, and (3) appropriate for the Medicare patients.

There are three main routes to coverage:

1. If the new technology and the indication for which it is approved fall under existing payment categories or codes, it will be added to the Medicare covered technologies by simply billing (coding) it under the existing payment category for the specific diagnosis/condition for which the technology is approved.
2. A local coverage decision (LCD) by the Medicare Administrative Contractor (MAC) which holds authority for processing Medicare reimbursements in each specific region. Most coverage decisions (more than 90%) are local and the manufacturer must work with each region in turn to gain coverage.
3. National Coverage Decision (NCD) by Medicare/CMS is usually a route taken by the manufacturer if the new technology represents a significant medical advance or if local coverage policies are inconsistent. Since the NCD process and decision is public, most private payers pay attention to the NCDs.

There are many benefits (e.g. a possibility to appeal a negative coverage decision) for companies to work through such local coverage decisions (LCD), even though it is more work than getting a single national coverage determination (NCD).

Currently, CMS is not required to and in fact is prohibited formally from taking cost-effectiveness of a product into consideration while making coverage decisions. The main criterion for coverage by Medicare is medical benefit to its beneficiaries, and the most commonly used gold standard to determine benefit is the randomized clinical trial. However, with the rising costs of healthcare weighing on the U.S. economy and on its budget, CMS informally continues to consider the impact of coverage decisions on healthcare costs (Box 8.2).

Box 8.2 Cost considerations and health economics
Payers do consider cost in making a reimbursement decision

In 2004, CMS was faced with a difficult choice, two drugs, recently approved by the FDA were up for coverage and reimbursement decision: Bexxar from GlaxoSmithKline and Zevalin from Biogen Idec. The drugs, which cost about $28,000 for a single dose (also the entire course), were approved as a third-line treatment for non-Hodgkin's lymphoma, only after other treatments had failed. However, some physicians were using these expensive drugs as first-line medicines in "off-label" or unapproved indications. In a *New York Times* article [Harris (2004)], Medicare officials were quoted as saying that cost was the program's primary concern in reviewing its reimbursement policy for Bexxar and Zevalin when they were used in non-approved indications (off-label use). Until then, Medicare was paying for all uses of approved drugs. It was apparent that the increased costs of these expensive drugs, and that they were being used by physicians in settings where their utility had not been proven by randomized clinical studies, led to a review by Medicare of its policy against these and other new expensive drugs used off-label in diseases like cancer. A lengthy review in early 2004 ended up making no change in national coverage policy for use of these drugs and passed the decision to the local coverage decisions. In general, as long as the drugs were approved for one indication, the reimbursement decision for off-label uses was left to the individual local contractors. In their written decisions (accessed on the CMS website), the CMS declared that cost was not a consideration.

In the UK, the National Institute for Health and Clinical Excellence (NICE) is an independent organization, responsible for deciding which medicines are paid for on the National Health Service (NHS). NICE has been issuing recommendations for taking drugs off the NHS list based largely on cost-effectiveness evaluations. These recommendations have generated significant public and industry discussion in the last few years, as many drugs which are being reimbursed in the United States and other European countries have been rejected as not being cost-effective enough by NICE.

Health economics

Health economics is a branch of economics concerned with issues related to scarcity in the allocation of health and healthcare and covers the various types of analyses and sub-disciplines discussed below. Other types of health economic analyses are cost-minimization analysis, cost-benefit analysis, cost-effectiveness analysis, and cost-utility analysis.

Cost-effectiveness, healthcare economics, and pharmacoeconomics

Cost-effectiveness (or cost-utility) is a measure of cost per unit of effective outcome ("effective outcome" varies with context and perspective) of a therapy, procedure, program etc. For example, a common measure of effectiveness of a

Box 8.2 *(cont.)*

medical intervention is through Quality-adjusted life years, or QALYs, based on the number of years of life that would be added by the intervention. The cost per QALY is typically used as a comparator or threshold to review a novel medical treatment or new intervention against the existing treatment paradigm. The UK agency (NICE) typically is said to use about US$30,000/QALY as the threshold to decide whether a new therapy is worth paying for by the National Health Service whereas the United States is reported to use a QALY figure of US$50,000/QALY (i.e. if a medicine costs more than $50,000 and only increases the life of a patient by 1 year on average, it may be judged to be not cost-effective enough and either the QALY count should increase or the cost must decrease for it to be covered for reimbursement). Various measures of cost-effectiveness are typically used by private payer organizations to make decisions on coverage. The CMS does not formally include cost-effectiveness in its decisions on coverage.

Cost effectiveness is only one factor considered by a broader field of study called Health Economics, which is a branch of economics concerned with issues related to scarcity in the allocation of health and healthcare.

ICER in the United States and others – Founded in 2006, the Institute for Clinical and Economic Review (ICER) is an independent and non-partisan research organization that objectively evaluates the clinical and economic value of prescription drugs, medical tests, and other healthcare and healthcare delivery innovations. It follows similar methodologies used by NICE. As of this writing in 2020, ICER's reports on "cost effectiveness analyses" or "value assessments" of specific drugs or devices are used by many insurance companies to determine whether or not to pay for specific treatments. As a private organization, it's not bound by federal laws restricting the use of QALYs to make recommendations for coverage or pricing. As part of the Affordable Care Act, Congress banned Medicare from using the QALY methodology out of concern that it could hurt the ability for the elderly and people with disabilities to access medical care and limit innovation in drugs and devices. However, as the cost of healthcare continues to spiral in the United States, there is increasing awareness of using cost-effectiveness metrics to make decisions on coverage and payment and ICER's reports have had an impact on the U.S. healthcare system, serving to frame the discussion on drug pricing and other similar discussions [Pizzi (2016)].

Some other examples given below of the widespread use of cost-benefits economics used to make decisions on coverage and payment, similar to NICE and ICER, are that of Germany, Japan and Taiwan.

Germany's Institute for Quality and Efficiency in Healthcare (IQWiG) is an independent advisory body to the German Health Authority that reviews the efficacy and quality of the healthcare to understand which therapeutic and diagnostic services are feasible and valuable.

Japan also has recently (in 2017) established guidelines for cost effective analysis of products by the Central Social Insurance Medical Council (Chuikyo),

Box 8.2 (*cont.*)

a sub-committee of the Ministry of Health, Labour and Welfare. Prior to this, while economic analyses on cost and pricing were carried out, reimbursement for granted for almost all approved products [Shiroiwa et al. (2017)].

In 2007, Taiwan's Department of Health established a Health Technology Assessment group within The Center for Drug Evaluation (CDE) to provide evidences on value of new healthcare technologies for decision makers in Bureau of National Health Insurance (BNHI). BNHI is the reimbursement body that now covers 99% of population and contracts with 91.5% of healthcare providers in Taiwan. The reports generated are not public and are used by the BNHI to review and make decisions on reimbursement.

Healthcare economics (distinct from "health economics") is a term usually used in the industry to refer to cost considerations for the purchasers and an assessment of economic benefit afforded to the purchasers and users through adoption of the new products/services. These calculations usually integrate facilities and other costs and typically cover the entire referral chain up to discharge of the patient or resolution of the illness, assessing the economic impact to the entire healthcare system and particularly to the purchaser/payer. Healthcare economic data is collected by the manufacturer of the product and is typically used to help influence decision-making for purchasers or payers. For example, Smith and Nephew (a large medical device firm) stated on its website in its 2005 sustainability report: "Healthcare economic considerations are integrated into the product development process to ensure that the benefits from the company's new products and line extensions not only seek to improve patient outcomes, provide better treatment and procedures for both clinician and patient but also contribute more cost effective solutions for healthcare services ... we also aim to deliver overall cost savings through such benefits [as] improvements in efficiency, such as reduced frequency of reduced dressing changes, shorter operating theatre times, reduced length of time spent in hospital stay and overall faster patient recovery rates. Healthcare economic benefits are a primary focus in our product development and marketing." In studies with Lipitor (cholesterol-reducing atorvastatin drug), Pfizer explicitly referred to the reduction in hospitalization costs that result from the use of Lipitor, showing the economic value of prescribing that drug for the patient groups studied. Baxter's department of Healthcare Economics, "has expertise in assisting healthcare providers, patients and payers with a variety of economic, insurance, and reimbursement-related issues." An increasing number of product development clinical trials are a priori including the collection of economic and clinical outcomes data in the trial design.

Pharmacoeconomics is a specific sub-discipline of *health economics* that compares the value of one drug therapy to another, considering costs and effects (discussed above). Pharmacoeconomic data is now regularly collected by manufacturers during clinical trials of their drug product and submitted to private payers

> **Box 8.2** (*cont.*)
>
> to help make the case for appropriate payment, positive coverage and formulary decision. However, cost-effectiveness economic analysis alone do not influence CMS coverage decisions as pointed out above.
>
> See References and additional readings list at the end of this chapter for more details on these topics.
>
> *A website with ratios used in cost-utility analysis publications is at* www.tufts-nemc.org/CEARegistry/.

Private insurers may include economic analysis of the resultant change in overall healthcare costs for their own coverage decisions. Competition among private insurers for these Medicare contracts will likely increase and will start reducing coverage in optional areas to show reduced costs. If this comes true, it could impact product development, requiring a higher threshold for efficacy with more statistical power in results, a better value proposition, and focus on a Medicare population in the clinical trials.

A key step in the coverage process for a new technology may be the *technology assessment* review. However, it is important to note that most devices (over ~95%) do not go through a formal coverage review process (technology assessment) as they are usually introduced and used as part of an already covered process. Most drugs also do not go through a formal coverage review process as the rigorous FDA approval process establishes the efficacy of the drug with pertinent endpoints. Private insurers or payers in other countries (e.g. UK) are more likely to conduct technology assessments. In the United States, the private payers either follow CMS' lead in coverage decisions or perform a technology assessment.

The following criteria could be used by a Technology Assessment panel at an insurer or CMS:

- The appropriate governmental regulatory bodies must approve the technology.
- Scientific evidence must permit conclusions about the effect on health outcomes.
- The technology must improve the net outcome.
- The technology must be as beneficial as any established technology.
- The improvement must be attainable outside the investigational setting.

However, it is important to note two points: (1) A positive review from a technology assessment organization like the Blue Cross Blue Shield is no guarantee of coverage from other insurers; (2) a negative technology assessment may occur even with an FDA approved product, and insurers may still decide to cover the product. Examples of issues raised during such technology evaluation are shown in Box 8.3. It is not unusual for such technology evaluations to reject a product that has passed FDA approval, especially as they require longer term follow-up (e.g. 3 years, instead of the

Box 8.3 Technology assessment for coverage determination

Technology assessments or evidence-based assessments are carried out by various groups in the United States, such as AHRQ, ICER, Hayes, California Technology Assessment Forum (CTAF), and others, giving an insight into the process of decision-making at major payers across the United States. Various payers contract with these private organizations to carry out assessments on multiple areas. These assessments could include first-hand interviews with patients, caregivers, administrators etc. as needed. These short cases below, highlight the point that approval is no guarantee of reimbursement coverage.

For example, the review summaries that are publicly available on the Hayes website (www.hayesinc.com) highlight key issues and concerns about specific technologies in the form of controversies and questions:

1. Artificial Pancreas with the MiniMed 670G for the Management of Diabetes Mellitus

    Controversy: Achieving glycemic control may be challenging for patients with type 1 DM.

    Questions raised in the summary indicate the following: Potential need for longer term assessment of the Minimed system with existing clinical alternatives, better identification of the patient selection criteria, safety in long term management of diabetes.

2. Occipital nerve stimulation (ONS) for treatment of chronic cluster headache

    Controversy: Some patients obtain limited or no relief from this treatment, and it is difficult to predict which patients with chronic CH will benefit.

    Questions raised in the summary indicate the following: possible need for further safety determination, need for better patient selection, comparison with alternate treatments and procedures such as deep brain stimulation.

3. Autologous fat grafting (AFG) for breast reconstruction following breast cancer surgery

    Controversy: concerns that filling a former tumor bed with regenerative cells could theoretically increase the risk of cancer recurrence and that the procedure may hinder breast cancer imaging. In addition, there is a lack of long-term strong comparative evidence demonstrating oncologic safety and patient satisfaction with aesthetic outcomes

    Questions raised in the summary indicate the following: Safety, effectiveness in cosmetic and aesthetic results, better comparison to alternatives, better patient selection criteria.

4. Amniotic Allografts for Tendon and Ligament Injuries

    Controversy: limited availability of amniotic membrane allograft (obtained from caesarean section deliveries) as it must be free of transmissible diseases such as human immunodeficiency virus and viral hepatitis. Use of this allograft may not be necessary due to other options such as percutaneous interventions

> **Box 8.3** (*cont.*)
>
> for treatment of refractory tendinitis and physical therapy protocols for prevention of adhesions after joint surgery.
>
> Other organizations listed above also sometimes publish public summaries of technology assessments or evidence-based assessments.
>
> A trend to note in these assessments is that they all seem to be looking at effectiveness and safety over longer horizons (2-3 years) than typically reported in regulatory registration (product approval) trials.

3-month follow-up data submitted for FDA approval) to ensure effectiveness for the patient's clinical outcome.

The rare technical assessment step to determine coverage for new technologies, either done internally at CMS or through one of several external technology review panels, is likely the most important reimbursement process step that can be directly influenced by the product development program and clinical study design. In particular, *the quality of the data given by the manufacturer to the payer will dictate the coverage decision.* The value of a product development (and clinical trial) plan that includes multiple inputs in planning becomes apparent here.

The data must address the above- described technology assessment criteria. Frequently, the clinical trial design that is sufficient for approval (for example, a trial for 510(k) clearance) may not have sufficient rigor or strength of evidence to convince CMS or other private insurers or commercial payers. The CMS guidance document for industry titled, *Coverage with Evidence Development*, can be found on the CMS website, www.cms.gov/medicare-coverage-database/details/medicare-coverage-document-details.aspx?MCDId=27. Companies should carry out rigorous randomized clinical trials to satisfy not only CMS coverage review, but also collect data for private insurers to prove the product's cost-effectiveness. This data could be collected (including the cost of ancillary services, radiological exams, lab and pathology tests, etc.) during the clinical trials, as increasingly, private payers will want to see some such data on outcomes and costs savings before offering coverage and payment.

In general, the strength of the evidence decreases in order with the following designs of clinical trials:

- Large double-blind, multi-center randomized control trial (patients are assigned randomly to treatment or control group, without the patients or the caregivers knowing which group is which – typical design for drug trials)
- Large, multi-center randomized control trial (patients randomly assigned with identification of control and treatment groups – typical in device trials)
- Meta-analysis of grouped data
- Smaller, single-site randomized controlled trials
- Prospective cohort studies
- Retrospective cohort

- Poorly controlled studies (historical controls)
- Uncontrolled studies (case-series or reports)

A manufacturer designs their trials to gain FDA approval and the various meetings with the FDA – pre-IND / pre-IDE meetings and pre-PMA / pre-Phase III – are important points at which clinical trial design is discussed and agreed on (see Chapter 6). Similarly, the manufacturer could also meet with CMS' Coverage and Analysis Group or other payers, early during product development to receive input on trial design and the data to be collected based on the projected coverage requirements. However, the decision to meet with payers should be a carefully considered decision and is not necessary for most products.

The threshold for coverage varies among the private and public payers. The value proposition for a coverage decision may be different among payers and may vary between countries. For example, Avastin and Erbitux are two cancer (infused) drugs that are approved and reimbursed in the United States by most payers but are not covered for payment in the UK health system.

8.4.2   Coding

Alphanumeric codes assigned to products and services allow for uniform and efficient processing of payments. Codes are used on insurance claim forms by purchasers/ providers to get reimbursed. Codes are important for healthcare providers as they facilitate submission of claims, and the use of codes enables payers to process and pay claims efficiently. However, getting a code is not a guarantee of coverage, nor is there any guarantee that the payment rate assigned to the code will be reasonable. Although manufacturers do not use these codes (they do not normally seek reimbursement from the payers), knowledge of which codes apply to their products is critical to helping position their products and to help purchasers obtain adequate reimbursement. However, it should be kept in mind that use of specific codes on claim forms is always at the sole discretion of the provider.

Two different types of codes are described briefly below:

Codes that identify specific procedures or products: CPT (+ PLA codes for
    diagnostics), HCPCS, and ICD-10/11 procedure codes
Codes that identify the related diagnosis or disease: ICD-10/11 diagnosis codes

Codes mentioned elsewhere are the Diagnosis Related Group (DRG) codes, which classify hospital cases according to certain groups, which are expected to have similar hospital resource use (cost). These codes are used for inpatient payment system calculations for reimbursement (Section 8.4.3).

CPT / HCPCS codes: The Current Procedural Terminology codes (CPT codes: a numeric coding system maintained by the American Medical Association [AMA]) *identify specific services carried out by physicians*. CPT codes are also synchronously published by the CMS, and in their system, CPT codes are called Level 1 HCPCS (Healthcare Common Procedure Coding System) codes. *Products and services* not included in the Level I HCPCS codes are identified in a group of Level II HCPCS codes. Some level II HCPCS code groups relevant to this book are:

B codes: Enteral (oral or through the gastrointestinal system) and parenteral (such as through intramuscular or intravenous injection) therapy

C codes: Transitional Pass-Through codes for reporting new technologies that are used in outpatient settings

D codes: Dental codes

E codes: Durable medical equipment (DME)

G codes: Procedures/Professional Services

J codes: Drugs and solutions (infused or injected; delivered by a caregiver)

L codes: Orthotics, prosthetics, and implant procedures

P codes: Pathology and laboratory services

Q codes: Temporary codes

S codes: Temporary local codes (used locally; non-Medicare)

The International Classification of Disease Ninth Edition, Clinical Modification (ICD-11) numeric codes classifies disease diagnoses (volumes 1 & 2) and procedures (volume 3). All hospitals and ambulatory care settings use this classification to capture diagnoses for administrative transactions. The ICD-11 codes go into effect in January 2022 and ICD-10 codes are used by the hospitals on their inpatient claim forms. The ICD codes are managed by the World Health Organization (WHO), and in the United States, the American Hospital Association (AHA) and CMS act as clearing houses and consultants on this international standard.

- ICD-10 diagnosis coding is matched up with and supports the CPT codes used for reimbursement of services/devices; these have to be consistently and correctly applied on claim forms.
- For almost all medical services and procedures there is a corresponding CPT code.
- Combinations of CPT and ICD procedure and diagnosis codes are packaged into the various context-dependent payment groupings (discussed in Section 8.4.3).

Other coding systems:

CDT: Current Dental Terminology (CDT) is used for reporting dental services. CDT codes are also included in alphanumeric HCPCS with a first digit of D.

NDC: National Drug Codes (NDC) are used for reporting prescription drugs in pharmacy transactions and claims by health claim professionals. The NDC is a unique three-segment numerical code that is generated by the FDA as a universal product identifier, and identifies the labeler, product, and trade package size.

Drugs that are self-administered (pills or self-dosed solutions) are usually claimed for reimbursement through their NDC codes by pharmacies. Drugs that have to be administered by the caregivers are also assigned HCPCS level II J-codes or C-codes that must be used on the claim forms for reimbursement. Diagnostics and devices get CPT / HCPCS codes for the procedure associate with the diagnostic test or devices.

If the procedure associated with the product is not described by any of the existing codes, a new CPT / HCPCS code might have to be obtained – this is not an easy task and can take up to 18 months. This delay of over a year (depending on timing of submissions and committee meeting dates) can really hurt a smaller company that was planning on a rapid uptake to generate sales revenue immediately after FDA approval or clearance. Assistance of the specialty physician group involved in the treatment is critical to getting approval for launching a new code (if one is needed) and manufacturers must work with the appropriate medical society from an early stage in the process. A written request is made by the manufacturer to the appropriate American Medical Academy (AMA) committee detailing the need for a new code, along with a complete description of the procedure or service and with a full list of references to peer-reviewed articles published in U.S. journals showing the safety and effectiveness of the procedure.

8.4.3   Payment

> Note: The payment processes discussed here are assumed to be under CMS/
> Medicare, unless specifically mentioned otherwise. Some of the text in this section
> is reproduced from the MedPac documents on the website www.medpac.gov.

Manufacturers set the price and get paid for their new technology by wholesalers, providers, or distributors as appropriate for each product type. However, understanding the system by which these purchasers get paid is critical to ensuring good sales and revenue growth. While not discussing pricing strategies here (pricing is a complex consideration that follows thorough review of not only reimbursement, but also many market conditions and parameters), it is clear that an inadequate reimbursement level for a high-priced item will keep the new product from reaching its potential maximum market penetration and revenue. The manufacturer can influence the reimbursement payment levels by strategically planning their product development in the context of a reimbursement strategy (see Section 8.5).

Drug reimbursement levels (for self-administered drugs; prescription pills, etc.)

These are based on pricing communicated by the manufacturers and negotiated by the payers (insurance companies, CMS), and payments are calculated from either AWP (average wholesale price) or ASP (average sales price) or other set formulae (drug acquisition costs to the retailer) reported by the manufacturer. Physician-administered drugs are usually reimbursed based on ASP data (Medicare reimbursed at 106% of ASP in 2005). Section 8.6 (Box 8.4) has more details on drug sales price terms. Other steps in the distribution chain may change the prices paid by insurers and retailers, such as rebates and discounts given through wholesalers and group purchasing organizations, pressure by state formularies, and negotiated discounted pricing or rebates to pharmacy benefit managers (PBMs). See Section 8.3.1 for some discussion on payment pathways for drugs.

Box 8.4 Companies should be proactive on government policy
Why companies involved in product commercialization should also be acutely aware of and proactive on government policy and policy makers
The healthcare industry is highly regulated through all aspects of the value chain

Regulations are laid down in response to laws that are passed by elected representatives. These representatives are sensitive to political pressures and public opinions.

As seen by the example below, laws and regulations can impact payment and reimbursement for many products and services, as the government (Centers for Medicare & Medicaid Services, Veterans Administration, and others) is the largest single payer for healthcare in the country.

Case example: Amgen, EPO, and the government in April 2007
The Drug: Amgen has been selling Epogen (erythropoietin; a stimulant of red blood cell production), a protein drug that stimulates the production of red blood cells in the body, reducing the need for transfusions.
The Application: The major market for this drug is dialysis patients who need regular transfusions. EPO was approved for the reduction of transfusions in patients undergoing dialysis. Physicians titrate EPO by more frequent or higher doses to get patients' hemoglobin content to between 10–12 mg/dl (normal range) or slightly higher for some to 13 mg/dl. Recent data have shown that dialysis patients with a hemoglobin target of 13 mg/dl or more are at higher risk for heart attack, stroke, or death.
The market: The 2006 worldwide sales of Aranesp and Epogen (EPO drugs) for dialysis and chemotherapy induced anemia were $6.6 B.
The Payers: The U.S. government (CMS) is the largest customer for Amgen as 93% of dialysis patients qualify for the Medicare End Stage Renal Disease program.
The story: Several news outlets reported in 2007 that private dialysis centers were making high profits from buying lower-priced drugs and getting higher reimbursement from Medicare as the drugs are reimbursed separately. These doctors and centers potentially had financial incentives to over-prescribe the use of EPO drugs and a House Ways and Means committee (of the US Congress) meeting in December 2006 noted that 40% of dialysis patients had hemoglobin counts greater than 12 mg/dl.
The politics: The House Ways and Means committee wants the CMS to eliminate separate payment for EPO drugs, bundling this payment into the overall cost of dialysis, thus forcing clinics to focus on overall profit margins and removing financial incentives to overuse EPO drugs. The chair of the committee said (as reported in the news articles) that Congress would have to act to safeguard the public from abuse from profit seeking centers, if CMS did not change the payment schedule, a clear demonstration of how politicians and regulations influence the biomedical markets.

Box 8.4 (*cont.*)

Note

While the EPO drugs are an unusual case in terms of government focus on a drug, elected representatives in particular are very conscious of their constituents' complaints on the high cost of healthcare. Manufacturers of biomedical products (drugs, devices and diagnostics) must engage with their representatives and industry associations to make sure that appropriate information is delivered to the decision makers who can influence the healthcare markets.

Payment structures for products delivered with physician administration or in a provider facility (devices, diagnostics, physician-administered drugs)

Reimbursement for services and products is usually made as a bundled payment, mainly composed of two or three parts:

1. Facility payment for services and supplies provided (includes product cost) paid by Medicare Part A.
2. Physician payment for services provided paid by Medicare part B.
3. Sometimes, a specific extra payment is made for (very few) new products that are deemed to result in "substantial clinical improvement" by Medicare part B.

The payment process starts with physician's notes in the medical record. The billing and coding department of a provider facility (hospital, physicians' office) reviews the notes and selects the appropriate diagnosis and procedure codes. Virtually all public and private payers require ICD-10/11 diagnosis codes in claim filings, to document the patient's illness and reason for the treatments. The ICD-10/11 diagnosis codes help justify the medical necessity for a given procedure or product (identified by CPT/ HCPCS codes). Compliance with accurate coding is a complex task and coders are trained on a regular basis.

The billing or claim form is then submitted to the payer (Medicare contractor), who reviews the coding and documentation for medical necessity and enters the various codes into grouping software. A group code is assigned by the software based on location of service (see Figure 8.9), the base payment rate is looked up, and adjustments are made on the basis of the provider's charges and local pricing in each market area. The operating and capital payment rates are adjusted to account for facility and case-specific factors. If the procedure is deemed medically necessary based on the ICD-10/11 diagnosis code(s), the calculated payment is sent to the provider facility and physician. All these steps are carried out through electronic data forms in computing systems. However, if the product is not covered or has been assigned a temporary procedure code, the provider is required to turn in a printed form which is manually processed by the payer, resulting in significant delays in receiving payments and the documentation required to justify the charges can be burdensome to the provider.

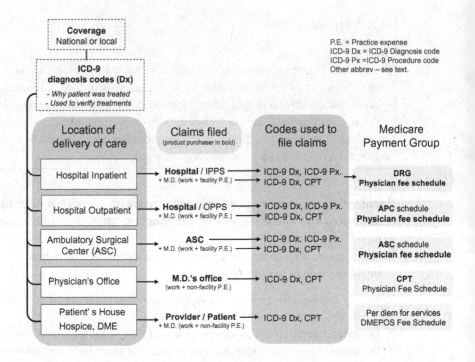

Figure 8.9 Payment codes and calculations of reimbursement amount vary by setting of healthcare delivery. DRG – Diagnosis Related Groups, APC – Ambulatory Procedure Classifications; DME – Durable Medical Equipment, DMEPOS – Durable Medical Equipment, Prosthetics, Orthotics and Supplies.

How are payment levels for facilities and physicians set for given procedures and treatments?

Each caregiver facility type (hospital inpatient locations, hospital outpatient locations, ambulatory surgical centers, or other clinics or doctors' offices) has a different set of parameters applied to calculate value, typically capital payment rates that cover depreciation, interest, rent, insurance and taxes, and operating payment rates that cover labor and supplies. The final calculated reimbursement level depends largely on the setting in which the product or service is applied (see Figure 8.9) and then on other case-specific factors such as the complexity of procedures. This reimbursement is a lump-sum (or bundled) payment that covers the cost of the product. The cost of the product may be a large or very small part of the overall payment, depending on each product and associated procedures. For example, the cost of an artificial hip joint might be about $6,500 which is a large portion of the $9,500 average Medicare reimbursement payment to the hospital for the entire process of implantation to discharge of patient. This does not leave much room for increasing the price of the artificial joint unless a significant improvement in clinical outcome can also be shown. On the other hand, a catheter used in the same surgery may cost about $50 and may face less pressure from the hospital for increases in price.

Prospective payments are pre-negotiated prices that Medicare uses to reimburse hospitals for inpatient and outpatient services, as well as skilled nursing facilities, rehabilitation hospitals, and home health services. Most insurers work with this type of payment process. Payments to hospitals for inpatient services are made in the *Inpatient Payment System* (IPPS), using the *diagnosis-related groups (DRG)* grouping codes. DRG grouping is based on the diagnoses for which the patient is admitted. The DRG codes work by grouping the 20,000+ ICD-10/11 diagnosis and procedure codes into a more manageable number of meaningful patient categories (~500). Patients within each DRG category are similar clinically and in terms of resource use. The payment amount is the same regardless of the length of the stay. Payments for medical devices and other physician-administered products are covered under DRG payment groups, which has significant implications on revenue, marketing and pricing considerations. Hospitals also can receive additional payments in the form of outlier adjustments for extraordinarily high-cost services or payments for qualified new technologies in the form of add-on ICD-10/11 procedure codes for the given DRG.

Another prospective facility payment system is the *Outpatient Payment System* (OPPS), based on the Ambulatory Payment Classifications (APC) which cover groups of procedures or services provided in outpatient settings. The unit of payment under the OPPS is the individual service as identified by HCPCS codes. CMS classifies services into ambulatory payment classifications (APCs) on the basis of clinical and cost similarity. The OPPS sets payments for individual services using a set of relative weights, a conversion factor, and adjustments for geographic differences in input prices. New technologies that meet the CMS criteria and are shown to deliver substantially improved clinical outcomes, become eligible for "transitional pass-through" payments. New services remain in these pass-through APCs for 2 to 3 years while the CMS collects data necessary to develop payment rates for them after which they are bundled into an APC payment rate. More than one APC payment may be made during one episode of care, depending on the services required. In setting the payment rates in the OPPS, CMS intends to cover hospitals' operating and capital costs for the services they furnish. CMS pays separately for professional services, such as physician services.

The *physicians' payment* is based on a relative value assigned to the specific procedure code and is based on the location and their specific practice. Medicare refers to the Physician Fee Schedule which is an elaborate price list that Medicare uses to pay physicians for their services (and also to pay other healthcare providers for items and services that are not bundled into prospective payment systems). Physicians are paid separately from the hospital for the procedures they performed. After a product is approved or cleared by the FDA for a specific indication, and a procedure code (or codes) is assigned or established, a relative value is assigned to that procedure code by an AMA subcommittee, starting from the recommendation of the medical society representing that practice or therapeutic area (e.g. American Academy of Dermatology, American Academy of Otolaryngology – Head and Neck Surgery, and others). A recent abrogation of this process, when Johnson & Johnson got CMS payment review started before FDA approval was obtained for their drug-coated stent (Cypher^{TM}), gained kudos from industry rivals but also raised a lot of controversy.

The manufacturer works collaboratively with the medical society from an early stage to establish the appropriate detailed procedures and costs involved with using that product, including establishing the clinical benefits to multiple stakeholders in the process. The relative value for a physician's service is calculated based on (1) cost to provide service – practice expenses, professional liability insurance, etc., (2) physician's labor – effort, skill, and time, and (3) practice expenses based on rent, supplies and equipment, and other staff costs. This value is then passed on to CMS, which usually accepts the AMA committee's recommendation.

If a procedure for which the device is used cannot be described by an existing CPT code, there are two choices for getting a code assigned. The more prevalent choice of temporary code today is to get a Category III CPT code assigned. The other choice, more common in past decades, is to have a "not otherwise classified" (NOC) or other similar generic code used. Pricing and reimbursement have to be worked out each time a procedure that uses the device is billed. If coverage is denied, either the provider will have to work with the insurer's process to help obtain some level of payment from the insurer or the patient may have to pay out of pocket (there are several more complicating issues in this patient liability, depending on the benefit package and contractual arrangements). Therefore, not having a CPT code assigned can hamper device sales, preventing a product from reaching its market potential. However, the case in Box 8.10 is an interesting example of a company that successfully got a higher reimbursement level for their new product by working the system aggressively with a temporary CPT code.

The process for establishing an appropriate payment and increasing access for new (and complex multi-biomarker algorithmic tests, called Advanced Diagnostic Laboratory Tests or ADLTs) diagnostics was made simpler with the Protecting Access to Medicare Act of 2014 (PAMA). Applicable laboratories report private payer (ie, commercial, Medicare Advantage, and Medicaid Managed Care) payment rates, including discounts, and volumes for existing tests, with those rates used to establish a weighted median for payment with different considerations established for ADLT tests (multi-biomarker tests with algorithmic analysis of results, or FDA approved tests) and general lab tests.

For *diagnostic products*, the clinical laboratory payment environment is largely based on Medicare published schedules that use CPT codes (now called PLA codes for advanced diagnostics, see Section 8.8) for each diagnostic step or set of processes. CMS has established the Laboratory Fee schedule, which is a schedule of CPT (PLA) codes and corresponding payment rates. Each CPT (PLA) code could cover individual components of laboratory tests or a panel of tests together as one bundle. This schedule is used as a benchmark for reimbursement and contracts between laboratories and private insurers. More details on pricing/payment for specific product categories are discussed in the following sections.

It is important to remember that when relative value units are assigned to procedure codes (CPT codes and ICD-10/11 procedure codes), the payment levels calculated in various payment systems (Figure 8.9) are fixed for a given year, with annual adjustments influenced by new data, grouping changes or policy changes. Government policies have a major influence in this area, and powerful manufacturers', hospitals', and physicians' lobby groups work to provide input to CMS and to exert pressure through Congress when changes in payment rules are proposed by CMS. In general, manufacturers should stay aware of specific policy changes (see Box 8.4) that could completely change market conditions or open new markets or close an existing product market.

Private insurers do not have to follow the CMS relative values and can assign their own values to the procedures and services associated with the new product. See Box 8.3 for value assessment of new therapies and Box 8.5 summarizing two recent paths laid out by Medicare for payment of new breakthrough technologies that have shown meaningful improvement in clinical outcomes.

The following are common payment systems used by payers:

Capitation fee structures are used by private insurers who contract with a healthcare provider to pay a fixed fee for a set of medical procedures, but unpredictable increases in costs are turning many groups away from this method of payment.

Carve-outs, less commonly used today, were payment contracts built to avoid the risk associated with capitated contracts, allowing for specific separate payment for new technologies or procedures. Thus, differing payment models have been utilized to incentivize improved outcomes and decrease costs, including unique payment models (e.g. capitation, bundled payments, and quality programs) and leveraging purchasing power (e.g. integrated delivery networks, accountable care organizations, group purchasing organizations).

Increasingly common in many countries are "managed entry agreements," which distribute the risk of uncertain benefit of the product on the country's population, between the payers (or budget holders) and the manufacturers. Between the simpler financial-based agreements, which may cap a budget spend for that product per patient or negotiate volume discounts or rebates, and the performance-based agreements, there is increasing use of the latter. As an example, beginning in 2017, the U.S. CMS reduces payment to hospitals when patients who have had a coronary artery bypass graft procedure are readmitted within 30 days of discharge. Performance-based agreements such as coverage with evidence development are used to manage the uncertainties on the impact of the new drug or device or diagnostic on utilization of healthcare resources or on clinical outcomes, by continuing to review prospective patient data. Giving their population access to novel technology while predetermining what evidence will be needed to ensure coverage, allows the payers (usually country health systems as budget holders in most European countries) to have greater certainty and greater value for their money spent by using managed entry agreements based on coverage with evidence development agreement. Manufacturers with truly breakthrough or beneficial products benefit by rapid access and sales, as do patients.

The UK launched a process in mid-2016 to allow patients to access selected new cancer drugs while collecting more data on effectiveness towards an agreed-on criteria and requirements for prospective evidence generation. Sweden and the Netherlands,

Box 8.5 Medicare offers New Technology Add-On Payment and Medicare Coverage of Innovative Technology coverage pathways for new technologies

Medicare Coverage of Innovative Technology (MCIT)

In January 2021, the Centers for Medicaid & Medicare Services (CMS) announced the launch of a new Medicare coverage pathway, Medicare Coverage of Innovative Technology (MCIT), for FDA-designated breakthrough medical devices. The MCIT rule provides national Medicare coverage as early as the same day as FDA market authorization for breakthrough devices and coverage lasts for 4 years. A breakthrough device must provide for more effective treatment or diagnosis of a life-threatening or irreversibly debilitating human disease or condition and must also meet at least one part of a second criterion, such as by being a "breakthrough technology" or offering a treatment option when no other cleared or approved alternatives exist.

Four years of Medicare coverage will allow manufacturers a chance to get coverage and work out payment structures and evidence to show improvements in clinical outcomes for their Medicare patient populations. This time period for coverage will allow clinical studies with Medicare patients to be completed while providing broad immediate access. When MCIT coverage sunsets, manufacturers will have all current coverage options available such as a National Coverage Determination (NCD), one or more Local Coverage Determinations (LCD), and claim by claim decisions.

New Technology Add-On Payment (NTAP)

Medicare also provides for additional payment for new device or drug technologies in the Inpatient Prospective Payment System (IPPS) through a program known as New Technology Add-On Payment (NTAP). Under the IPPS, Medicare pays for a patient's inpatient hospital stay under one bundled payment, which covers all costs for acute care services performed. Some examples of costs include room and board, operating room, nursing, supplies, laboratory services, radiology and also drugs, devices, and supplies. NTAPs provide for an exception to the bundling and establish a process of identifying and ensuring adequate payment for new medical services and technologies. The technology or product that requests such add-on payment has to satisfy criteria that are reviewed rigorously by Medicare. Specifically, the three criteria that must be satisfied for a new medical technology or service to receive an add-on payment are the following (1) the medical service or technology must be new; (2) the medical service or technology must be costly such that the DRG rate otherwise applicable to discharges involving the medical service or technology is determined to be inadequate; and (3) the service or technology must demonstrate a substantial clinical improvement over existing services or technologies. In addition, certain transformative new devices, Qualified Infectious Disease Products (QIDP) and Limited Population Pathway for Antibacterial and Antifungal Drugs (LPAD) may qualify under an alternative inpatient new technology add-on payment pathway. The manufacturer must provide substantive evidence

> **Box 8.5** (*cont.*)
>
> of clinical improvement over existing services or technologies and hence this pathway will require collection of specific clinical outcomes and in many cases the clinical outcome for the treatment may only be assessed by long term (3 years) follow up.
>
> Manufacturers that do not have Medicare (65 years or older) patients in their treatment population will have to work with private insurers to establish any additional payments as discussed elsewhere in this chapter.

for example, have been executing these types of evidence development agreements since 2008 to collect additional evidence over periods of 3 to 4 years in order to finalize coverage decisions.

In the United States, the CMS in 2017 had 22 such Coverage with Evidence Development agreements, largely on cardiology drugs followed by cancer drugs. Under this program, a technology may be approved and paid for by a patient on the condition that they are participating in a registry or clinical trial. In addition, in September 2020, CMS released guidelines for states' Medicaid directors to consider shifting away from traditional financial fixed fee-for-service arrangements which have incentivized providers to pursue volume over quality and outcomes, to alternative value-based payment structures (performance-based). A September 2020 letter on this topic from Medicaid Director to State Medicaid Directors, is found at www.medicaid.gov/Federal-Policy-Guidance/Downloads/smd20004.pdf.

8.5   Reimbursement planning activities

Reimbursement planning activities begin in the very early stages of product development when market assessments are made (Figure 8.10). Most smaller companies do not explicitly consider reimbursement planning as it is a complex area of analysis, but implicitly look at it through market research, potential pricing and revenue forecasts and competitive assessments of the market. These smaller companies that do not have experienced reimbursement personnel in house are best advised to consult with reimbursement and regulatory experts at an early stage to assess the specific steps (and timing of those steps) of reimbursement activities to achieve maximum value of their products.

The first review of reimbursement issues arises when a patient population is identified as being the most appropriate (target) for the new product (Figure 8.10). Based on the population demographics, the payer mix can be identified and reimbursement strategy planning can start in earnest. During product development, the data collected in clinical trials should typically satisfy the value proposition that will be evaluated by the payers and should be able to meet at least the technology evaluation criteria discussed in Section 8.4.1. However, it is useful to note that most devices and

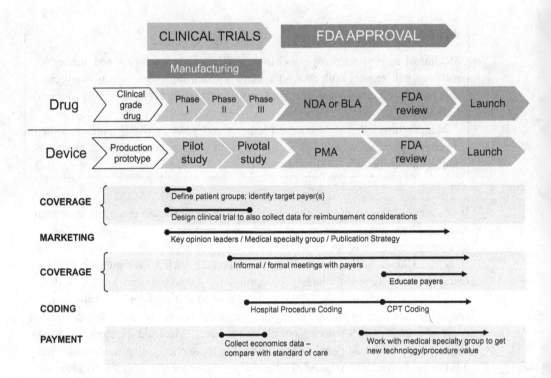

Figure 8.10 Timing of activities related to reimbursement planning.

diagnostics can be described by existing codes, are covered, have appropriate payment, and will not go through a detailed and formal technology evaluation by the payers. The challenge is in designing the clinical studies to meet regulatory challenges.

As explained further in Section 8.10, the marketing functions of a company are also active in reimbursement planning. Market research is used during product development to assess market perceptions of pricing, acceptance, awareness of product and possibility of reimbursement/coverage by payers. Marketing efforts at the early product development stages are focused around awareness, aiming to influence the coverage decisions by increasing awareness of the product technology or novelty and its value proposition (see Section 8.10). Educating payers is also another vital component of marketing efforts.

Just as there are meetings with the FDA during product development, companies can meet with payers in informal and formal meetings to convey information and data on clinical usage of their products and to get guidance on development parameters or product characteristics. In particular, economic usage or additional clinical outcome data requested by payers, if collected during clinical trials (Figure 8.10), may increase the development costs, but these costs may be recouped in the faster rise in sales due to quicker acceptance of coverage, coding and appropriate pricing. Economic and clinical outcomes data could also help in achieving a suitable pricing level for the product. *Note:* These efforts must be done with some advanced understanding of the field, otherwise there is a risk that the product could end up with more scrutiny than needed and greater burden; the company might do better in some cases to not be too aggressive in these efforts, and the judgement needs to be made thoughtfully and with some consultation among people experienced in the reimbursement area.

8.6   Reimbursement path for self-administered drugs (mostly pills)

Drugs that are taken directly by the patients (self-administered) have a fairly straight-forward reimbursement path. The manufacturer has to approach the various providers (payers, hospitals, etc.) with the FDA approval letter and then demonstrate with existing or additional clinical trial data, the value proposition of their drug and establish a price point for their drugs. On approval, the drug is automatically assigned an NDC code and coverage for approved indications is almost automatic. Most FDA-approved drugs are covered and paid for by insurers directly, as the clinical trials required for drug approval are more rigorous than for devices, and the efficacy and safety balance is usually clearly established by the FDA approval and label. However, patients may have one or more levels of co-payments and pharmacies/providers may have different reimbursement levels depending on the tiers in the formulary (see Section 8.3.1). But if the drug is not covered by the formulary or has not got "preferred" status, and the doctor has not been able to get prior authorization (a process within each health plan) for coverage, the patients may have to pay the full price for the drug at the pharmacy. Therefore, the bulk of efforts by manufacturers relate to positioning their drug in the preferred group in the formularies. These efforts may take several months after approval depending on the payer mix. Matching pricing and reimbursement levels for drugs is a complex issue, as multiple rebates and discounts offered to various parties impact on the actual cost of the drug to the purchasers, as illustrated in Figures 8.6 and 8.7. Payers use their own standards to calculate reimbursement rates and various pricing calculations are described in Box 8.6.

In the process of reimbursement planning, once an indication is chosen for the new product, the patient demographics can be determined and subsequently the specific insurer mix is identified. Since private payers and Medicare Part D and Medicaid/VA payers have different approaches to managing their formularies and pricing, the company can start to bring this reimbursement feedback into the portfolio planning discussions. In particular, once the specific insurers are identified, various points of emphasis in the clinical trial – from the data to be collected, to trial design or patient recruitment strategies – become clear and a reimbursement strategy can be developed hand in hand with the product development plan.

8.7   Reimbursement path for devices and infused drugs

In many cases, early-stage biotech drug or device companies whose product has reached this stage and who plan to take their product to the market themselves, will want to contract with a third-party logistics provider or consultant to help navigate

---

Box 8.6 Terms used in calculating actual drug pricing and reimbursement levels

Average Manufacturer Price (AMP): The average price paid to a manufacturer by wholesalers, for drugs distributed to retail pharmacies. Section 1927(b)(3) of the Social Security Act requires a participating manufacturer to report quarterly to CMS the average manufacturer price (AMP) for each covered outpatient drug. Section 1927(k)(1) defines AMP as the average price paid to the manufacturer by wholesalers for drugs distributed to the retail pharmacy class of trade, after deducting customary prompt pay discounts. AMP data is not publicly available.

Average Sales Price (ASP): Defined in the Social Security Act [42 U.S.C. 1395w-3a], ASP is reported by the manufacturer as the weighted average of all non-Federal sales to wholesalers net of chargebacks (*see below for definition*), discounts, rebates, and other benefits tied to the purchase of the drug product, whether it is paid to the wholesaler or the retailer. The basis for reimbursement for products covered under Medicare Part B changed under the Medicare Modernization Act of 2003 from AWP to ASP.

Average Wholesale Price (AWP): average price at which wholesalers sell drugs to physicians, pharmacies, and other customers. It is a figure reported by commercial publishers of drug pricing data and is used by payers to make reimbursement payments (AWP minus some percentage). AWP is sometimes referred to as a "sticker" price and is not reflective of the true market price as it does not include the discounts or rebates given by the manufacturers. The basis for reimbursement for products covered under Medicare Part B changed under the Medicare Modernization Act of 2003 from AWP to average sales price (ASP).

Chargeback: When the manufacturer has negotiated discounts with PBMs or large pharmacies that brings the price to be paid by the pharmacies to the wholesaler lower than the AMP (price paid by the wholesaler to the manufacturer), the wholesaler charges the manufacturer to make up the difference in payment – this is called a chargeback.

Wholesale Acquisition Cost (WAC): The price paid by a wholesaler for drugs purchased from the wholesaler's supplier, typically the manufacturer of the drug. Publicly disclosed or listed WAC amounts may not reflect all available discounts.

Thus, the actual price paid by a provider for a particular drug product can be difficult to measure accurately, giving rise to fairly complex averaging calculations carried out by various payers.

---

these complex steps. Clearly, the providers/consultants need to be hired at an early stage of development, as seen in Figure 8.10.

- Reimbursement planning begins in the preclinical phase, where the target patient population affected by the disease and treatment/diagnosis is identified

- The mix of patient demographics affected by the targeted disease will dictate the mix of payers. For example, Product A targets a patient population with the following demographics (payer mix):
  10% – Medicare
  70% – Under 65 years of age, private payers (e.g. Cigna, United Health Care, Blue Cross/Blue Shield)
  10% – Medicaid
- As public and private insurers pay differently for the same treatment, the information about payer mix expected for a particular product could be used as a component input into the portfolio selection process and also into product pricing and development planning activities.
- The manufacturer could approach representatives from the payers to discuss the planned clinical benefit from their technology and also familiarize the payers with the novelty and advantages offered by the technology. *Note:* Most products will not need to approach the payers, as payments can be generated by simply selling the approved product to the buyers (care providers or patients) with a matching code that fits into the existing reimbursement plans. Meeting with the payers could create additional (negative) ramifications and this decision should be made in consultation with someone familiar with the reimbursement systems and processes.

The path diverges slightly here for physician-administered drugs and devices.

8.7.1 Reimbursement path for physician-administered drugs

- While drug marketing approval is being obtained from FDA, meet with payers to familiarize them with technology, benefits etc. (see cautionary note above)
- When a drug is approved by the FDA, reimbursement coverage for the approved use is usually assured as soon as it is sold. Reimbursement levels can be influenced by the manufacturer as described in Section 8.6. Physician-administered drugs are typically reimbursed under a J-code in the CPT (procedure codes), or the company could get one of the temporary J-codes such as J9999 or a Category III CPT code.
- The growth in sales will be hindered with a temporary code, because a temporary code requires a physician or provider to first purchase the drug up front, fill out a paper form, send it in to the insurance company/payer, address any specific questions from the payer, and then finally receive payment a few weeks later. On the other hand, an established code can be entered electronically and payment is received rapidly once the payment amount is established.
- Getting a permanent J-code for the physician-administered drug is a process that can take from 8 months to 18 months, depending on the timing of FDA approval and the CPT panel bi-annual meeting date. A submission that does not get in for the first panel meeting at mid-year, will take 18 months to be assigned a new permanent code.
- The submission can be made with some assistance from the medical specialty group, but the involvement of the medical specialty group is not as important to this

process, as it is for the devices, which typically involve much more procedural work by the medical specialty group.

- Payment for physician-administered drugs provided in a physician's office is primarily based on the price of the drug as the work component of the procedure is a small part of the total reimbursement. Drugs administered in the hospital to inpatients are paid (reimbursed) under a DRG (IPPS system) payment which reflects the total cost of administering the drug in the inpatient system. Similarly, payments for drugs administered in the outpatient setting are paid through the APC (CPT J-codes) schedule in the OPPS system. If the drugs are administered in an ambulatory surgical center, the payments are based on the prospective ASC fee schedule. With the existence of significant discounts from the manufacturer flowing through the distribution and payment system (see Figures 8.6 and 8.7), the provider typically makes a profit from the reimbursement payment they receive (see Box 8.7).

8.7.2   Reimbursement path for devices

- While device marketing approval is being obtained from the FDA, meet with payers to familiarize them with technology, benefits etc. (see Box 8.8, Box 8.9, and Box 8.11). However, as noted above, a sophisticated understanding of the overall process should be used to decide on the need or timing of such meetings to make sure it does not create additional hurdles to coverage and payment.
- A major difference between approved infused drugs and marketed devices is that FDA approval or clearance for marketing a new device does not always mean reimbursement will be provided by the payers.
- Describe the procedure and the caregiver setting (inpatient, outpatient, physician's office) in which the device will be used/delivered. Identify if an existing CPT code will apply to this procedure (if applied in the inpatient setting, identify an ICD-10/11 procedure code also). Simultaneously determine the ICD-10/11 diagnosis code used to describe the patient indication/problem. Combinations of these various codes will point to the specific payment grouping code that will be used to pay the providers in the various payment systems (see Figure 8.9). The various groupings are discussed separately below.

IPPS: Inpatient Payment System

- The combination of the diagnosis code and the procedure codes will be used by the payers' grouping software to determine the DRG group code. From the DRG code, the manufacturer can determine the specific payment that the hospital receives for carrying out the procedure. Payment for the device is included in the DRG payment level but is bundled as a single payment that includes hospital costs. Thus, if the DRG for a new device is not adequate to cover the increased cost of a new improved device, there is a disincentive for the hospital to purchase and continue to use the device unless it helps in improving efficiency and increases savings in some

Box 8.7 Same disease, similar drugs, different reimbursement paths

Written with assistance from Jayson Slotnik, JD, MPH, currently a Partner at Health Policy Strategies, Inc. and past Director of Medicare Reimbursement and Economic Policy at the Biotechnology Industry Organization (BIO).

Differing reimbursement structures impact reimbursement planning, clinical trial design, marketing strategies and patient assistance programs. The example discussed here shows how differing reimbursement paths between two competing drugs can impact drug development and clinical decision making.

Etanercept (Enbrel) and infliximab (Remicade) are protein therapeutics that bind to TNF-alpha (a naturally occurring protein) and inhibit the inflammation caused by the release of TNF-alpha at sites of rheumatoid arthritis or other pathologies. The drugs are both approved for treatment of rheumatoid arthritis but are administered differently creating distinct reimbursement issues and considerations. Both drugs are expensive, with 100 mg of each costing ~$650 to ~$700 and with an annual cost of between $10,000 and $16,000 to treat a patient with rheumatoid arthritis.

Etanercept is a protein that mimics the receptor of TNF-alpha and inhibits TNF-alpha. This drug is self-administered by the patient with a single injection below the skin twice every week and is therefore covered by either the patient's pharmacy benefit or Medicare Part D. In this case, the specialty pharmacy or regular pharmacy gets reimbursed by both, the insurance company (or Medicare) under drug prescription benefits, and by the patient's co-pay.

Infliximab is a monoclonal antibody that is part human and part murine, requiring co-administration with methotrexate. The drug must be administered in a physician's office or hospital setting by IV infusion over 2 hours every 6 to 8 weeks with the provider buying the drug and receiving reimbursement under either the patient's medical benefit or Medicare Part B. The physician's office or hospital charges the insurer using procedure codes covering the work done for the infusion and cost of drug.

Because of the differing reimbursement structures, each therapy must overcome certain challenges to gain market share and manage pricing pressures. For example, because infliximab is IV administered by a provider and covered under Medicare Part B, the hospital or doctor's office (provider) must first purchase the therapy and then collect part of the payment from the government and the remaining part from the patient. The expense of this drug obviously hinders cash flow for the provider as they have paid for the product but may have to wait for reimbursement. The payment situation also poses certain cost challenges for both the patient and the provider. This dynamic creates pricing pressure on the manufacturer. On the positive side, since there is no formulary in this outpatient setting for the physician-administered drug, coverage is essentially guaranteed.

Etanercept has contrasting positives and negatives. Because Etanercept is self-administered, it is generally covered under Medicare Part D. However, the manufacturer must negotiate for formulary placement with each of the hundreds of

Box 8.7 (*cont.*)

Medicare plans and the other private insurers. On the positive side for the sales environment, the provider is not involved in the cash flows and the patient liability is capped at their annual deductible, reducing the pricing pressure on the manufacturer.

Both companies likely have to make different types and levels of investments in patient assistance programs and would also likely adjust their marketing mix to the specific purchase-influencers (physicians or patients) for each drug based on mode of administration. Reimbursement planning in this case offers significant insight into pricing pressures and issues around market acceptance or sales projections.

This example shows that an understanding of the reimbursement dynamics, in addition to studying potential product competition and market acceptance issues, should be used to make better decisions or plan better early on in the manufacturer's drug development process. Drug developers should ask the following questions as part of their early product and reimbursement planning:

What is the targeted population (demographics, etc.) and resulting payer mix?
Do we want to focus on CMS or private plans for coverage and reimbursement?
Are the patients healthy enough to self-administer?
Do we want to market to patients or physicians?
What is our pricing and discounting strategy?
What is our patient assistance strategy?

Box 8.8 The importance of planning for device reimbursement early in the process

Written with assistance from Mitchell Sugarman, Director of Health Economics, Policy and Payment at Medtronic Vascular, and fellow associates.

During a surgical procedure known as angioplasty, a balloon is inflated inside an artery at the site of occluding plaque deposits to re-establish the lumen and increase blood flow through the artery. These procedures are usually performed on patients in hospital operating rooms or catheterization imaging labs by interventional cardiologists. To get the balloon into the artery (typically the coronary arteries) a guide wire has to be put into the artery at a distant site like the groin or arm and the guide wire has to be worked past the site of the lesion. Then a catheter with the balloon is threaded on this guide wire to reach the site. The biggest risk during this process was that part of the plaque could get loose as an embolus or thrombotic plaque and flow into the circulation, potentially occluding a smaller vessel completely and causing a heart attack, or flowing into the brain and causing a stroke.

Medtronic, a large medical device maker, acquired a company (Percusurge Inc.) that had just gotten FDA approval for the first emboli-protection device that would work with these interventional procedures (1999). The only indication approved

Box 8.8 (*cont.*)

for use of this device was in angioplasty procedures in saphenous vein grafts (which would frequently get occluded post transplantation). The PercuSurge device was a hollow guide wire with a balloon at its tip. When initially positioned past the lesion, the small balloon at the end was inflated distal to the lesion site, blocking the lumen completely and preventing any loose emboli from flowing further. The angioplasty catheter could be threaded onto the guide wire as usual, and after angioplasty, the occluded pool of blood between the lesion site and the distal occluding balloon would be suctioned up and irrigated, removing all emboli or loose debris from the plaque that had entered the blood stream. The balloon at the end of the PercuSurge guide wire was then deflated, resuming blood flow and the guide wire could be withdrawn. This device was the first one on the market that would reduce the significant risk associated with the angioplasty procedure with other solutions following at least two years behind (in 2007, most companies marketed mesh baskets at the end of guide wires that allow blood but not emboli to flow, instead of completely blocking blood flow, which was the mode of operation of the PercuSurge device). The device had the potential to be a high-revenue product as it was the first solution on the market addressing a grave problem with an increasingly popular interventional procedure. The Percusurge device was priced significantly higher than the existing guide wires that were being used.

However, the device did not have adequate reimbursement. The small company that Medtronic bought (for $225 million in December 2000) had been focused only on getting FDA approval and there had been no reimbursement planning put in place during product development.

In the inpatient setting (payment system – IPPS), where almost all angioplasty procedures were done, Medtronic assessed the specific existing ICD procedure codes and ICD diagnosis code for saphenous vein graft angioplasty and ascertained the payment DRG group code. Looking up the current payment level, it was clear that this payment would not allow for adequate reimbursement to cover the hospital costs and a higher price for the enhanced Percusurge guide wire device. Medtronic could appeal for a new DRG to be created to reflect the new procedure and device. In order to do that, they would have to collect additional clinical data and outcomes data (likely not done by the PercuSurge venture) and then appeal to the CMS. Medtronic assessed that their chances of getting a new DRG created were low even after this process. At that time (1999), DRG add-on payments were not yet implemented (only available after October 2001) where an extra payment for this device might have been available to a new technology used under an existing DRG (and the data was likely not available even if DRG add-ons had been available). The outcome of this review meant that the hospital would have to bear the costs of the new technology ($1,500) under the existing level of DRG payment, creating a significant disincentive for a hospital to purchase and use this PercuSurge device.

Box 8.8 (*cont.*)

Reviewing the outpatient setting (payment system – *OPPS*) for the outpatient setting (where very few angioplasty procedures were performed), a similar situation existed with inadequate APC group payment available (see Figure 8.9). However, in the APC groups, a transitional pass-through code that covered the cost of the device could be obtained. Medtronic applied for and received approval for a pass-through transitional code that would reimburse adequately for use of the device in the outpatient setting. While a good outcome, it would have negligible impact on the overall sales of the device.

The *physician fee schedule*, based on CPT codes was also reviewed by Medtronic. Current CPT procedure codes for use of guide wires in angioplasty were assigned relative values (used for payment calculation by payers) based on two main components – the physician work and the practice fee. The work fee component of the existing CPT code did not recognize the extra work that the PercuSurge device required, as inserting the guide wire past the lesion site was a difficult and risky process. Therefore, Medtronic worked with the physician specialty group (vascular interventional cardiologists, American College of Cardiology [ACC]) to apply for a new CPT code to reimburse them adequately. However, the ACC learnt that in order to review this petition, the committee would likely have to review all the existing codes for these associated interventional cardiology procedures. Not wanting to risk a reduction in reimbursement value levels that might result from additional reviews of existing codes, the ACC preferred to withdraw the new CPT code petition for the procedure associated with the PercuSurge device. Physicians could use a temporary 99 code for this device to get reimbursed for the extra time it took them to use the PercuSurge, but no values were defined for this code and each usage would have to be followed by extensive discussion with payers to get the physician appropriate reimbursement for their extra effort. Medtronic put together a template to help physicians with forms and information to guide them through this process.

These steps resulted in a less than optimum outcome for reimbursement for the device. There was little that could be done in this situation except continue to educate the payers.

The device is still available, and the price has come down significantly to allow it to compete with existing new technologies (mesh baskets that open like umbrellas at the end of the guide wire to catch emboli larger than the mesh openings) but the device has never quite reached its potential revenues. Medtronic had the first such device on the market 2 years before its competitors, but inadequate planning for reimbursement early on by the startup company certainly hampered the revenue growth for this device.

In hindsight, if the startup company had started the reimbursement planning before starting clinical trials, they might have realized the reimbursement challenges ahead. They might have tried a few things listed below.

Box 8.8 (*cont.*)

Physician fee: Begin dialogue with the physicians early on, who might have helped to get the procedure cost bundled into the physician CPT codes or ICD-10/11 procedure codes as they have now done with carotid stenting. When carotid stenting started becoming popular in recent years, the ACC obtained two codes, one for the procedure with embolic protection devices and one without embolic protection devices (the difference between the two reimbursement levels was only $50).

Indication expansion: The PercuSurge device was approved with indication only for use with occluded saphenous vein grafts in 2000, not in native carotid arteries and the company might have also started planning to expand its indications early on as it would have ensured high sales volume and possibly a better reimbursement review.

Collect appropriate clinical data and educate payers early: The startup company could also have lobbied CMS and the CPT committee early on to educate them about the value of using this embolic protection device, and built support by collecting data on the significance of this advancement in care and the resultant reduction in post-operative complications.

Update note February 2021 (additional comments from Jo Ellen Slurzberg of JR Associates): For endovascular (and many other) procedures, there has been a steady move toward outpatient and ASC billing. Patients that usually meet criteria for inpatient only procedures are usually in an acute setting or major surgery. The two-midnight rule released by the CMS (Medicare) has pushed more procedures to be done in outpatient or ASC settings. The CMS rule states: "For stays for which the physician expects the patient to need less than two midnights of hospital care (and the procedure is not on the inpatient-only list or otherwise listed as a national exception), an inpatient admission may be payable under Medicare Part A on a case-by-case basis based on the judgment of the admitting physician. The documentation in the medical record must support that an inpatient admission is necessary, and is subject to medical review."

Payments for these settings are also "global "or episodic in most cases, similar to DRGs.

---

part of the overall procedures covered under that DRG. For example, if a new and improved device leads to reduced hospital stay, the hospital might be able to save enough (the current DRG pays for the longer hospital stay in its bundled single payment) to justify the higher purchase price of the new device.

- On the other hand, if the procedure for the device falls into a DRG that does not pay adequately then the manufacturer has a couple of options (outlined below), but these options will take time to implement and revenues will likely suffer in the

Box 8.9 Reimbursement pathway for in vitro diagnostics (IVDs)

Diagnostics are generally reimbursed similarly to infused drugs and devices – the primary purchaser is the service provider (central labs services, hospitals, physician's offices) – and reimbursement is claimed using an ICD-10/11 procedure code (or DRG grouping code) or PLA code (alphanumeric CPT code) for the test. Proprietary Laboratory Analyses (PLA) Codes are an addition to the CPT code set approved by the AMA CPT Editorial Panel. They are alphanumeric CPT codes with a corresponding descriptor for labs or manufacturers that want to more specifically identify their test. Tests with PLA codes must be performed on human specimens and must be requested by the clinical laboratory or the manufacturer that offers the test. The PLA Code section includes (but is not limited to) Advanced Diagnostic Laboratory Tests (ADLTs) and Clinical Diagnostic Laboratory Tests (CDLTs) as defined under the Protecting Access to Medicare Act of 2014 (PAMA).

Diagnostics that have emerged in the past decade, such as molecular diagnostics (or nucleic acid testing [NAT]), have had a particularly difficult time getting reimbursed at appropriate levels, because of the outdated Clinical Laboratory Fee Schedule (CLFS) used by Medicare to set payment rates. However, the Protecting Access to Medicare Act of 2014 (PAMA) has updated the outdated payment structure to meet current market trends and allow patients quicker access to new, more complex tests (Advanced Diagnostics Laboratory Tests [ADLTs]) and specific CLA codes have been designated as above. All rates for tests on the CLFS are now valued on market-based payment and volume data, somewhat similar to the average sales price (ASP) methodology used for outpatient drugs and biologics.

The CMS has also established the *cross-walk and gap-fill processes* to determine payment amounts for new codes added to the fee schedule. *Cross-walking* is the mapping of new codes to technologically or clinically similar codes on the fee schedule and payments are made based on the existing fee schedule. Cross-walk determinations are made internally at the CMS with no formal process for correction currently in place, leading to protests from manufacturers. The *gap-fill* process requires local determination of payment rates for new codes, with no formal process in place to help make that determination, leading to confusion and variation in the first year of payment for the new code. In the second year of a new code, the CMS adopts a national limitation amount (NLA) in its fee schedule for the new code, based on the local carrier rates. The lower of the local rate or the NLA is then used to pay for the newly coded test.

In order to prepare for reimbursement, the diagnostics manufacturer has to

- Collect appropriate clinical effectiveness data.
- Work with main regional/local bodies of CMS to obtain coverage. Approach major identified private payers for coverage determination.
- Identify procedure codes
  - CPT-4 (level I HCPCS codes) – physician procedure codes

Box 8.9 (*cont.*)

- level II HCPCS codes – material/device or procedure codes
- ICD-10/11 procedure codes (hospitals' procedure code)
- Pick existing applicable procedure codes (preferred) or establish new code (1–3 years) through American Medical Association – CPT board
- Identify Medicare benefit category – "diagnostic test"
- Submit application to HCPCS working group at CMS for coverage and reimbursement level determination
- Determination by CMS in 3–6 months by cross-walk or approach local carriers by gap fill method. Or under new rules, work with private payers to gain appropriate reimbursement payment and in nine months, the clinical laboratories will report these payment rates to their local Medicare Administrative Center (MAC) which will help determine prospective payment by Medicare for the Advanced Diagnostic Lab Test.
- Work with other major third-party payers (Blue Cross, Cigna, etc.) dependent on patient demographics.

---

meantime (few years). Therefore, the reimbursement assessment must be done at an earlier time so that the manufacturer can appropriately make portfolio investment decisions.

- Insurers use past data on claims submitted for similar procedures to carry out annual reviews and to set appropriate payment levels for a procedure or group of procedures. The manufacturers can ask the hospitals to place a charge for the new device on the claim forms submitted to insurers as this may eventually register in the system as a DRG that needs increased payment. If a significant number of sites starts using a more expensive new technology, the hospital charges will initially get reimbursed at the set DRG group rate, but then after a couple of years, the review will show that this grouping code needs to be raised to account for these consistently higher charges. The problem is that several years could pass before a higher payment is recognized for that DRG. During that time hospitals will likely progressively reduce the purchase of items for which they lose money due to inadequate reimbursement and the product never gets used by enough sites to show up in the national review of DRG charges, leaving the payment rate unchanged.
- An argument could be made to CMS for higher-valued payment by grouping into a new DRG – this is done by an initial appeal (with assistance from the medical specialty group) to the CMS, followed by review by CMS - and could take several years for them to collect the data to justify the right level of payment and the new grouping.
- In the interim, in the DRG (inpatient) setting, if the device meets CMS criteria (or other payers' criteria), an add-on payment may be made to the hospital. Evaluation is based on criteria of newness, clinical benefit, and cost.

Box 8.10 Reimbursement challenge for new nucleic acid diagnostic tests
Written with assistance from Dr. Kim Popovits, currently Chief Operating Office and President of Genomic Health.

Genomic Health Case Study

Genomic Health has developed a novel cancer diagnostic test that predicts the likelihood of recurrence of breast cancer by genomic analysis of breast cancer biopsy samples. In order to gain market acceptance and establish clinical validity of their test, they used over $50 million (invested up to 2005) since inception in 2000 and carried out retrospective clinical trials with good statistical rigor. The studies were successful, and results showed that physicians and patients could use the results of their OncoTypeDX test to make decisions in management of their breast cancer, by giving people some quantifiable risk of recurrence and the likelihood that they might benefit from chemotherapy treatment. Their OncoTypeDX Test was qualified under homebrew or ASR diagnostic testing (not requiring to be passed through either 510(k) or a PMA process) and their central reference laboratory, which received samples from around the nation, was granted a CLIA certification with rating sufficient to carry out the most complex tests.

The business model for Genomic Health challenged the existing diagnostic business model of high volume, (relatively) low margin operations of most successful companies like LabCorp or Quest Diagnostics. The biggest challenge the company faced (and that was the biggest impedance to getting further investment into the company to continue to develop new diagnostic tests) was the lack of adequate reimbursement based on existing CPT codes. Stacking up all existing CPT codes that could apply to their multi-step test, a cross-walk approach would get them to a reimbursement level of about $1,300 which was less than half of the target amount for the company to sustain its model of investing in intensive R&D for new tests. It was imperative for the future survival of Genomic Health to establish a higher price point for their test, closer to a therapeutic rather than a diagnostic. They chose to break ground by classifying their product under a Miscellaneous CPT Code category. Seeing this CPT code on a reimbursement request would prompt a manual review of each request, requiring appeals to the payer and submission of validation data for the test. Genomic Health took another inventive step of organizing a reimbursement support team that made up to the maximum allowed three steps of appeals on behalf of the patient, who had to sign an authorization for Genomic Health to carry out the appeals. If the payer did not agree to the pricing, Genomic Health would have to bill the patient. This process could potentially create significant delays of 6 to 14 months in collecting revenues from a given test, creating another level of risk, but thus far Genomic Health has seen good success in this strategy, getting full reimbursement at the right level. This example points to a creative and aggressive approach to the current reimbursement hurdle faced by developers of new diagnostic tests.

Box 8.11 Early-stage startup's considerations for reimbursement and payment
Written with assistance from Jonathan Romanowsky, co-founder and Chief Business Officer of Inflammatix.

Inflammatix is building novel molecular diagnostics based on their abilities to interpret, using advanced bioinformatics techniques, how the immune system responds to diseases. The company's first test (the InSep™), an acute infection and sepsis test, informs on the presence, type, and severity of acute infections for patients seen in hospital emergency departments.

The product consists of (1) an instrument (called Myrna™; note: the name is a play on "My RNA," since the technology focuses on human immune response gene expression analysis), (2) specific assay cassettes (microfluidics, all analytes and reagents contained in the cassette), and (3) the software in the instrument used to analyze the assay readings. Specifically, the software applies machine learning–based algorithms on the expression levels of multiple genes and generates results that are clinically actionable. Based on conversations with FDA to date, their InSep test and Myrna instruments are likely to be Class II and will require FDA clearance through the 510(k) route.

Their Myrna instruments will be sold to the hospitals, and these facilities will purchase cassettes based on assays they want to run (the InSep test will be the first one). The software is included within the instrument and future updates will be loaded at no charge. The company will get paid directly by the buyer.

The reimbursement strategy for the InSep test varies based on the setting of the testing and the patient type. Approximately half of expected patients in the United States are covered under Medicare.

For inpatient scenarios (when patients are admitted into the hospital), the hospital cannot request Medicare to be reimbursed individually for an InSep test run, as the hospital gets one lump sum bundled payment (DRG grouping code) for treating that admitted patient for that indication. So, the cost of the test (the InSep cassette cost) will have to be taken out of that bundled payment. The company can work with hospital customers to obtain a new technology add-on payment from Medicare to the specific DRG code group. The company and the hospital will collect data over time to provide justification to the payers to get the DRG payment eventually increased to cover the added InSep test.

In pricing the instrument, Inflammatix considered the manufacturing cost and the average budgetary thresholds for purchase of similar point of service testing equipment. Since this instrument is not going to be an individually reimbursable item, it was important to arrive at a price point that would make the adoption and purchasing decision easy. The business model Inflammatix used is one where the goal is to gain profits from the sale of many razor blades not from the razor handle.

Another consideration for pricing the InSep test is the cost of the currently used methods for diagnosing acute infections. For example, the cost to hospitals for tests that diagnose respiratory viruses is approximately $150. Of course, if there is a significant savings seen on the treatment of the patient, for example, due to early

> **Box 8.11** (*cont.*)
>
> and accurate rule out of bacterial infection that reduces the treatment costs including shortening the stay of the patient, then the hospital will be better incentivized to purchase and pay for the diagnostic instrument and assay cassettes.
>
> In certain Medicare patient scenarios (when patients tested are not admitted into the hospital, aka "outpatients"), the hospital buyers can seek reimbursement for each test by health insurance companies for the InSep tests (cassettes). So, in this case, even though the company does not interact with the payers for reimbursement of their products directly, the company has to take the steps needed to gain coverage with appropriate level of payment to make it economically attractive for the hospitals. The company will need to conduct appropriate interventional trials (with and without the InSep test) to measure clinical outcomes and health economics. They will need to get a new PLA code as an ADLT human genetic test. For getting an appropriate payment level, they will need to circulate the economic and clinical outcomes data to the payers and make their case on getting appropriate payment assigned to each test. This process will take a couple of years and will be done in parallel with early sales.
>
> For non-Medicare patients, the company will need to engage with numerous private payers and Medicaid health plans to obtain specific test reimbursement, when applicable.

- If neither of these options work, under an existing set of codes, then a new set of ICD-procedure codes must be applied for

OPPS: Outpatient Setting

- The CPT codes and the ICD diagnosis codes are grouped into the appropriate APC codes (see Figure 8.9).
- The CPT codes for the procedures involving the device are used to identify the APC group code and thus determine the APC payment level. If the payment is not adequate to cover the new device, and the device qualifies as a new technology with significantly improved clinical outcomes or benefits, an add–on payment at cost of the new device may be approved as a "transitional pass-through" payments. The additional reimbursement is based solely on invoice cost of the new device. This is a transitional code, which is used to collect claims data and in a couple of years this payment level is rolled into an existing or new APC. See Box 8.10 for one example of how a new diagnostic technology for cancer was able to get higher reimbursement levels.

Physician's Fee Schedule: Based on CPT procedure codes

- Payment is made by accounting for specific physician work in the procedure and their practice fee (facility practice fee if carried out in a hospital and non-facility practice fee if in their own offices).

- If no suitable CPT code currently exists to pay the physician for performing additional services with the new technology, collaborate with the appropriate Medical Specialty society and file an application for new code with appropriate data, with the AMA (American Medical Association) CPT Editorial Committee (see Section 8.4.2 above). Work with the appropriate medical specialty society/ AMA committee to get a value assigned to the CPT code for reimbursement basis. Assistance from a medical specialization association is critical at this stage. A reimbursement relative value scale is used to assign values to the new CPT code depending on the complexity of the procedure (attested to by the professional specialty association). For clarity, note that the cost of the device to the hospital is covered under the DRG or APC payments.
- Wait for 18 to 24 months for a new permanent code to become active. Until that time, a temporary procedure code or transitional or add-on code can be applied for as described above. Without permanent codes and approved coverage, the risk of the payer refusing reimbursement or extra efforts required by the physician to get reimbursement could reflect in their choice to recommend purchase or use of new devices.

Final Steps

- Set up customer service at the company to assist providers / patients with reimbursement process. Publish educational materials describing appropriate billing codes, coverage rules, and payment rates and policies. This educational material and any such services must have content that adheres to federal guidelines on appropriate educational activities by manufacturers.

8.8   Major differences among selected national healthcare and reimbursement systems

A few national healthcare systems are profiled here in brief and data are drawn from the World Bank (http://data.worldbank.org) and the Commonwealth Fund (www .commonwealthfund.org/international-health-policy-center/countries).

Canada has a publicly financed, privately delivered healthcare system, which provides universal comprehensive access to healthcare. Hospitals are operated as private non-profit entities run by community boards or provincial health authorities. Doctors are private practitioners paid on a fee-for-service basis. About 10.57% of the GDP is spent on healthcare (2017).

The French healthcare system provides universal coverage to all citizens through a model that integrates public and private insurance. Healthcare expenditures were 11.31% of GDP in 2017. The public system is financed by employer payroll deductions and federal tax. End users are covered with a fee schedule based on socioeconomic status. Private plans reimburse treatment costs above the public system level for out-of-plan expenses. Although the government owns and operates most of the hospital beds, most ambulatory and outpatient care is provided by private professionals who are paid on a per-diem basis. Prescription drugs are reimbursed at prices set nationally.

The German public-private healthcare system covers about 90% of the population, and private healthcare insurance covers the rest. Primary funding comes from payroll taxes and is augmented by federal matching funds for certain costs. Employers organize sickness funds, the primary purchasers of government healthcare and unemployed and indigent are provided membership in sickness funds by the government. The government relies on negotiated spending caps on ambulatory care and pharmaceuticals with individual target hospital budgets. Germany spends about 11.25% of its GDP on healthcare (2017).

China has a two-pronged healthcare system that distinguishes rural versus urban healthcare. About 10% of the rural population is covered by the Cooperative medical system while the remainder is financed by out-of-pocket payments in a fee-for-service system. In the urban healthcare system, about 60% of the population is covered by a nationally paid, regionally administered system. However, out-of-pocket fees remain the predominant mechanism of payment for healthcare through most of the countryside. Healthcare spending is about 5.15% of its GDP (2017).

India relies on the private sector to finance nearly 80% of its healthcare system. The central government plays a limited role in the financing and delivery of healthcare with widespread regional variations in health status and outcomes due to varying abilities of state and municipal governments. The central government provides some insurance coverage through the Employees State Insurance Scheme which covers healthcare for organized labor and is funded through mandatory employer and employee contributions; the scheme covers some 28 to 30 million workers and retirees. The central government also runs the majority of rural hospitals and clinics that offer ambulatory, neonatal, family planning and maternal services. The government spending covers 25% of healthcare spending with the remaining 75% coming from household out-of-pocket expenditure. At the state level, the Directorates of Health Services and the Departments of Health and Family Welfare are responsible for organizing and delivering healthcare services to their populations, including all medical care, from primary care and pharmacies to secondary and tertiary hospital care. Expenditures for healthcare are about 3.53% of GDP (2017).

Australia offers both public and private health insurance, achieving universal coverage under a two-pronged system. Public insurance covers about 75% of the population and private insurance covers the rest. Free access to public hospitals is provided through the public system (Medicare) and patients who opt to go to private hospitals pay out-of-pocket or through private insurance. Strict regulations on private healthcare keep private fee rates low and the healthcare provision across public and private resources is not too different. Healthcare expenditures were 9.21% of GDP in 2017.

8.9   Marketing

The purpose of marketing is to convince someone to buy a product and, once they have bought it, to keep them buying it again. The four key parameters used by marketers, known as the four Ps of marketing, are also called the "marketing mix." These four Ps are used to create a marketing plan and strategy.

The four Ps of marketing are:

- *Product*: The product management and product marketing aspects of marketing deal with the specifications of the actual good or service, and how it relates to the end-user's needs and wants.
- *Pricing*: This refers to the process of setting a price for a product, including discounts.
- *Promotion*: This includes advertising, sales promotion, publicity, and personal selling, and refers to the various methods of promoting the product, brand, or company.
- *Placement* or distribution refers to how the product gets to the customer; for example, point of sale placement or retailing. This fourth P has also sometimes been called *Place*, referring to "where" a product or service is sold, e.g. in which geographic region or industry, to which segment (young adults, families, business people, women, men, etc.)[1]

In a technologically complex environment such as healthcare, where patients are not fully informed about choices, it is difficult to communicate and market the products accurately with full disclosure, as the general public may not know how to use the information. In fact, mass marketing to consumers of pharmaceuticals is banned or restricted in every western country except the United States and New Zealand. The U.S. FDA does oversee marketing messages by reviewing advertising materials and regulates the specific claims that can be made while marketing a product or service.

Marketing and reimbursement go hand in hand, as perception of value over the risk can drive demand and can also drive reimbursement. A clear example of this was seen in the J&J Cordis stent case discussed earlier (Section 8.4.3). According to an industry reimbursement executive, "Demand makes reimbursement challenges go away." If you can create a strong perception of value and need from the consumer, then the demand from the consumer can drive positive decisions on reimbursement. An example of use of this strategy is described in Box 8.12.

The marketing mix in the United States is complicated by the multipayer system, as each stakeholder needs different information to make purchasing decisions or selections for healthcare. For the large part, marketing efforts from pharmaceutical and device companies are focused around the need to educate all the stakeholders about the benefits and applications of the new product that uses advanced technology product. Adoption of a new technology in the existing healthcare paradigm is a key problem due to the conservative nature of the practice of medicine among the bulk of practitioners. Examples of various efforts made by a medical device company, Aspect Medical Systems, to improve adoption of a cutting-edge technology are summarized in Box 8.13.

In the biomedical industry, marketing is involved fairly early in the product development stage. The role of marketing is to ensure rapid growth in market adoption, as quickly as possible after approval, not just by the early adopters and

---

[1] Original text from http://en.wikipedia.org/wiki/Marketing.

Box 8.12 Diagnostic company's novel market strategy ramps sales rapidly

Excerpts from "Marketing Diagnostic Tests Directly to the Consumer", an article by Sue Auxter in Clinical Laboratory News, *April 2000, vol. 26, no. 4, reproduced here with permission.*

A new Pap test for cervical cancer, called the ThinPrep test, was approved in 1996 by the FDA. The manufacturer, Cytyc (Boxborough, MA), invested heavily in evaluating the new test against the standard Pap smear, and got FDA approval for a label stating that the ThinPrep test "was significantly more effective" than the conventional Pap smear. The company initially marketed the test to physicians, labs and managed care companies as typical diagnostic tests are marketed. In the markets where it was well accepted by payers and providers, it then proceeded to launch a direct-to-consumer marketing campaign using advertising in women's magazines and TV spots which resulted in many women asking for this improvement in their annual Pap testing. "We gained a ten percent market share during the period of time when we ran the DTC campaign, over and above what we would have expected to increase the market share without DTC" said Levangie [Dan Levangie, senior vice president of Cytyc Corporation]. "We've been fairly successful in directly educating *payers* [emphasis added] about the value of a better Pap test but I don't think there is any question that consumer pressure also gets their attention," explained Levangie, "and to some extent, a direct to consumer campaign by a company like us is indirectly influencing the insurance coverage for the test." The cost of the ThinPrep test varies considerably because of the complex laboratory contracts and regional differences, but the test generally costs $15 to $20 more than a traditional Pap smear...". The combination of astute regulatory, reimbursement, and advertising strategy helped the company succeed in gaining rapid sales and record over $140 million in revenues in 2000.

---

cutting-edge teaching hospitals, but by the larger group of practitioners who make up most of the market. Marketing efforts also include reimbursement specialists who advise on strategies and timing to educate the various payer groups. Product positioning efforts thus focus on different stakeholders throughout the development process. Key opinion leaders (KOLs) can greatly influence early and widespread adoption. Typically, the company will want to involve the opinion leaders in carrying out and publishing results of the clinical studies. Marketing efforts also include developing a publishing and presentation (at conferences) strategy and feedback on competitive landscape. Competitor information is useful in planning appropriate head-to-head comparison trials in clinical studies, not required by regulatory bodies but very important to show clinical benefit of a new treatment. Key scientific results of clinical studies are used to influence a broader segment of potential prescribers as the product gets close to approval. As the product (e.g. a drug) moves to patent expiration, patients

---

Box 8.13 Aspect Medical Systems climbs the adoption curve

Adapted from HBS case (Atkins, 2000) and RPI class interview with the founder and CEO Nassib Chamoun.

Aspect has developed a controversial device that measures electrical signals from the brain during anesthesia to evaluate some measure of consciousness – the BIS device. Clinical use of this device has shown that anesthesiologists that use this device are able to better judge optimal dosage, frequently resulting in reduced dosage of drugs and faster functional recovery of patients after surgery, sometimes reducing the length of post-operative hospital stay.

However, after early adopters accepted the device, its sales stagnated (even after FDA approval and reimbursement assignment) as anesthesiologists felt their judgment and skills were being questioned by a machine and many did not want to take chances on a machine or to change their established practice. In order to increase market adoption, the company undertook the following multiple approaches:

- Get the BIS endorsed as a standard of care by the medical society (American Society of Anesthesiologists).
- Carry out multiple studies in various groups of patients to generate more data.
- Recruit key opinion leaders to use the device.
- Create public awareness of consciousness during surgery. Manage public and professional relations during this process.
- Obtain coverage and reimbursement from most major payers.
- Obtain FDA approval for more indications so as to increase applications of equipment – thus making it more economical for hospitals.
- Innovative sales strategy to encourage hospitals to put one in every surgery room in the hospital – thus facilitating widespread usage.
- Capture market channels of distribution to keep off competition.
- Going through the PMA approval process created barriers to entry for others.

Adoption of a new technology can be rapid at first and then typically slows down once the "early adopters" are won over. The BIS device had not been accepted by the second tier of end users, the (typically) larger base that makes up 50 to 70% of the "mass" market. Pushing the BIS device into that market would see the device gain "blockbuster" status, but it also required a significant change in practice of anesthesiology. Thus the determinants that would sway the larger audience had to be carefully understood and market acceptance issues approached on multiple levels for the device to truly attain blockbuster status. The above elements had to be put in place almost simultaneously.

---

exert more influence on the decision to purchase the branded drug over the generics and the focus of marketing efforts shifts in concert (Table 8.3).

Market research (see Chapter 2), one of the functions of marketing departments in companies, specifically serves an important and useful role in product development as

Table 8.3 The focus of marketing changes during development

Phase of development	Purchase influencer targeted
Clinical development	Key opinion leaders
Pre-launch	Physician stake holders, Payers
New Indication	Prescribers, Payers
Patent expiration	Patients

described in Chapters 2 and 5. Marketing efforts to position the product can help steer the clinical trials of the product, just as reimbursement considerations influence clinical trial design. As an example, Pfizer's marketing department saw a new market possibility in the side effects of Viagra (sildenafil) when it was being developed in early clinical trials for blood pressure reduction. Similarly, GlaxoSmithKline's Paxil (paroxetine) was originally developed as an antidepressant but marketing saw more indications of interest and it has subsequently been approved for social anxiety disorder and panic disorder indications.

8.10   Sales

Depending on the route of distribution to the marketplace, salespeople are hired to approach buyers and close a purchase contract for the goods. When dealing with wholesalers or PBMs or with payers such as private insurers, the salespeople use a combination of incentives or discounts to the buyers, usually depending on the volume of sales expected. Sales to the patient are rare and salespeople almost always are dealing with institutional purchasing departments (materials management department in a hospital) or third-party payers to ensure coverage and inclusion in formulary or inventory.

In the *drug industry*, sales personnel are extensively trained by the company and are expected to be able to explain complex clinical results to physicians. Another tier of salesperson in the drug industry called the Medical Science Liason (MSL), typically a trained physician or scientist, engages thought leaders. An MSL is hired to explain the benefits of the new drug by discussing the data and clinical relevance, usage, patient concerns, etc. with the practicing physician at a scientific and practical level. These MSLs play an important role in between marketing and sales.

Sales positions in the *device industry* are also highly specialized jobs and most salespeople have a bachelor's degree in sciences or engineering and many have master's degrees in biomedical engineering or related fields. Many sales personnel will need to have intimate knowledge of human anatomy and physiology in addition to product information (depending on their specific role and the product itself). These salespeople often serve as consultants for the physicians or surgeons, standing by to address questions or problems and offer tips in the surgical suite as the surgeon is implanting or using the new device, until the surgeon is comfortable with the new device and associated procedure. Often the sale of a new device will include one or more training sessions, as a sound procedure of usage of the device is at least as important as the quality of the device itself. In the device industry in particular, salesforce feedback can play a vital role in new product design and development as they come back with insightful observations from their complete participation in the product application and usage process.

8.11   Product liability

Once a medical product is approved by the FDA for sale in the market, the company is generally protected against wrongful use of the product – e.g. a physician who uses or prescribes the product for any indication other than the one approved is putting themselves at risk for a lawsuit from the patient if something goes wrong, but the manufacturer cannot generally be sued in this instance. For example, if a surgeon implants a device incorrectly, it is the surgeon at risk for malpractice or negligence rather than the device manufacturer. As long as the product is used in the manner and indication for which it is approved, and it works as claimed, the manufacturer in general is not liable for faulty usage.

Specifically, product liability lawsuits have as their basis one of three claims – *negligence*, *strict liability*, or *breach of warranty*. Lawsuits against biomedical product manufacturers typically claim one or more of these legal causes as the basis for recovery of damages.

Negligence is based on the following fact: Manufacturers have a duty to care for the patients who are the end users of their products; the standard for carrying out the duty was breached and as a result a compensable injury resulted to the plaintiff.

In these *negligence cases*, the burden is on the plaintiff (patient) to prove negligence. However, manufacturers of products that can have serious risk to the patient should be even more careful during product development than manufacturers of products that may carry lower risk.

Strict liability or product liability cases focus on the product. Does the inherent design of the product have a defect that makes it unreasonably unsafe? Did the product leave the manufacturer's factory in a condition that causes unreasonable danger to the user or consumer? The critical focus in a strict liability suit is whether the product is defective and unreasonably dangerous or whether the manufacturer failed to adequately warn of hazards. The latter is then also a case of negligence, and the majority of cases are of this nature (failure to warn). The criterion of "unreasonably dangerous" is based on a risk/benefit analysis.

Breach of warranty cases assert that the manufacturer breached the warranty or representations made about the product.

Some hallmark *negligence and liability* lawsuits against drug manufacturers include those against GlaxoSmithKline's Plavix (a claim that GSK did not report studies that showed the drug increased suicidal tendencies), American Home Products' Fen-Phen (diet drug that led to heart valve problems; AHP set up a trust fund to settle class action lawsuits without admitting wrongdoing), Merck's Vioxx (lawsuits claim that Merck did not reveal increased risk of heart attacks found in early studies), and Bayer's Baycol (a cholesterol-lowering drug that caused rare muscular disease; all cases were settled after Bayer won first two cases). Such lawsuits against medical device manufacturers include those against Guidant (covered up knowledge of defects that would cause its Ventak defibrillator to short and fail) and its recently acquired subsidiary Endovascular Corp (misled FDA and did not report majority of malfunctions in its Ancure stent-graft device). In cases involving gross negligence, where the manufacturers have been shown to have been at fault, the manufacturers usually settle the case with payments made to the plaintiffs. The FDA marketing application approval and FDA inspections of the manufacturing facilities provide no protection against these types of lawsuits. All of the lawsuits above had claims resulting in millions of dollars of payments (reaching billions in many cases).

Along with these product-liability lawsuits and claims, companies usually have to face shareholder lawsuits that claim damages for fall in share price due to mismanagement.

Some of these cases are driven by a product recall announced by the company or the FDA, rather than on specific proof of wrongdoing or misconduct. In one of the most famous cases of this type, after the FDA issued recall of silicone-filled breast implants, the large number of ensuing lawsuits and claims led the Dow Corning company to file for bankruptcy in order to reorganize and settle the claims through a trust fund. However, subsequent analysis of the medical data has shown no clear link between the reported health problems (silicone adjuvant disease where silicone was claimed to trigger some unknown factor in the immune system) and the silicone breast implants.

In the *product liability cases* that typically blame faulty or unsafe design, the cases have to be introduced into the state judicial system, and often come up against the fact that state law cannot preempt (a device approved under) federal law, thus providing some liability protection for FDA-approved devices. In one case, *Thorn* v. *Thoratec*, Mrs. Thorn alleged that a defect in the design of HeartMate (a ventricular heart assist device) caused the device to fail in her husband and led to his death. A suture that was required to position a tube came off and a resulting embolism traveled to the brain causing a fatal brain hemorrhage. The judges in the U.S. 3rd Circuit Court of Appeals rejected her bid, as the ruling would conflict with the federal law and PreMarket Approval (PMA) regulated process and state law cannot conflict with federal law under which products were approved for sale.

However, the position that FDA approval provides some shield to manufacturers from certain types of product liability lawsuits is shifting, with several ongoing court cases and proposed bills addressing this issue, arguing for consumer protection over manufacturer protection.

8.12   Summary

Start early in product development to understand the reimbursement path, definitely before starting clinical trials. The increasing emphasis by payers on real-world clinical outcomes and cost-benefit economic analysis will impact the product development strategy. Payment and coverage are often driven by local considerations, and product developers should be prepared to work step by step through the markets to achieve coverage and adequate payment. New technologies will have to justify their value proposition with clinical outcomes and work to obtain adequate reimbursement from payers, and products that meet a real unmet medical need will usually have an easier path to adoption and adequate reimbursement.

Exercises

8.1.   What is the patient mix for the product?

8.2.   Who are the payers for these types of patients?

8.3.   What is the setting(s) in which your biomedical technology will be used? Based on that answer, in which payment group (DRG, APC, etc.) will it be categorized?

8.4.   What is the coding (CPT/ICD-10/11) for your technology and its applications? (If a drug product, what is it that determines whether it needs a CPT code?)

8.5.   How will your product be distributed to get it to its proper setting for use?

8.6.   Prepare a distribution/reimbursement flow chart for your product.

8.7.   Highlight the key issues that will deliver a positive coverage decision for your product showing value over the competition. Have you planned to collect economic data for analysis in your clinical trials?

8.8.   Do you need to rethink your product characteristics based on the above questions?

8.9.   What is the value proposition of your product for insurers? For physicians? Can you summarize the risk vs. benefits of your product?

8.10. What levels of payment can you expect right after approval and then over time for your product?

8.11. Based on answers to the above questions, you should be able to plan the reimbursement path and predict possible revenue cash flows for your product.

References and additional readings

Atkins, N, and Bohmer, R. (2000). *Aspect Medical Systems. Case Study no. 9-600-076.* Cambridge, MA: Harvard Business School Publishing.

Auxter, S. (2000). Marketing diagnostic tests directly to the consumer. *Clinical Laboratory News* 26(4), 4.

Campbell, J. (2005). *Understanding Pharma*. Raleigh, NC: Pharmaceutical Institute. ISBN 0-9763096-0-2

Drummond, MF, Sculpher, MJ, Torrance, GW, O'Brien, BJ, and Stoddart, GL. (2005). *Methods for the economic evaluation of health care programmes*. Third edition. Oxford: Oxford University Press. ISBN: 0-19-852945-7

Folland, S, Goodman, A, and Stano, M. (editors). (2007). *Economics of Health and Health Care*. Upper Sadle River, NJ: Prentice Hall Publishing. ISBN: 0132279428

Harris, G. (2004, January 30). US weighs not paying for all uses of some drugs. *New York Times*.

Kerr, NF, and Recupero, T. (2015). *Product Launch: Practical Guide to Launching Medical Device Products*. Kerr Consulting Group LLC. ISBN: 978-0990908104

Ohsfeldt, RL, and Schneider, JE. (2006). *The Business of Health: The Role of Competition, Markets, and Regulation*. AEI Press. ISBN: 0-8447-4240-6

Pekarsky, BAK. (2015). *The New Drug Reimbursement Game: A Regulator's Guide to Playing and Winning*. Adis/Springer International Publishing. ISBN: 978-3-319-34920-6

Phelps, CE. (2003). *Health Economics*, 3rd ed. Reading, MA: Addison-Wesley, 2003. ISBN-13: 978-0321068989

Pizzi LT, (2016). The Institute for Clinical and Economic Review and its growing influence on the US healthcare. *American Health & Drug Benefits* 9(1), 9–10.

Provines, CD. (2012). *Strategic Pricing for Medical Technologies: A Practical Guide to Pricing Medical Devices & Diagnostics*. Christopher D Provines. ISBN: 978-0615661896

Shiroiwa, T, Fukuda, T, Ikeda, S, et al. (2017). Development of an official guideline for the economic evaluation of drugs/medical devices in Japan. *Value in Health* 20(3), 372–378.

Walker, TP. (2018). *Insights: 33 Lessons Learned in Medical Device Marketing*. Utah: Life Catalyst Publishing. ISBN: 978-0997635829

Walley, T, Haycox, A, and Angela Boland, A. (2004). *Pharmacoeconomics*. Churchill Livingstone Press. ISBN: 044307240X

Websites of interest

CPT code-making process:	www.ama-assn.org/practice-management/cpt
CMS Centers for Medicare & Medicaid Services	www.cms.gov
MedPAC – advisory to Congress on Medicare (summary documents on Medicare processes; look under "other publications" link)	www.medpac.gov
Medicare Coverage Database (search for National and Local coverage decisions)	www.cms.gov/Medicare/Coverage/CoverageGenInfo/index
HCPCS Codes at CMS	www.cms.gov/Medicare/Coding/MedHCPCSGenInfo
CPT codes and how they are managed American Medical Association	www.ama-assn.org/practice-management/cpt/cpt-overview-and-code-approval
Useful list of acronyms used in healthcare and by CMS	www.cms.gov/acronyms

Outpatient Prospective Payment system at Medicare	www.cms.gov/Medicare/Medicare-Fee-for-Service-Payment/ASCPayment/index
Medicare physician fee schedule	www.cms.gov/Medicare/Medicare-Fee-for-Service-Payment/PhysicianFeeSched/index
Kaiser Family Foundation	www.kff.org
Leonard Davis Institute of Health Economics	www.upenn.edu/ldi
National Bureau of Economic Research (has many working papers on health-related topics)	www.nber.org
National Health Service (UK)	www.nhs.uk/
National Center for Health Statistics (USA)	www.cdc.gov/nchs/
WHO (World Health Organization) Statistical Information System	www.who.int/whosis/
World Bank	data.worldbank.org

Visit https://shreefalmehta.com/csbtbook for additional enriching readings around the topics covered in the book, topical updates on the content and for industry viewpoints and news.

Glossary

Abbreviated New Drug Application (ANDA) Contains data that, when submitted to FDA's Center for Drug Evaluation and Research, Office of Generic Drugs, provides for the review and ultimate approval of a generic drug product. Generic drug applications are called "abbreviated" because they are generally not required to include preclinical (animal) and clinical (human) data to establish safety and effectiveness. Instead, a generic applicant must scientifically demonstrate that its product is bioequivalent (i.e., performs in the same manner as the innovator drug). Once approved, an applicant may manufacture and market the generic drug product to provide a safe, effective, low-cost alternative to the American public. (Text from www.fda.gov/cder/drugsatfda/Glossary.htm)

ADMET Acronym for parameters used to understand drug behavior in a living system: Absorption, Distribution (among the tissues), Metabolism, Excretion, and Toxicology

AI Artificial Intelligence (AI), also machine learning (ML) or neural networks; all of the these represent adaptive programming methods wherein the software program heuristically trains and statistically selects input data patterns to associate with desired outcomes. This type of software adaptively learns to improve its analysis as more data and validations are fed to it. The adaptive nature of the analysis algorithms often means that there is no clearly defined analytical formulation, and results to the same inputs may differ over time as the program learns.

Antisense A nucleic acid sequence that is complementary to the coding sequence of DNA or mRNA.

API Active pharmaceutical ingredient in a drug (biological or synthetic chemical). Any substance or mixture of substances intended to be used in the manufacture of a drug product and that, when used in the production of a drug, becomes an active ingredient in the drug product. Such substances are intended to furnish pharmacological activity or other direct effect in the diagnosis, cure, mitigation, treatment or prevention of disease or to affect the structure and function of the body (as defined by the FDA).

ASP Average sales price (ASP): a new system created by federal and state governments to ensure more accurate price reporting for drugs. ASP is the weighted average of all non-federal sales to wholesalers and is net of chargebacks, discounts, rebates, and other benefits tied to the purchase of the drug product, whether it is paid to the wholesaler or the retailer.

ASR Analyte-specific reagents (ASR; also known as home-brew tests): the active ingredient of an in-house diagnostic test. The FDA defines ASRs as: "antibodies, both polyclonal and monoclonal, specific receptor proteins, ligands, nucleic acid sequences, and similar reagents which, through specific binding or chemical reaction with substances in a specimen, are intended for use in a diagnostic application for identification and quantification of an individual chemical substance or ligand in biological specimens."

AWP Average wholesale price (AWP): a national average of list prices charged by wholesalers to pharmacies. AWP is sometimes referred to as a "sticker price" because it is not the actual price that larger purchasers normally pay. For example, in a study of prices paid by retail pharmacies in eleven states, the average acquisition price was 18.3% below AWP. Discounts for HMOs and other large purchasers can be even greater. AWP information is publicly available.

Bioinformatics The use of extensive computerized databases to solve information problems in the biological sciences. These databases generally contain protein and nucleic acid sequences, genomes, etc. Bioinformatics also encompasses computer techniques such as 3D molecular modeling, statistical database analysis, data mining, etc.

Biologics Biological macromolecules or large molecular (weight) entities (proteins such as monoclonal antibodies, or enzymes and other molecules such as glycoproteins, hormones, etc.) developed through biotechnology for therapeutic intervention as drugs for specific diseases. Counterparts are synthetic chemical small molecule drugs that make up the traditional pharmaceutical industry.

CFR Code of Federal Regulations (CFR): the book in which all regulations are codified as promulgated by various executive agencies. The CFR is the codification of the general and permanent rules published in the Federal Register by the executive departments and agencies of the Federal Government. Title (volume) 21 contains regulations pertaining to food and drugs. Latest text is available at www.govinfo.gov/app/collection/cfr.

CMO Contract manufacturing organization (note: When referring to personnel, CMO = Chief Medical Officer or Chief Marketing Officer).

CMS Centers for Medicare & Medicaid Services administer the Medicare and Medicaid programs, which provide healthcare to about one in every four Americans (~145 million people). Medicare provides health insurance for elderly and disabled Americans. Medicaid, a joint federal-state program, provides health coverage for low-income persons, including children, and nursing home coverage for low-income elderly. CMS also administers the State Children's Health Insurance Program. Established as the Health Care Financing Administration (HCFA) in 1977. Headquarters: Baltimore, Maryland. Website: www.medicare.gov, www.cms.gov. CMS had a budget of $826 billion in 2020.

CPT codes Current Procedural Terminology codes (CPT-4): numeric coding system maintained by the American Medical Association (AMA). The CPT codes are used primarily to identify medical services and procedures furnished by physicians and other healthcare professionals. These codes are also known as Healthcare Common Procedure Coding System (HCPCS) level I codes.

CRO Contract research organization (CRO): can range from a one-person biostatistical consultant to a large multinational organization that coordinates all aspects of the clinical trial. CROs also work on preclinical research, most commonly on GLP toxicology studies.

Device Any product that achieves its primary functions in the body through mechanical or physical action (rather than chemical).

DME Durable medical equipment (DME): can usually withstand repeated use, is useable at home, and is not beneficial to a person without an illness or injury. Examples include hospital beds, wheelchairs, and oxygen equipment.

Drug A product that achieves its primary functions in the body by chemical action. In this book this term covers small molecular weight chemical compounds and also biologics.

Enantiomers Stereoisomeric compounds that are non-superimposable mirror images of one another. For more details and implications of enantiomers in pharmaceutical product development, see http://en.wikipedia.org/wiki/Enantiomer

Evidence-based medicine Principle of making individual treatment decisions by consideration of the results of many scientific studies.

FDA The Food and Drug Administration assures the safety of foods and cosmetics, and the safety and efficacy of pharmaceuticals, biological products, and medical devices. Website: www.fda.gov. Headquartered at Rockville, Maryland, the agency had 17,686 employees and a budget of $5.9 billion in 2019.

FISH Fluorescent in situ hybridization (FISH): a process that vividly paints chromosomes or portions of chromosomes with fluorescent molecules. A fluorescent hybridization probe is created for the DNA segment of interest. The probe, and attached fluorescent molecule, will combine with any complementary DNA or RNA it encounters. (Text from www.wikipedia.org)
This technique is useful for identifying chromosomal abnormalities and gene mapping. It can also be used to identify microorganisms.

Formulary A select menu of drugs that a managed care organization or insurer will cover under its health plan.

GCP Good clinical practice (GCP): an international quality standard for clinical trials involving human subjects. The GCPs are provided by the International Conference on Harmonization (ICH) and issued as regulations by the FDA. GCP Guidelines include standards on how clinical trials should be conducted, define the roles and responsibilities of clinical trial sponsors, clinical research investigators, and monitors.

Gene The fundamental physical and functional unit of heredity that is made up of tightly coiled threads or polymers of deoxyribonucleic acid (DNA). A DNA molecule consists of two strands that wrap around each other to resemble a twisted ladder or double helix. DNA is an informational molecule and is made up of four distinct nucleotides: deoxyadenosine (A), deoxyguanosine (G), deoxythymidine (T), and deoxycytidine (C). It is the nonrandom order of these individual "bases" that results in DNA being an informational molecule. However, in and of itself, DNA has no functional property. It is a chemical that, when placed in an appropriate environment, will direct the synthesis of particular and specific proteins, which make up the structural components of cells, tissues, and enzymes (molecules that are essential for biochemical reactions). This environment is found within the cell. Organisms, from single-celled protozoans to far more complex human beings, are made up of cells containing DNA and associated protein molecules. The DNA is organized into structures called chromosomes, which encode all the information necessary for building and maintaining the organism. A DNA molecule may contain one or more genes, each of which is a specific sequence of nucleotide bases. It is the specific sequence of these bases that provides the exact genetic instructions that give an organism its unique traits.

Gene therapy Treatment that alters genes (the basic units of heredity found in all cells in the body). In early studies of gene therapy for cancer, researchers were trying to improve the body's natural ability to fight the disease or to make the tumor more sensitive to other kinds of therapy. This treatment may involve the addition of a functional gene or group of genes to a cell by gene insertion to correct a hereditary disease.

Generic drug The same as a brand-name drug in dosage, safety, strength, how it is taken, quality, performance, and intended use. Before approving a generic drug product, the FDA requires many rigorous tests and procedures to assure that the generic drug can be substituted for the brand- name drug. The FDA bases evaluations of substitutability, or "therapeutic equivalence," of generic drugs on scientific evaluations. By law, a generic drug product must contain the identical amounts of the same active ingredient(s) as the brand name product. Drug products evaluated as "therapeutically equivalent" can be expected to have equal effect and no difference when substituted for the brand-name product.

Genomics The identification and functional characterization of genes.

GLP Good laboratory practice (GLP): guidelines on record-keeping and procedures to be followed in conducting nonclinical studies. These guidelines are regulations issued by the FDA. Nonclinical data required by the FDA as part of product approval applications must be carried out under GLP.

GMP Good manufacturing practices (GMP) (see also QSR) (Also cGMP = current GMP): guidelines that are developed by the World Health Organization (WHO) and the ICH are incorporated into the FDA regulations. GMP regulates the manufacturing and laboratory testing environment and processes. An extremely important part of GMP is documentation of every aspect of the process, activities, and operations involved with drug and medical device manufacture.

HCFA See CMS

HCPCS codes The Healthcare Common Procedure Coding System (HCPCS) is divided into two principal subsystems, referred to as Level I and Level II of the

HCPCS. Level I of the HCPCS is comprised of Current Procedural Terminology (CPT) codes, which are used to describe and identify services and procedures furnished by physicians and other healthcare professionals. Level II of the HCPCS is a standardized coding system that is used primarily to identify products, supplies, and services not included in the CPT codes, such as ambulance services and durable medical equipment, prosthetics, orthotics, and supplies (DMEPOS) when used outside a physician's office.

HHS The Department of Health and Human Services (HHS) is the principal U.S. agency for protecting the health of all Americans and providing essential human services. The department includes more than 300 programs, including the NIH, FDA, CMS. Many HHS-funded services are provided at the local level by state or county agencies, or through private sector grantees. It administers more grant dollars than all other federal agencies combined (representing nearly a quarter of all federal outlays). The HHS Medicare program is the nation's largest health insurer, handling more than 1 billion claims per year. Medicare and Medicaid together provide healthcare insurance for one in four Americans. HHS programs are administered by 11 operating divisions. Website: www.hhs.gov/ HHS had a budget of $1.2 trillion for FY 2020.

Home-brew tests See ASRs for more details

HRSA The Health Resources and Services Administration (HRSA), an agency of the U.S. Department of Health and Human Services, is the primary Federal agency for improving access to healthcare services for people who are uninsured, isolated, or medically vulnerable.

IACUC Institutional Animal Care and Use Committee (IACUC): this committee reviews and approves all experimental work that involves animals with a view to preventing unnecessary.

ICH International Conference on Harmonization (ICH): A unique project that brings together the regulatory authorities of Europe, Japan, and the United States and experts from the pharmaceutical industry in the three regions with the goal of making the international regulatory processes for medical products more efficient and uniform.

ICD-10/11 International Classification of Diseases (ICD): the international standard for reporting diseases and health conditions. It is the diagnostic classification standard for all clinical and research purposes. ICD defines the universe of diseases, disorders, injuries and other related health conditions, listed in a comprehensive, hierarchical fashion. ICD-10 was approved in 1990 and is used by more than 150 countries around the world. The World Health Organization (WHO) is responsible for maintaining and updating this classification. In the USA, the American Hospital Association (AHA) serves as a clearing house for all ICD issues. ICD-11 has been adopted by the Seventy-second World Health Assembly in May 2019 and came into effect on January 1, 2022.

In silico biology The use of computational algorithms to create virtual systems that emulate molecular pathways, entire cells, or more complex living systems. The use of computers to simulate or analyze a biological experiment.

In vivo Inside the living organism

In vitro In the laboratory or outside the organism

Indication An indication for a drug or device refers to the particular disease or stage of that disease that the drug or device is intended to treat.

Incidence The incidence of a disease is defined as the number of new cases of disease occurring in a population during a defined time interval. (Also see prevalence)

IPPS Inpatient payment system (IPPS): Medicare's payment system for reimbursing hospitals for inpatient services and products provided.

IRB Institutional Review Board: a committee that is required to review ethical principles of biomedical studies and approve any protocol that involves human subjects. The role of the IRB is to protect the rights and welfare of subjects.

ISO International Organization for Standardization (ISO): an independent, non-governmental, international organization that develops standards to ensure the quality, safety, and efficiency of products, services, and systems. ISO standards are in place to ensure consistency. For example, ISO 9001:2015 refers to the standard number 9001 and the (year) version that is being met by the certified facility.

IVD In vitro diagnostics (IVD): products used to diagnose disease by analyzing or reacting with samples of human tissues, blood or extracts thereof.

Label The FDA-approved label is the official description of a drug product which includes indication (what the drug is used for); who should take it; adverse events (side effects); instructions for uses in pregnancy, children, and other populations; and safety information for the patient. It defines the specific market for the product. Labels are often found inside drug product packaging.

LCD / NCD Local coverage decision (LCD; made by local Medicare contractors for each region) / national coverage decision (NCD; made by CMS; applies to all local providers).

Lead These are typically leading compounds that are being developed towards a drug candidate. Next development step is to optimize the lead compound with respect to multiple product characteristics.

Microbiology A branch of biology dealing especially with microscopic forms of life.

Monoclonal antibody (mAB) Highly specific, purified antibody that is derived from only one clone of cells and recognizes only one antigen.

Novel chemical entity (NCE) A compound not previously described in the literature. Any new molecular compound not previously approved for human use, excluding diagnostic agents, vaccines and other biologic compounds not approved by the FDA's Centers for Drug Evaluation and Research (CDER). Also excluded are new salts, esters and dosage forms of previously approved compounds.

New molecular entity (NME) An active ingredient (usually a synthetic small molecule chemical) that has never before been marketed in any form. (The terms NME and NCE are used interchangeably.)

NIH National Institutes of Health – NIH is the world's premier medical research organization. It invests over $40 billion annually in medical research, supporting research projects in cancer, Alzheimer's, diabetes, arthritis, heart ailments, AIDS, and many more. Includes 27 separate institutes and centers. Website: www.nih.gov

NCD / LCD National Coverage Decision (made by CMS; applies to all local providers) / Local Coverage Decision (made by local Medicare contractors for each region).

NPD New product development.

Off-label The practice of prescribing drugs for a purpose outside the scope of the drug's approved label.

OPPS Outpatient Payment System (OPPS): Medicare's payment system to reimburse hospitals for outpatient procedures.

Pass-through codes Section 1833(t)(6) of the Social Security Act provides for temporary additional payments or "transitional pass-through payments" for certain innovative medical devices, drugs, and biologicals for Medicare beneficiaries, even if prices for these new and innovative items exceeded Medicare's regular scheduled OPPS payment amounts. For drugs and biologicals, the pass-through payment is the amount by which 95% of the average wholesale price exceeds the applicable fee schedule amount associated with the drug or biological. For devices, the pass-through payment equals the amount by which the hospital's charges, adjusted to cost, exceeds the OPPS payment rate associated with the device.

PBM Pharmaceutical benefit managers (PBMs) manage pharmacy benefits, maintain formularies and obtain discounts for bulk purchases of drugs for their clients – employers, insurance companies, and unions. A handful of large national companies and many small regional PBMs influence more than 80% of prescription drug coverage. The original purpose of PBMs was to offer cost-effective services such as reliable claims information and issuance of drug cards for easy ID and account tracking. Over time, however, PBM's functions have evolved to include large-scale "block purchases" of drugs and medical products that dramatically lower their wholesale costs.

PCT Patent Cooperation Treaty (PCT): originally formed in 1970 (and modified several times) over 130 states are signatories to the PCT. The Treaty makes it possible to seek patent protection for an invention simultaneously in each of a large number of countries by filing an "international" patent application.

Pharmacogenomic test Pharmacogenomics is the study of the stratification of the pharmacological response to a drug by a population based on the genetic variation of that population. This assay is intended to study interindividual variations in whole-genome or candidate gene, single-nucleotide polymorphism (SNP) maps, haplotype markers, or alterations in gene expression or inactivation that may be correlated with pharmacological function and therapeutic response. In some cases, the pattern or profile of change is the relevant biomarker, rather than changes in individual markers.

PhRMA The Pharmaceutical Research and Manufacturers of America (PhRMA): an organization that represents the country's leading pharmaceutical research and bio-technology companies. Website: www.phrma.org

Pharmacokinetics (PK) The study of the process by which a drug is absorbed, distributed, metabolized and eliminated by the body over time. PK is often called the study of what the body does to the drug, whereas PD is the study of what the drug does to the body.

Pharmacodynamics (PD) Pharmacodynamics is the study of the biochemical and physiological effects of drugs and the mechanisms of drug action and the relationship between drug concentration and effect. PD is often called the study of what the drug does to the body, whereas PK is the study of what the body does to the drug. (From wikipedia.org)

Phase I Clinical testing phase for new drugs with the main aim to determine drug safety. Drugs are typically tested in a small group of healthy volunteers. (Exceptions to this exist – for example, cancer drugs are usually tested in cancer patients in Phase I studies.)

Phase II Clinical testing phase for new drugs, that is aimed at identifying the optimal dose to be used in Phase III trials and tests for proof of efficacy of the drug with statistically significant results on the endpoint. Typically, Phase II trials are double-blinded and have placebo controls.

Phase III Clinical testing phase for new drugs, that is aimed at definitively determining the drug's effectiveness and its side-effect profiles in significantly large (thousands of patients) trials. These studies are also typically double blinded and placebo controlled. After the study is closed and while data are being readied for presentation, a Phase IIIb may allow continued "compassionate use" of drug by patients who have been in the Phase III study.

Phase IV After a drug has been approved, pharmaceutical companies may conduct further studies of its performance, often examining long-term safety or expansion to other indications.

Prevalence The number of cases of a disease at a specified time divided by the number of individuals in the population at that specified time. The prevalence may be reported as a percentage of the population or as the prevalence per 100,000 people. (Also see incidence)

Proteomics The study of gene expression at the protein level, by the identification and characterization of proteins present in a biological sample.

QSR Quality system requirements (QSR): Manufacturers must establish and follow quality systems to help ensure that their products consistently meet applicable requirements and specifications. The quality systems for FDA-regulated products (food, drugs, biologics, and devices) are known as current good manufacturing practices (CGMPs). QSR for medical devices is detailed in 21 CFR 820.

RBRVS and RVUs Resource Based Relative Value Scale (RBRVS) and Relative Value Units (RVUs): A relative value system that is used for calculating national fee schedules for services provided to Medicare patients. Physicians are paid on relative value units (RVUs) for procedures and services. The three components of each established value are: work RVU, practice expense RVU, and malpractice expense RVU.

Recombinant DNA (rDNA) A combination of DNA molecules of different origin that are joined using recombinant DNA technologies.

SOP Standard operating procedure (SOP): documented so that the SOP document reflects actual practice of procedures in various manufacturing, design review, testing, assays, laboratory procedures, etc. SOPs are an integral part of cGMP and QSR.

siRNA Silencing or short-interfering RNA (ribonucleic acid); control mechanism for controlling production of specific proteins by interfering with gene transcription.

Surrogate endpoint A marker – a laboratory measurement or physical sign –- that is used in clinical trials as an indirect or substitute measurement that represents a clinically meaningful outcome, such as survival or symptom improvement. The use of a surrogate endpoint can considerably shorten the time required prior to receiving FDA approval. Approval of a drug based on such endpoints is given on the condition that post marketing clinical trials verify the anticipated clinical benefit. The FDA bases its decision on whether to accept the proposed surrogate endpoint on the scientific support for that endpoint. The studies that demonstrate the effect of the drug on the surrogate endpoint must be "adequate and well controlled" studies, the only basis under law, for a finding that a drug is effective.

Target Usually a protein, an enzyme or a receptor in a cell or tissue that has been discovered to play a central role in the development or diagnosis of a disease.

USC United States Code (USC): the codification by subject matter of the general and permanent laws and statutes of the United States. Website: https://uscode.house.gov/ (In contrast, the CFR contains all of the regulations promulgated by executive agencies.)

Wholesale acquisition cost (WAC) The price paid by a wholesaler for drugs purchased from the wholesaler's supplier, typically the manufacturer of the drug. On financial statements, the total of these amounts equals the wholesaler's cost of goods sold. Publicly disclosed or listed WAC amounts may not reflect all available discounts.

Wholesaler A company that serves as a bridge between a drug manufacturer and a covered entity. This means any entity (including a pharmacy or chain of pharmacies) to which the labeler sells the covered outpatient drug, but that does not relabel or repackage the covered outpatient drug.

Index

Printed in the United States
by Baker & Taylor Publisher Services